Dentition of Living Primates

Dentition of Living Primates

DARIS R. SWINDLER

Department of Anthropology,
The Regional Primate Research Center,
University of Washington,
Seattle, Washington, USA

1976

ACADEMIC PRESS
London · New York · San Francisco
A Subsidiary of Harcourt Brace Jovanovich Publishers

ACADEMIC PRESS INC. (LONDON) LTD
24–28 Oval Road,
London NW1

US Edition published by
ACADEMIC PRESS INC.
111 Fifth Avenue,
New York, New York 10003

Library of Congress Catalog Card Number: 75-19680

ISBN: 0-12-679250-X

Text set in 10/12 pt. Linotron Times Roman, printed by offset litho, by J. W. Arrowsmith Ltd., Bristol BS3 2NT

This book is dedicated to the memory of two fine colleagues and friends, the late Bertram S. Kraus and Lawrence Oschinsky.

Preface

The idea for this book goes back over a period of years and owes its conception, as is so often the case in science, to another source. In *The Jaws and Teeth of Primates*, (published in 1960) W. Warwick James presented a systematic description of one specimen, of one species, of one genus of the majority of extant primate genera. I have long wanted to add substantively to this work by presenting both qualitative and quantitative statistics of the parameters of dental variability in extant primates. To do this, my graduate students and I prepared dental casts of over 2,000 primate dentitions representing all of the principal genera and many of the species of living primates. We have studied these dentitions over the past several years and the results of this combined effort are presented here.

The degree of dental diversity observed in extant primates underlines the value of obtaining as correct an estimate as possible of dental variability within and between the genera and species. One thing is patently clear from the present work: many more detailed and systematic investigations of adequate samples of each species (even subspecies) are required before the true nature of the primate dentition becomes known. It is hoped that this reference book will provide the necessary background and stimulation for these studies.

The organization of the book is taxonomic, beginning with the tree shrew and ending with man. A total of 55 genera were studied and described. In addition, a general description of the dentition is presented for each primate family. One point must be made clear: the taxonomic organization of the book is not intended to suggest in any way that the dentition of one group gave rise to that of another group. The organization is solely heuristic, not evolutionary, since there is no temporal aspect to the data presented in the book.

The primate classification used in the book is, for the most part, that presented by Napier and Napier (1967). Certain emendations are included regarding the genetic status of *Nasalis larvatus* and *Simias concolor* as well as

Pygathrix nameus and *Rhinopithecus roxellanae*. In the former situation the genus *Simias* is suppressed and the single species becomes *Nasalis concolor* while in the latter, *Rhinopithecus* becomes a subgenus of *Pygathrix (Rhinopithecus) roxellanae* (Groves, 1970). The primate affinities of the tree shrew are dubious and the accumulating evidence of both paleontology and neontology strongly suggests that these interesting creatures be excluded from the primates (Van Valen, 1965; Hill, 1965; Jane, Campbell and Yashon, 1969). Notwithstanding, tree shrews are included here since, in our opinion, their dental morphology affords a valuable underpinning for understanding the dental anatomy of the living primates.

There is a great deal of original statistical material presented in this book. Several persons have assiduously proofread the data several times, particularly the tables in the Odontometric Appendix, but some mistakes are almost inevitable. I take full responsibility for these and hopefully no major error has found its way into the final publication.

Seattle, Washington
November, 1975

Daris R. Swindler

Acknowledgements

I wish to thank the following museums and staff for the generous use of collections and facilities: American Museum of Natural History, New York; Field Museum of Natural History, Chicago; National Museum of Natural History (Smithsonian Institution), Washington, D.C.; and the Cleveland Museum of Natural History, Cleveland. Also, Drs Neil C. Tappen, Department of Anthropology, University of Wisconsin/Milwaukee and Henry C. McGill, Department of Pathology, Louisiana State University School of Medicine, permitted us to make casts of their private collections of Old World monkeys. I am also grateful to Drs James A. Gavan, Department of Anthropology, University of Missouri and the late Melvin H. Knisely, Professor Emeritus, Department of Anatomy, Medical University of South Carolina College of Medicine for encouraging and supporting some of my initial studies of primate dentitions.

The following individuals have assisted us in so many different ways over the past several years that it becomes difficult to thank each one separately for their many contributions to the final publication: Dr Ann McCoy Beck, Dr Frank J. Orlosky, Dr Robert Simmons, Dr Joyce E. Sirianni, Dr Arno W. Weiss, Jr and Ms Susan Frost. I must single out the many individual contributions by Drs Orlosky and Sirianni over the past several years. Their interest in primate odontology has provided a stimulating background from which I have freely drawn. In addition, I benefitted greatly from many conversations with Mr Kenneth Byrd, Mr Lewis Tarrant and Ms Jane Robinson, all graduate students at the University of Washington. I also wish to thank Mr David Gantt, Washington University, St. Louis and Ms Carla Culver for their assistance in the odontometric portions of the project.

I wish to extend my sincere thanks and appreciation to three excellent artists, Ms Beverly Marshall, Medical Art Student, University of Texas Southwestern Medical School, Dallas; Mr Keith Smith, Laboratory Technician, Regional Primate Research Center, University of Washington and Ms

Phyllis Wood, A.M.I., Medical Illustrator, Regional Primate Research Center and Faculty, Art Department, University of Washington. Their unflagging industry and cooperation is gratefully noted.

Special thanks are due to Dr Glenn Short, Department of Anthropology, Central Washington State College, for making available dental casts of several rare prosimians.

I am grateful to Dr F. Clark Howell, Department of Anthropology, University of California, Berkeley, for lending me an excellent specimen of *Theropithecus*. Also, Mr Gerald Eck, Department of Anthropology, University of Washington, generously permitted the use of his unpublished measurements of the teeth of *Theropithecus*.

Dr Maurice Zingeser, Oregon Regional Primate Research Center, Beaverton, Oregon gave me the dental casts of *Brachyteles* and I wish to acknowledge his generosity at this time. Also, Dr D. Carl Johanson, Department of Anthropology, Case Western Reserve University sent me dental casts of *Pan paniscus* as well as original data for which I am most grateful.

The assistance of Ms Roberta Swindler with typing, lead-lining and labeling is gratefully noted.

I wish to express my appreciation to Ms Kathleen Schmitt, Editor at the Regional Primate Research Center, for proofreading the manuscript. She made many helpful suggestions and corrections.

Finally, I wish to acknowledge the support of the National Institutes of Health from 1962 to 1968. Without their grant, DE-02955, this book would not have been written. NIH grant RR-00166 also supported portions of the project during 1974.

Contents

1

Materials and Methods

The basic data presented in this book were taken from plaster dental casts which were made from alginate impressions. Both impressions and casts were made of the permanent teeth of primate skulls residing in the following museums: American Museum of Natural History, The National Museum of Natural History (Smithsonian Institution), Chicago Field Museum and the Cleveland Museum of Natural History. Additional casts were procured from the collections of Drs Neil C. Tappen and Henry C. McGill.

The casting technique is relatively simple and requires little time, yet provides permanent material for detailed study in the laboratory. All casts were poured within five to ten minutes after the impressions were made, thus minimizing the possibility of any dimensional change (Skinner, 1954). Two separate studies have demonstrated that measurements taken on dental casts using modern dental materials are directly comparable to measurements of the original teeth (Warrer, 1952; Swindler, Gavan and Turner, 1963). The observed differences are more likely due to instrumentation than to dimensional change resulting from the dental materials. All casts were made by my assistants and myself.

The original collections were made by collectors from the respective museums, and specimens came from many different geographic areas of the world. In the majority of cases, species are represented from a wide range within their normal geographic range, although in certain groups, e.g. *Papio cynocephalus*, the animals were collected from a more limited area and may well approximate an interbreeding population. And, of course, we used *Macaca mulatta* specimens from Cayo Santiago, Puerto Rico, which certainly represent animals with more restricted breeding opportunities. Unfortunately, several species were represented by only a few specimens, or in one or two cases, by a single specimen. This usually meant that these species were rare in museum collections and because of money and time, we were unable to increase the sample. Also, the manner in which specimens were collected influenced the randomness of a sample and anyone who has ever used museum collections is quick to realize this fact. There are obviously other biases in such a collection of specimens (2,000) as studied in this book. However, since the principal objective of this work was to describe the normal dentition and present a statement of the range and magnitude of dental variability obtaining within the major genera of extant primates, we believe the influence of such unavoidable biases to be slight.

The sex of the animals was determined in the field at the time of collection and any specimen of doubtful sex was excluded from the study. In the analytical descriptions of each section of the book the sexes were pooled unless otherwise mentioned.

The primitive mammalian dental formula was $I^3_3\ C^1_1\ P^4_4\ M^3_3$. The majority of early primates had lost one incisor, and by the Eocene premolar reduction had begun. The first incisor is usually considered the missing member of the group and premolar loss also occurred mesiodistally, i.e. P^1_1 was lost first, followed by P^2_2. As we shall see, extant primates have lost the first premolar and some have lost the first two premolars.

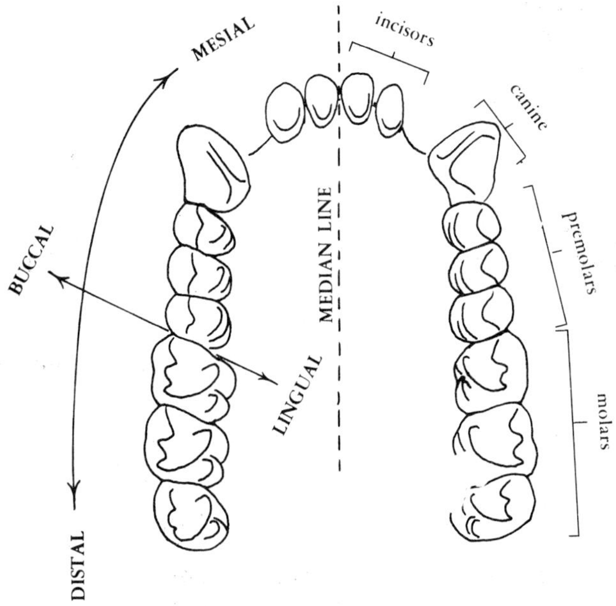

Fig. I. Maxillary dental arch. Terms of position within the oral cavity.

The incisors and canines are known as anterior teeth; premolars and molars are called posterior teeth (Fig. I). The tooth surfaces facing toward the cheek are called buccal surfaces (some students distinguish between buccal and labial tooth surfaces, thus labial is limited to the incisor and canine surfaces facing the lips). All surfaces facing the tongue are referred to as lingual. The mesial (anterior) surface of a tooth faces toward the front of the oral cavity while those most distant are called the distal (posterior)

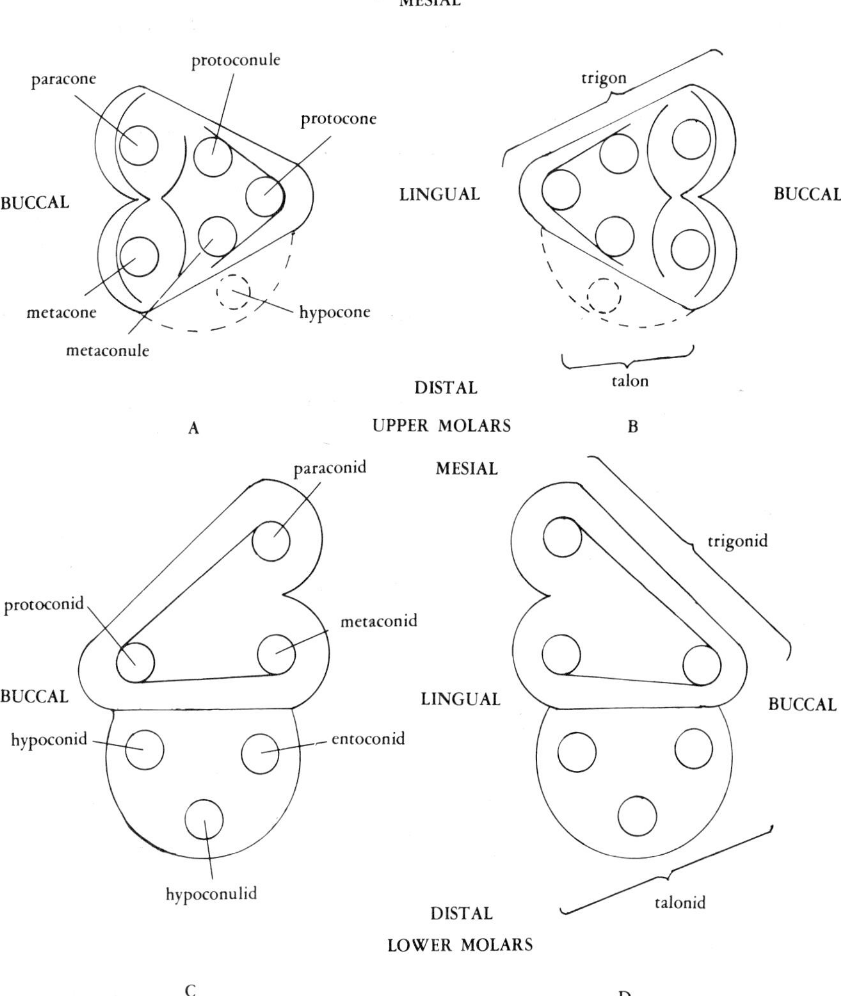

Fig. II. Cope–Osborn mammalian molar nomenclature. (After Day, 1965.)

surfaces. A cusp or cusplet is defined as a tooth structure having structural or functional occlusal areal components delimited by developmental grooves and having independent apexes (Hornbeck and Swindler, 1967).

The Cope–Osborn nomenclature of mammalian molar crown morphology is used to designate the molar cusps of the premolars and molars (Fig. II). This terminology has been entrenched in the world literature for nearly a century and its weaknesses and strengths are recognized and understood by (almost) all practicing odontologists. To change it at this late date seems of somewhat doubtful value, but one should read Vandebroek (1961) and Hershkovitz (1971) for opposite views. We have, for the most part, followed the nomenclature suggested by Van Valen (1966) and Szalay (1969) for naming the different crests connecting the cusps and cuspules of the crown surface of the premolars and molars (Figs III and IV).

Since there is some confusion regarding terminology in the odontological literature we present the terms used here as well as a list of the more common synonymies. The suffix "id" is added to terms to designate the lower dentition from the upper. In many cases this permits the same terms to

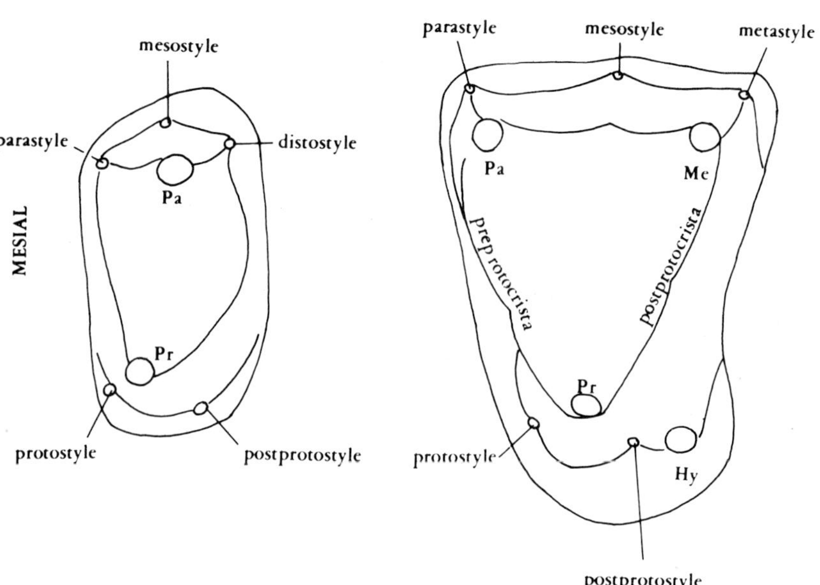

Fig. III. Cusp and style nomenclature, maxillary, premolar and molar. Pa = paracone; Pr = protocone; Me = metacone; Hy = hypocone. (Modified from Kinzey, 1973.)

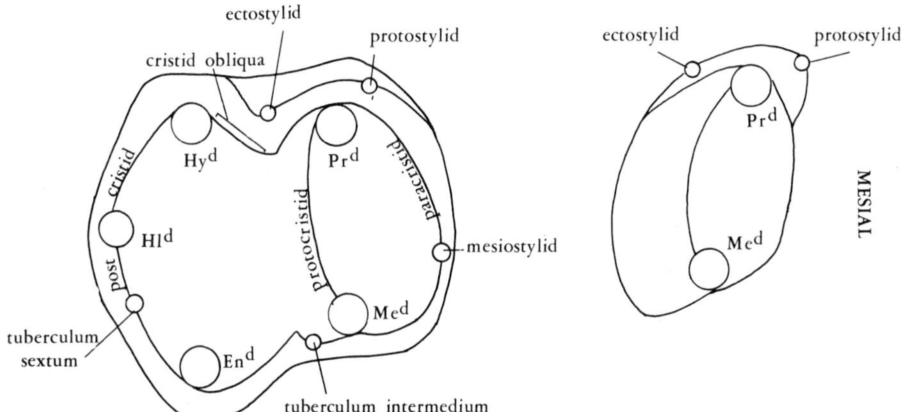

Fig. IV. Cusp and style nomenclature, mandibular premolar and molar. Pr^d = protoconid; Me^d = metaconid; Hy^d = hypoconid; En^d = entoconid; Hl^d = hypoconulid. (Modified from Kinzey, 1973.)

be used in both the upper and lower dentitions and yet remain easily distinguishable.

Tooth Nomenclature

This book	Synonymy
Upper teeth	
paracone (O)	eocone (Vb)
protocone (O)	epicone (Vb)
metacone (O)	
metaconule (O)	plagioconule (Vb)
protoconule (O)	paraconule (VV)
distoconulus (R)	postentoconule (H)
parastyle (O)	mesiostyle (Vb)
mesostyle (O)	ectostyle-1 (H)
metastyle (O)	distostyle (Vb)
distostyle (P^{2-4} K)	
protostyle (O)	Carabelli cusp
postprotostyle (K)	interconule (R)
preprotocrista (VV)	protoloph (O)
postprotocrista (VV)	crista obliqua (R)
entocrista (H)	
premetacrista (S)	
postmetacrista (S)	
trigon basin (S)	protofossa (VV)

Tooth Nomenclature—*cont.*

This book	Synonymy
Lower teeth	
paraconid (O)	eoconid (Vb)
protoconid (O)	epiconid (Vb)
metaconid (O)	
entoconid (O)	
hypoconid (O)	distostylid (Vb)
hypoconulid (O)	parastylid
mesiostylid (Vb)	
ectostylid (K)	
protostylid (K)	postmetaconulid (H)
tuberculum intermedium (R)	postentoconulid (H)
tuberculum sextum (R)	metalophid (O), protolophid (VV)
protocristid (S)	premetacristid (H)
cristid obliqua (S)	
postentocristid (H)	paralophid (VV)
paracristid (S)	
postmetacristid (S)	prefossid (VV)
trigonid basin (S)	postfossid (VV)
talonid basin (S)	

(H) = Hershkovitz, 1971; (K) = Kinzey, 1973; (O) = Osborn, 1907; (R) = Remane, 1960; (S) = Szalay, 1969; (Vb) = Vandebroek, 1961; (VV) = Van Valen, 1966.

The fundamental structure of a tribosphenic molar is represented in Fig. IIA–D. The trigon (Fig. IIA,B) is a three-cusped triangle formed by the paracone, metacone and protocone. With the distal expansion of the cingulum a heel or talon is added which commonly presents a new cusp, the hypocone. The trigonid (Fig. IIC,D) is also triangular and is made up of the paraconid, metaconid and protoconid. In the majority of extant primates (*Tarsius* excepted) the paraconid is absent. The talonid develops on the distal aspect of the trigonid and frequently bears three cusps, the hypoconid, entoconid and hypoconulid. The molar of all extant primates is derived from the basic tribosphenic pattern depicted here, and for a lucid, readable presentation of the basic transformations leading to the primate molar the student is referred to Le Gros Clark (1971). Of course, the most comprehensive work on the evolution of primate teeth still remains *The origin and evolution of the human dentition* (1922) by W. K. Gregory.

The principal cusps are connected by crests, which occasionally display minor cusps usually designated by the suffix "conule (conulid)", e.g. metaconule or hypconulid. A cingulum (cingulid) girdled the primitive tooth and portions of it may still be present on the teeth of extant primates. When present these structures are noted by the suffix "style (stylid)" and are named for the related cusp, e.g. protostyle or protostylid.

The tooth measurements were taken with a Helios caliper. The arms were ground to fine points for greater accuracy. Mesiodistal and buccolingual dimensions of maxillary and mandibular teeth were taken. In all odontometric calculations presented in this book, the sample size (n) refers to the number of animals measured. We believe this procedure is more realistic than presenting the number of teeth measured since it is well known that there are very few significant differences between the dimensions of right and left teeth. The right side is presented here; however, if a tooth was badly worn or absent its antimere was used. Also, teeth exhibiting noticeable wear were excluded. It should also be mentioned that in many cases the n presented in the Odontometric Appendix differs from the number of animals studied in the morphological section for a given species. This is due to the fact that in many cases the teeth could be examined for a particular trait, yet were too worn for odontometry, or vice versa.

Repeated measurements taken on the same teeth revealed an average difference between measurements of 0·2 mm. The teeth were measured by the author and research assistants, each of whom were trained by the author.

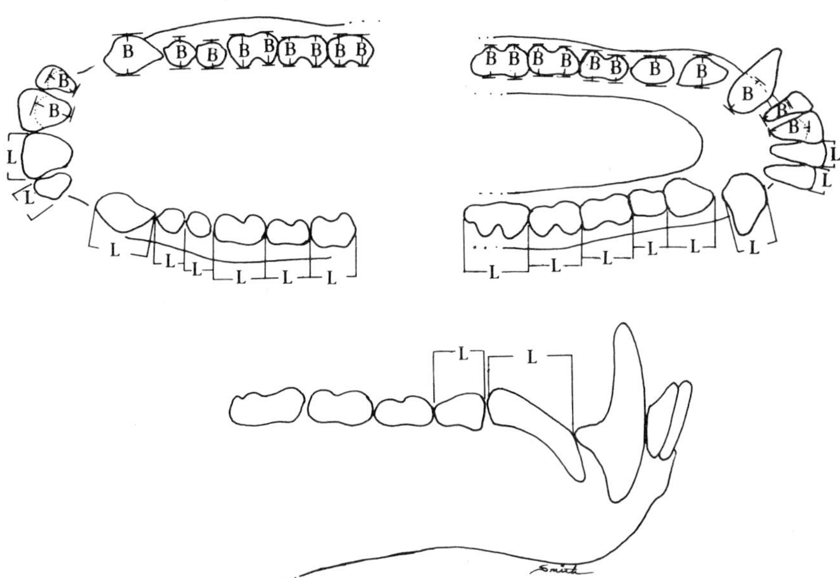

Fig. V. Odontometric landmarks. B = breadth; L = length.

The measurements appear in Fig. V and are defined as follows:

Incisors

Length. Mesiodistal diameter taken at the incisal edge of I^{1-2}_{1-2}.

Breadth. Buccolingual diameter taken at the cementoenamel junction at a right angle to the mesiodistal diameter.

Canines

Length. C^1, maximum diameter from the mesial surface to the distolingual border. C_1, mesiodistal diameter measured at the level of the mesial alveolar margin.

Breadth. Buccolingual diameter taken at the cementoenamel junction at a right angle to the mesiodistal diameter.

Premolars

Length. Maximum mesiodistal diameter taken between the contact points. If mesial contact is lacking on P^3_3 due to a diastema between it and the canine, the maximum horizontal distance is measured from the distal contact point to the most maximum point mesially.

Breadth. Maximum buccolingual diameter taken at a right angle to the mesiodistal diameter.

Molars

Length. Maximum mesiodistal diameter on the occlusal surface between the mesial and distal contact points.

Breadth. Maximum buccolingual diameter measured at a right angle to the mesiodistal dimension. The breadths of both the trigone (trigonid) and talon (talonid) were taken in this manner.

Statistical calculations for means, standard deviations (s.d.) and standard error (s.e.) of means were performed for the dental measurements of each species by sex. Hypotheses of equality of means between sexes of each species, where the samples were large enough, were tested using the appropriate small sample t-test statistic (Sokal and Rohlf, 1969). The results of the t-tests for sexual dimorphism are presented for each species in the Odontometric Appendix.

2

Family Tupaiidae

Present Distribution and Habitat

Tree shrews range throughout South-East Asia: from eastern India on the west, to Mindanao in the Philippine Archipelago on the east; from southern China on the north, southward to Java and the small islands dotting the coast off south-western Sumatra. Of the currently recognized genera, none exist throughout the entire range. The genus *Tupaia* has the widest distribution. Tree shrews are found in tropical rain forests as well as montane forests, and the genus *Dendrogale* lives in the moss-covered mountains of north-east Borneo from 915 to 3,100 m (3,000–11,000 ft) (Napier and Napier, 1967).

The majority of tupaiid species appear to live in both arboreal and terrestrial niches, the exceptions being *Ptilocercus*, *Tupaia minor* and *Tupaia nicobarica* (arboreal) and *Tupaia* (*Lyongale*) *tana* and *Urogale* (terrestrial) (George, 1973). It will take much more field work to achieve an understanding of the true pattern of tupaiid ecological stratification.

Dietary Habits

The tupaiid diet of insects and fruit may be considered a primitive marsupio-placental feature (Steuerwald, 1969). In addition, many species supplement this basic diet with earthworms, lizards and birds' eggs. In contrast to other species, *Tupaia tana* has been observed using the claws of its forefeet to dig and scratch into the ground for earthworms and other tidbits (Davis, 1962).

General Dental Information

Permanent dentition: $I_3^2\ C_1^1\ P_3^3\ M_3^3$

Deciduous dentition: $i_3^2\ c_1^1\ m_3^3$

Sequence of eruption of permanent teeth, *Tupaia glis* (Lyon, 1913):

	M^1		M^2 M^3 P^4	P^2		P^3		C	I^2	I^1
M_1		M_2	M_3 I_3 I_1 P_4		I_2	P_2	P_3	C		

The permanent dental formula of living tree shrews has been reduced by three teeth from the generalized Eutherian dentition of $\frac{3\text{–}1\text{–}4\text{–}3}{3\text{–}1\text{–}4\text{–}3}$, by eliminating one upper incisor and P^1_1. It should be noted that the possession of two upper and two lower incisors is frequently mentioned as a characteristic feature of the dentition of living primates. The eruption of *all* permanent molars prior to the elimination of the deciduous teeth is found in only one primate species, *Aotus* the small South American night monkey.

The three lower incisors are usually well developed and functional, although the third one is small and slightly distal to the second incisor. The first and second incisors are set close together and are more procumbent than the third, giving them a lemurine appearance. These teeth are used for combing the fur in a manner similar to that observed in lemurs. The major anatomical difference between tree shrew and lemur anterior teeth is that the lower canines of the former are not aligned with the incisors but are set apart from these teeth and are only slightly procumbent. The genus *Anathana* has small, incisiform and procumbent lower canines which are not functionally part of the dental comb. In *Ptilocercus* the upper canine is double-rooted and premolariform in shape. Two-rooted upper canines are occasionally found in other tupaiid genera, but they are not premolariform (Lyon, 1913).

The premolars in both jaws increase in size and complexity mesiodistally. Both second premolars are very small and practically functionless, whereas the fourth upper and lower premolars are molariform in most living genera. In addition, P^4 has three roots as do the molars.

The upper molars are distinguishable between the two subfamilies by the presence or absence of the mesostyle. The mesostyle is found in the Tupaiinae but not in the Ptilocerinae. In both groups the upper molars display the "W" or dilambdodont pattern when viewed from the occlusal surface. That is, the paracone and metacone are V-shaped and slightly separated from each other near the middle of the tooth (Plate 1). This configuration is better developed on the first and second molars. Also present on these two teeth is the hypocone which is rarely observed on the third molar. The upper molars of *Ptilocercus* possess a complete cingulum which is absent in *Tupaia*, who only have a lingual cingulum (Lyon, 1913). The lower molars have paraconids and thereby display the primitive trigonid arrangement of paraconid, protoconid and metaconid on the mesial moiety

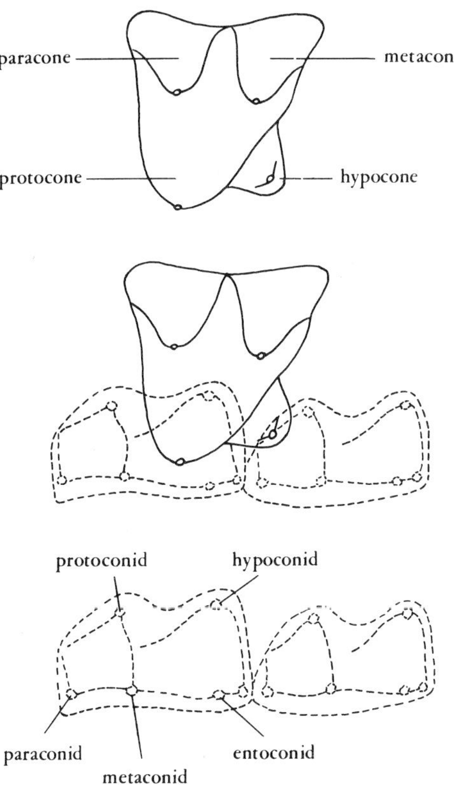

Fig. VI. Basic molar relations in *Tupaia.* (Modified from Mills, 1955.)

of each molar. Distally, the talonid has a hypoconid and entoconid on the first two molars while the third has a distinct hypoconulid. *Ptilocercus* has a well defined buccal cingulum on the lower molars which is absent in the Tupaiinae.

According to the recent work of Hiiemae and Kay (1973), during mastication the teeth of tree shrews are used in the following manner. The incisors do little if any biting, rather they are used for grasping and holding the food. The premolars and molars are primarily responsible for trituriting the food, or what is referred to as "ingestion by mastication" by Hiiemae and Crompton (1971). Thus, in occlusion the trigons of the upper molars fit between the trigonids of the lower molars and occlude with the talonid. Such an arrangement permits the protocone to contact the center of the talonid basin while the hypoconid nestles into the center of the trigon, between the paracone and metacone (Fig. VI). This is a basic mammalian dental pattern of normal occlusion which is present throughout the primates and, in some ways, it may be a more reliable criterion of molar normality than Angle's classification (Mills, 1955). It is an excellent functional alignment for puncturing, crushing and chewing food. Finally, it should be noted that the tree shrew enjoys a rather wide mediolateral excursion of the lower jaw near the end of the preparatory stroke during mastication (Hiiemae and Kay, 1973). This is a much different interpretation of the dynamics of tree shrew mastication than that of many earlier students who believed it to be strictly orthal.

Genus *Tupaia*

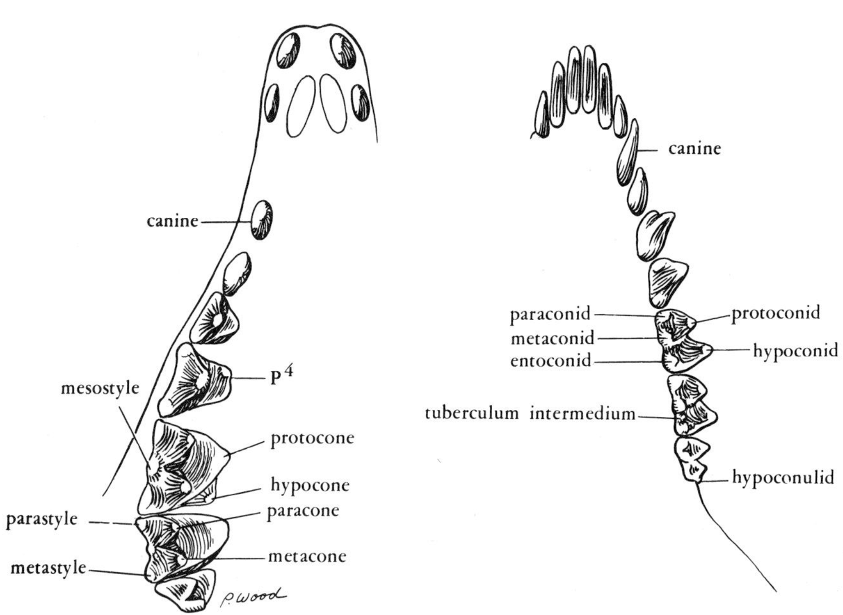

Plate 1. *Tupaia glis* female, occlusal view ×3.

Morphological Observations

	Sample Male	Sample Female
Tupaia glis (common tree shrew)	26	21
Tupaia javanica (small tree shrew)	13	17
Tupaia minor (pygmy tree shrew)	8	6
Tupaia tana (terrestrial tree shrew)	5	10

Incisors

Upper. The upper incisors are very similar in the specimens studied. They are caniniform and I^1 is larger than I^2. The difference is usually quite marked. In the genus *Urogale*, however, I^2 is much larger than I^1 and according to Lyon (1913) they function as canines. There are no discernible cusplets or any suggestion of lingual cingula in *Tupaia*. In *Ptilocercus*, however, I^2 is somewhat premolariform and possesses a distinct distal cuspule (Lyon, 1913; Le Gros Clark, 1971). There is a pronounced diastema between the central incisors as well as a large space between them and the second incisors.

Lower. There are three lower incisors, of which the third is much smaller than the others, while the second is somewhat larger than the first. Indeed, the third is frequently much reduced and may be functionless or even absent if the lower canine is large, as it often is in *Urogale* (Lyon, 1913). The lower incisors are procumbent, especially I_{1-2} which form a dental comb much as in lemurs with the notable exception that the lower canines are not functionally involved.

Canines

Upper. The upper canine is of moderate size in all living tree shrews and it projects above the occlusal level of the premolars. It displays a single cusp with no additional ornamentations. It may possess two roots as in *Ptilocercus.*

Lower. The lower canine is larger than the upper canine and always projects well above the occlusal line of the adjacent incisors and canines. In several taxa it is quite large when the upper central incisor is hypertrophied and the lower third incisor is diminutive, as mentioned above. According to the same author, the lower canine is larger in *Tupaia nicobarica* (Lyon, 1913). It is interesting to note that of living tree shrews, *Tupaia nicobarica* is usually considered to be one of the most arboreal (George, 1973).

Premolars

Upper. The second upper premolar is small and practically functionless, possessing a single, long paracone. The premolars grade in size and complexity mesiodistally culminating with the molarization of P^4. The third premolar is only slightly larger than the second and often possesses small but distinct buccal and lingual cingula. On rare occasions, 7–8% of *Tupaia glis*, there are incipient protocones present on the lingual cingulum and the percentages are similar in both sexes. According to Steele (1973) a protocone on P^3 is typical only in *Dendrogale*. Protocones are absent in the other species studied. The fourth premolar has two buccal and one lingual cusp as well as distinct buccal and lingual cingula.

Both P^{3-4} possess para- and distostyles along their buccal surfaces. The styles are better developed and more frequent on P^4. Mesostyles are not found on P^3, and on P^4 they are present only in *Tupaia javanica* and *Tupaia glis* (less than 1%). Of the three styles, the parastyle is always large on P^{3-4}. The para- and distostyles are slightly more common on P^4. There is no discernible sexual dimorphism in the frequency of the styles between P^{3-4}, or among the taxa studied.

Lower. The second lower premolar, having a single cusp, is the smallest of the series. The third premolar is larger and may have two cusps and a small talonid. A paraconid may be present in some species (*Tupaia gracilis, Tupaia splendidula* and *Urogale everetti*) Steele (1973). A slight lingual cingulum is present on P_3 in all species examined; however, buccal cingula are found only in *Tupaia javanica* and *Tupaia glis.* P_4 frequently has three cusps (paraconid, protoconid and metaconid) plus a small talonid. The paraconid has disappeared in most living primates and this is usually attributed to the increased complexity of the crown. It is consistently present in *Tarsius* and has also been reported as a vestigial element in both New and Old World monkeys, especially in the mesial premolars and more often in the deciduous than in the permanent teeth (Hershkovitz, 1971; Orlosky, 1973). P_4 closely resembles the single lower premolar of *Erinaceus* (European Hedgehog), a similarity first noted by Mivart (1867). As on P_3 the lingual cingulum is present in all species on P_4 while the buccal cingulum is only sporadically represented in the taxa examined, i.e. it is absent in *Tupaia tana* and *Tupaia minor* and present in 1% of *Tupaia javanica* and 8% of *Tupaia glis.*

Molars

Upper. The first and second upper molars are typically dilambdodont in all living tree shrews, and in addition, each possess well marked para-, meta- and mesostyles along their buccal cingula. The mesostyles are frequently bifid, a condition also noted by Steele (1973).The hypocones on these two teeth are small and positioned low on the crown near the neck of the tooth. The hypocones on M^2 are extremely diminutive but are always present in the species sampled. At the same time, the mesostyles are largest on M^2. Buccal and lingual cingula are present on M^{1-2}. The third molar is quite small, lacks the hypocone and if the mesostyle is present it is much reduced as are the cingula. Thus, in occlusal outline, M^3 is triangular.

These three molars display certain differences among the various genera of which the following are the most notable. The development of hypocones on M^{1-2} of *Anathana* is larger than in any other member of the family though nearly equalled in *Urogale* and *Ptilocercus* while they are absent in *Tupaia minor, Tupaia gracilis* and *Dendrogale* (Lyon, 1913; Steele, 1973). In *Ptilocercus* the cusps are blunter and more rounded than in the Tupaiinae and they lack a mesostyle (Lyon, 1913). According to the same author, all three upper molars of *Ptilocercus* are surrounded by a distinct cingulum.

Lower. All three lower molars have maintained the basic tribosphenic pattern consisting of an elevated trigonid and a low talonid. The trigonid consists of the paraconid, metaconid and protoconid while the talonid has the entoconid and hypoconid on M_{1-2} and an additional cusp, the hypoconulid, on M_3. A buccal cingulum is lacking in the species sampled but this structure is found in *Ptilocercus* (Lyon, 1913).

A small but distinct tuberculum intermedium (entoconulid) may be seen arising from the entoconid on M_{1-3} in *Tupaia glis.* This is the only species displaying the trait in the present investigation. The molar frequencies for the trait are: M_1 8%; M_2 10%; and M_3 6% and are similar for both sexes. This feature is not mentioned by Lyon (1913); however, Mivart (1867) noted a division as "sometimes" occurring in the posterointernal cusp (entoconid) of *Tupaia.* Steele (1973) also mentions its presence in *Tupaia nicobarica.*

Odontometry (Tables 1–16, Appendix)

The anterior teeth were not measured because we felt it was impossible to obtain accurate measurements. In addition to being small and difficult to measure, they were frequently broken or absent.

T-tests were calculated to determine whether significant sexual size differences were present within each of the four tupaiid species studied. There are no significant sex differences in tooth size of *Tupaia glis* or *Tupaia minor.* In *Tupaia javanica* the mesiodistal dimension of P_4 is significant ($P<0{\cdot}05$), while in *Tupaia tana* the same diameter is significant in M_2 ($P<0{\cdot}05$). In both cases the male has the largest teeth. It is apparent from these findings that there are few significant size differences between the sexes in these four species. This is true irrespective of mode of locomotion since these species span the locomotor categories outlined by George (1973).

Of the mandibular and maxillary molars, the first has the greatest mesiodistal diameter in both males and females and in all species with few exceptions. The third molars are always the shortest. The usual pattern in

Tupaia is $M_1^1 > M_2^2 > M_3^3$. In general, the upper second molar is wider than the other molars although there is somewhat more variability in this diameter than in the length dimension. The relations between the breadths of the trigonid and talonid are also quite variable (see Steele, 1973). Thus, in M_1 the talonid is equal to, or slightly wider than, the trigonid in *Tupaia glis* and *Tupaia javanica*, whereas in the other two species there is no discernible pattern. In M_2 the dimensions are equal, or the trigonid is slightly wider, in *Tupaia tana* and narrower in *Tupaia minor.* The maxillary first molar generally displays equal breadth dimensions while the second is always broader across its mesial moiety. Thus, molar size relations demonstrate a great amount of sequential variability in *Tupaia.*

3

Family Lemuridae

Present Distribution and Habitat

Lemurs inhabit the island of Madagascar and are also found on a few islands in the Comoro chain. In earlier geological periods they were more widely distributed, occurring in Europe and North America during the Eocene, disappearing from these continents before the Oligocene. When they arrived on the third largest island in the world is not exactly known but they have probably been isolated for more than 40 million years (Tattersall, 1973). They most likely arrived by raft from Africa since Madagascar is usually thought to have been adrift from its mother continent since the late Cretaceous. There is some evidence that Madagascar was still part of Africa during the Paleocene, and perhaps did not separate much before the Eocene (Cooke, 1968; Fooden, 1972). If this proves to be correct it means that prosimians *may* have been residents on this piece of real estate prior to its departure from south-eastern Africa. However, current evidence indicates that the earliest African prosimians came from a few East African locations of Miocene age. Until recently there were some 16 genera living on Madagascar; today there are only ten.

The climate of Madagascar is tropical with temperatures ranging between 73 and 88° F. The island is essentially mountainous with most species inhabiting the lower forested areas, although *Indri* is found from sea level to 1,790 m (5,900 ft). Most species are arboreal although *Lemur catta* is well adjusted to terrestrial foraging (Jolly, 1966). Locomotor adaptations are varied, ranging from quadrupedalism to vertical clinging and leaping (Napier and Walker, 1967). The lemurids display a complete spectrum of activity rhythm from nocturnal to diurnal. Of the two subfamilies, the Cheirogaleinae are virtually nocturnal while the Lemurinae are crepuscular, with one exception: *Lepilemur* is nocturnal (Petter, 1962).

Dietary Habits

The prosimian diet is quite varied: in some species it is considered specialized, while in others the animals appear to eat a rather wide range of food. *Lemur* and *Hapalemur* are vegetarians preferring leaves and flowers to insects or small animals, while the mouse lemur (*Microcebus*) shows a distinct taste for insects and honey (Petter, 1962). *Phaner* and *Cheirogaleus* are predominantly frugivorous. *Lepilemur* is more specialized, possessing a diet of some fruit, but mainly leaves, buds and barks; indeed, in captivity they require some bark to keep healthy (Petter, 1962).

General Dental Information

Permanent dentition: $I^2_2\ C^1_1\ P^3_3\ M^3_3$

Deciduous dentition: $i^2_2\ c^1_1\ m^3_3$

Sequence of initial eruption of permanent teeth (Schwartz, 1974a):

*Lemur mongoz**

M^1		I^2	I^3	C		M^2	P^2	M^3	P^4		P^3	
M_1	I_2	I_3	C		M_2		P_2	M_3		P_4		P_3

Lemur fulvus

M^1			I^2	I^3	M^2	C	P^2	P^4	M^3	P^3
M_1	I_2	I_3	C		M_2	P_2		P_4	M_3	P_3

Lemur macaco

M^1			I^2	I^3	M^2	C	P^2	?	M^3	P^4	P^3
M_1	I_2	I_3	C		M_2		P_2	?	M_3	P_4	P_3

Lemur catta

M^1	I^2	I^3		M^2	C	M^3	P^4	P^3	P^2
M_1	I_2	I_3	C	M_2		M_3	P_4	P_3	P_2

Lemur variegatus

M^1	I^3		I^2	M^2	?	C		P^2	P^4	M^3	P^3
M_1	I_2	I_3	C	M_2		?	P_2		P_4	M_3	P_3

Hapalemur griseus

M^1				I^2	I^3	M^2	P^4	M^3	P^3	C	P^2
M_1	I_2	I_3	C		M_2		P_4	M_3	P_3		P_2

* Schwartz (1974a,b) uses I^2_2 and I^3_3 instead of I^1_1 and I^2_2 to designate the central and lateral incisors since he believes the original mesial incisor (I^1_1) was lost phylogenetically.

Lepilemur mustelinus

M^1	M^2				M^3	P^4	C	P^3	P^2
M_1	M_2	I_2	I_3	C	M_3	P_4		P_3	P_2

Cheirogaleus major

(M^1	I^2	I^3		M^2)	P^2		C	M^3	P^4	P^3
(M_1	I_2	I_3	C	M_2)		P_2		M_3	P_4	P_3

Microcebus murinus

(M^1	I^2	I^3)	M^2		C	P^2	M^3	P^4	P^3
M_1	M_2	I_2	I_3	C	P_2	M_3		P_4	P_3

Phaner furcifer

M^1				I^2	M^2	?	I^3	C	P^2	P^4	M^3	P^3
M_1	I_2	I_3	C		M_2				P_2	P_4	M_3	P_3

The permanent dental formula listed above obtains in all living genera of Lemuridae with one exception. In *Lepilemur* the two maxillary incisors have disappeared completely, giving a formula of $\frac{0\ \ 1\ \ 3\ \ 3}{2\ \ 1\ \ 3\ \ 3}$, but they are present in the deciduous dentition. There are three premolars in each jaw, the first one having been lost in all lemurs after the Eocene. The second permanent molar in *Lepilemur* erupts before the incisors—a primitive condition. Also, it is worth noting that the sequence P4, P3, P2 is found only in *Lepilemur, Hapalemur* and *Lemur catta,* in fact, these three groups are the only prosimian taxa displaying this pattern (Schwartz, 1974a).

The modern genera of lemurs, with the exception of the aye-aye, display a curious specialization of the incisors and lower canines. The upper incisors are small, or absent as noted in *Lepilemur.* The two central incisors are separated by a wide diastema while being set close to the lateral ones. Both incisors are compressed buccolingually, a condition first noted in *Notharctus* (Gregory, 1922). The mandibular incisors are long, narrow teeth projecting nearly straight forward from the anterior portion of the jaw. The incisiform canines are in line with the incisors and together these six procumbent teeth form the dental comb (Plate 2). The term dental comb may be a misnomer; however, the fact that the apparatus is used in grooming is indisputable (Buettner-Janusch and Andrew, 1962). It should be mentioned, however that a recent article by Martin (1975) shows that these lower anterior teeth are used by some of the lemurs (*Phaner, Microcebus* and *Euoticus*) to scoop fresh gums and resins from trees. Indeed, he believes that the use of the tooth scraper for obtaining gum was an important factor in the development of these procumbent teeth and that using them for grooming could have been developed as a secondary characteristic. This specialization is found in all recent lemurs and may date to the Early Miocene genus *Progalago* (Le Gros Clark, 1971). Fossil lemurs prior to this period possessed vertically implanted lower incisors and the canines were not incisiform.

The upper canines are long, trenchant teeth with distally curving crowns. There is a slight diastema between them and the second premolar.

The premolars increase in size and complexity from front to back. In both jaws P^2_2 is caniniform, much more so in the upper jaw than the lower. In *Phaner*, P^2 is hypertrophied and is more caniniform than in the other genera. During mastication they function somewhat as canines, piercing and tearing the food, and in occlusion, P_2 glides against the distal surface of the upper canine. The third premolars are less caniniform, but each is still essentially a unicuspid tooth. The fourth premolars are broadened buccolingually, particularly the upper one. Indeed, the latter tooth displays a well developed protocone connected to a high paracone by a concave developmental crest in most genera. In *Hapalemur* and *Cheirogaleus* P^4 is molariform, and in the former genus, P_4 also resembles the molars, a unique condition among the extant members of the family.

The upper molars grade from tritubercular to quadritubercular in shape. The lingual cingulum is usually prominent and in some species a protostyle is developed. Likewise, the hypocone is variably exhibited among living lemurs, and Schwarz (1931) was able to identify different species of the genus *Lemur* expressing these two traits in different combinations. The evolutionary history of the hypocone among primates is of great taxonomic importance and the student should see Gregory (1922). Briefly, the hypocone is classified either as a true or pseudohypocone depending on whether it develops from the cingulum or the protocone respectively. Among the Northarctinae it appears as a pseudohypocone, whereas among the Adapinae it appears as a development from the cingulum, a true hypocone. Thus, members of the former subfamily of Eocene prosimians are considered aberrant, having no phylogenetic connections with modern lemurs. However, more recently Simpson (1955), Butler (1956) and Hershkovitz (1971) have rejected this distinction and the hypocone is considered to be a derivative of the lingual cingulum independent of the protocone. This seems to be the most reasonable explanation for the development of the hypocone.

The lower molars possess a trigonid and talonid. The paraconid is absent in all genera, while the hypoconulid is variable. The protoconid and metaconid are set close together and connected by an oblique transverse crest, the protocristid. Indeed, James (1960) describes the lower molars of *Hapalemur* as bilophodont, although I do not agree (see p. 28 for further discussion). In most cases the protoconid is mesial to the metaconid.

Occasionally a cingulum is found along the buccal surface of the molars, it is especially well developed in *Lepilemur.*

In summary, the dentition of the Lemuridae displays the following features: (a) procumbent lower incisors, (b) incisiform lower canines incorporated into the incisor series to form the "dental comb", (c) caniniform anterior premolars, (d) wide separation between upper central incisors and (e) molarization of P^4.

As noted in tree shrews, the incisors are of little or no use in biting, but are mainly employed in grasping and restraining food. The animals either take food in their hands and bite off pieces of it, or take food directly into their mouths and quickly shove it back to the premolar and molar teeth (Buettner-Janusch and Andrew, 1962). The mesial end of the oral cavity is mainly occupied with grooming. Thus, the non-nutritive functions of the anterior teeth among prosimians have probably played a more important role in their evolutionary history than have the food grasping and holding ones; however, see Martin, 1975. The tooth rows of lemurs are somewhat V-shaped and present an uneven occlusal plane. In most forms, the trigonid is slightly higher than the talonid. Also, the protostyle and hypocone probably function as a sort of "chopping block" for the sharp cutting edges of the trigonid of the mandibular teeth (Mills, 1973).

The actual chewing occurs in the premolar and molar area and this is true for soft as well as hard food (Buettner-Janusch and Andrew, 1962; Hiiemae and Kay, 1973). The primate molar is particularly well adapted for shearing, crushing and grinding. These various aspects of mastication are considered fully in Kay and Hiiemae (1974).

Genus *Lemur*

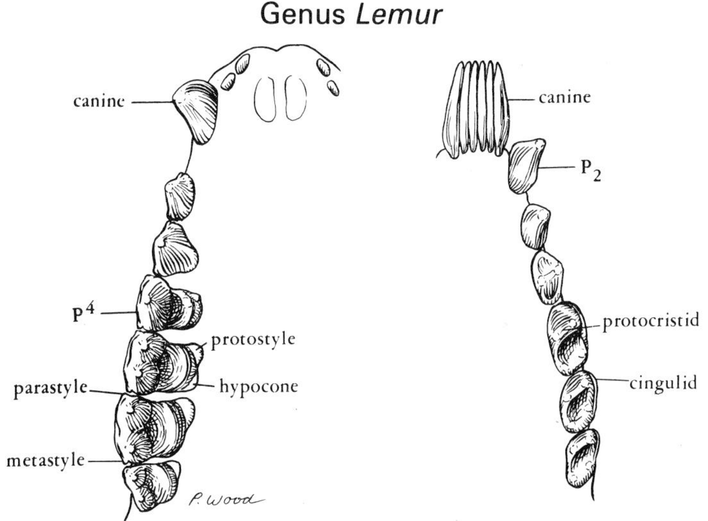

Plate 2. *Lemur macaco* male, occlusal view ×1·5.

Morphological Observations

	Sample	
	Male	Female
Lemur macaco (black lemur)	7	8
Lemur rubriventer (red-bellied lemur)	1	3
Lemur variegatus (ruffed lemur)	7	1
Lemur mongoz (mongoose lemur)	1	2
Lemur catta (ringtailed lemur)	1	2

Incisors

Upper. The upper incisors are very similar in the species studied. They are compressed buccolingually and show little if any differentiation. The central ones are separated by a wide, median diastema while being closely aligned with the laterals.

Lower. The four lower incisors are narrow, long teeth set closely together. They are procumbent and with the canines show the characteristic lemuriform arrangement.

Canines

Upper. The upper canine is a long, blade-like tooth with a wide base. A slight median lingual ridge is present mesial to which is a lingual groove. The canine is separated from P^2 by a small diastema.

Lower. The lower canine is procumbent and larger than the incisors. It presents the curved outer border which is characteristic of lemurids.

Premolars

Upper. The second upper premolar is caniniform and is the smallest of the three. It possesses a slight but complete lingual cingulum which displays small mesial and distal enlargements. There is also a slight median lingual ridge passing from the base to the tip of the crown. P^3 is unicuspid. A median lingual ridge extends from the crown tip of the tooth to the enlarged lingual base. From here, a distal lingual cingulum is visible passing to the hinder part of the tooth. The fourth premolar has two cusps, a large paracone and a smaller, lower protocone. The apices of these two cusps lie opposite each other and are connected by a transverse ridge. A lingual cingulum is present. In *Lemur macaco* ($n = 15$) a small but distinct protostyle is present on P^4 in 13% of the specimens. It is not present in the other species studied.

Lower. The second premolar is distinctly caniniform, possessing a broad base which tapers to a rather sharp crown tip. There is a complete but very narrow lingual cingulum. In occlusion, the tooth articulates with the upper canine, thus assuming the function of a true canine. The third premolar is unicuspid and somewhat caniniform, although not as compressed as P_2. A lingual cingulum is present. The fourth premolar possesses a large protoconid which is usually connected to a small but perceptible metaconid by an interrupted protocristid. In fact, the two-cusped P_4 is present in all specimens of each species examined except *Lemur variegatus* and *Lemur mongoz*. In the former it has an incidence of 28% ($n = 8$), in the latter 33% ($n = 3$). A small talonid basin extends distally, surrounded by an uninterrupted distal marginal ridge.

Molars

Upper. All three molars are essentially tribosphenic, possessing three well formed cusps, the paracone, protocone and metacone. The paracone is connected to the broad, slightly elevated protocone by a transverse enamel crest which sweeps distally along the summit of the protocone to terminate at the distal marginal ridge. In so doing, the crest delineates a shallow depression between the distal aspect of the protocone and the mesiolingual surface of the metacone. The depression receives the hypoconid when the teeth are in occlusion, and thereby becomes an important mechanism for triturating food. A prominent lingual cingulum is present on M^{1-2}, absent from M^3. Two variable structures are associated with the cingulum, the protostyle and hypocone. The former is found in all lemurs studied except *Lemur catta* while the latter is present in all species except *Lemur variegatus* (also see Schwarz, 1931). The other species express these traits in the following manner. In *Lemur macaco* ($n = 15$) both structures are always present on M^1; however, on M^2 the protostyle is found but the hypocone is present in only 20% of the sample. Neither trait is found on M^3. In *Lemur mongoz* ($n = 3$) the protostyle is present on all three molars while the hypocone is found only on M^1 and M^2. The protostyle is present on M^1 and M^2 and the hypocone is located only on M^1 in *Lemur rubiventer* ($n = 4$).

Lower. These teeth are compressed buccolingually and are divisible into the trigonid and talonid regions. The trigonid is slightly higher than the talonid and in living taxa, the trigonid lacks the paraconid. The latter cusp is present in fossil lemurs, at least on M_1 and M_2 (Gregory, 1922). The protoconid lies mesially to the metaconid and both cusps are set close together and connected by a protocristid. The talonid basin is shallow and presents a well defined marginal ridge circumventing the entire area. The hypoconulid is present on M_3. Accessory cusps are not present in this sample, although Bennejeant (1936) reported the presence of a tuberculum intermedium in one specimen of *Lemur catta* from the collection at the Musée d'Histoire Naturelle de Bâle.

Odontometry (Tables 17–18, Appendix)

Sample sizes are small for each species and only measurements for *Lemur variegatus* are presented in the Appendix. A few comments are necessary regarding tooth size.

The upper incisors are small and subequal. The canine is large and blade-like. Of the three premolars, P^4 is wider and longer than the others, although its crown height is slightly less than that of P^3. The second molar is the largest of the series while M^3 is very small.

The lower incisors and canines are procumbent and of the three, the canine is the largest. The premolars are well formed teeth; the caniniform P_2 is the largest mesiodistally while P_4 is widest buccolingually. The molar size formula is $M_1 > M_2 > M_3$.

Genus *Lepilemur*

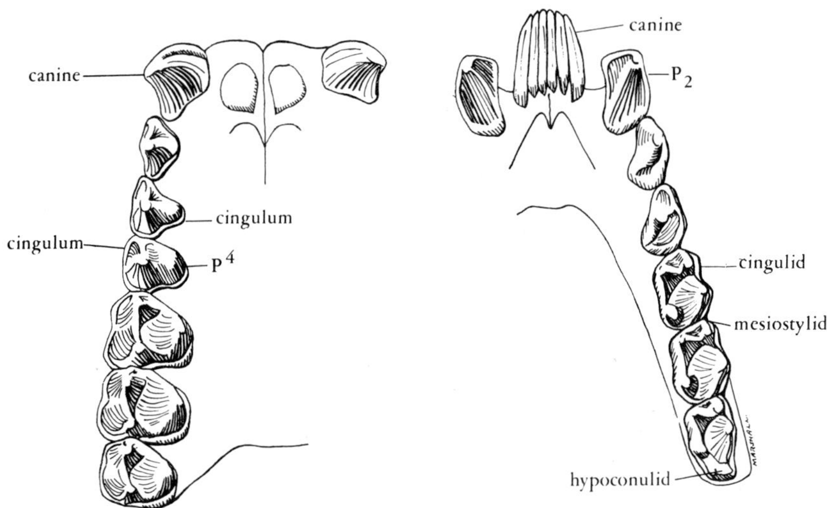

Plate 3. *Lepilemur mustelinus* male, occlusal view ×3·2.

Morphological Observations

	Sample	
	Male	Female
Lepilemur mustelinus (sportive lemur)	6	11

Incisors

Upper. The upper permanent incisors are absent; however, incisors are present in the deciduous dentition. Of the Lemuridae, *Lepilemur* is the only genus lacking upper incisors, although Colyer (1936) found incisors in three of more than 100 specimens.

Lower. The lower incisors are procumbent and quite regular in length and breadth.

Canines

Upper. The upper canines are long, trenchant and recurved distally. They are compressed buccolingually and present deep lingual grooves. A well formed distal heel or step is in contact with P^2.

Lower. The lower canines are in line with the lower incisors but are somewhat larger.

Premolars

Upper. P^2 has a single buccal cusp which is higher than the crowns of the other two premolars. The tooth is compressed buccolingually, and is therefore caniniform. A complete cingulum skirts the lingual aspect of the crown; there is no indication of a protocone. Both P^3 and P^4 are bicuspid and display lingual and buccal cingula. The paracones are larger than the protocones and lie opposite each other.

Lower. P_2 is broad, flat and its single buccal cusp is extremely high. A lingual ridge courses vertically from the cusp tip to end as a distal enamel flange that touches P_3. All three premolars are unicuspid and possess the lingual ridge described above. P_4 has a buccal cingulum.

Molars

Upper. The upper molars are essentially tribosphenic teeth. The paracone, protocone and metacone are well developed. Whether or not a hypocone is present is difficult to decide. James (1960) said a hypocone existed on M_1 and M_2, while Hill (1953) denied the presence of the cusp. A distolingual cingulum is present on the molars, particularly well developed on M_1 and M_2; however, in our opinion, it has not attained the size and function to warrant the title hypocone. A complete buccal cingulum is observable on all three molars.

Lower. The first two molars have four cusps and the last one has a fifth, the hypoconulid. All three are somewhat compressed buccolingually. The protoconid and metaconid are set close together and connected by a protocristid. The protoconid is always mesial to the metaconid. There is a complete buccal cingulum on all three molars. The protoconid is connected by the paracristid to a slight elevation on the cingulum as it passes onto the mesial surface of the tooth. This we believe to be the mesiostylid. A cristid obliqua is present. This crest creates a depression between the two buccal cusps and when viewed from above, the outline of the buccal half of each molar appears W-shaped.

Odontometry (Tables 19–22, Appendix)

The upper incisors and lower incisors and canines were not measured: the former teeth are absent while the latter constitute the dental comb.

The upper canines are about equal in size in both males and females. In females they are slightly larger in the buccolingual dimension, although the difference is not significant at the 0·05 level. Of the maxillary premolars P^2 displays the largest mesiodistal dimension, while P^4 has the greatest breadth diameter. The size relation of the upper molars is $M^1 > M^2 > M^3$.

In the mandible, P_3 is the longest and narrowest of the series while P_2 and P_4 are practically subequal in their dimensions. The molar formula is $M_1 < M_2 < M_3$.

If one looks closely at the means in Tables 19–22, it is obvious that for all teeth and for all dimensions, the size difference between the sexes is small. T-tests were performed and there were only two instances where sexual dimorphism was statistically significant: the buccolingual diameters of M^1 ($P < 0·04$) and M^2 ($P < 0·04$). Interestingly enough, in both cases the females were larger than the males. Also, of the 26 measurements taken, 17 revealed a slightly higher mean for females suggesting that, on average, the teeth of the female sportive lemur are slightly larger than her male counterpart.

Genus *Hapalemur*

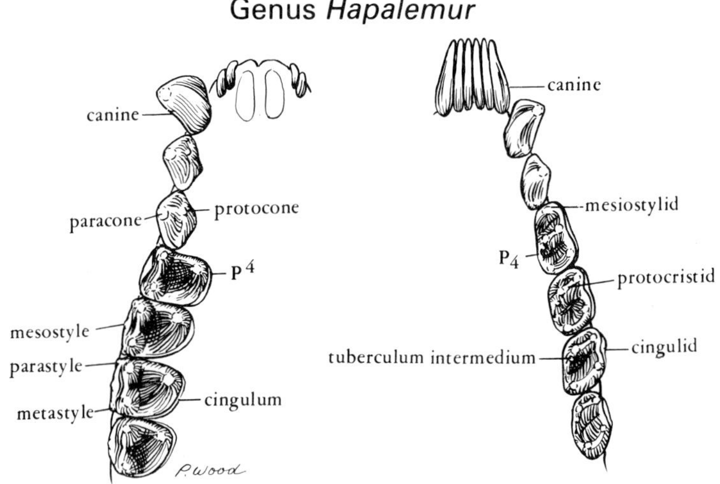

Plate 4. *Hapalemur griseus* female, occlusal view ×2.

Morphological Observations

	Sample Male	Sample Female
Hapalemur griseus (grey gentle lemur)	4	2

Incisors

Upper. Both upper incisors are small, subequal teeth exhibiting few topographic details. They are set close together and I^2 is located to the lingual side of the canine with which it frequently makes contact. This relationship is characteristic for the genus *Hapalemur.*

Lower. The four lower incisors are very narrow and procumbent.

Canines

Upper. The upper canine is well developed, being broadly based and presenting an elongated and slightly recurved crown. A small distal heel is observable. A vertically running crest separates the lingual surface into two almost equal portions. There is little if any space between the canine and P^2.

Lower. The lower canines are in series with the incisors.

Premolars

Upper. P^2 is less caniniform than in most lemurine species for it possesses an incipient protocone which increases its buccolingual diameter. The cusp is connected to the larger paracone by a prominent occlusal ridge which divides the lingual surface into a mesial and distal aspect. A very narrow cingulum circumscribes the lingual surface. In P^3 the lingual cusp is larger and more distinguishable than in P^2. An occlusal ridge connects the two cusps and a lingual cingulum is present. In addition, a narrow but perceptible buccal cingulum is present. The last premolar is decidely molariform, possessing a paracone, protocone and metacone. Buccal and lingual cingula are both prominent structures on the tooth. In size and morphology P^4 is a molar; indeed, its surface area is as great as M^2 and greater than M^3. Thus, this genus displays a definite meristic series from premolars to molars.

Lower. P^2 is large and high-crowned. There is a prominent lingual ridge dividing the lingual surface into two facies. A well developed heel projects distally to touch P_3 on its lingual side. P_3 in turn, presents a single sharply pointed cusp from which a distal ledge can also be seen. P_3 is much smaller than P_2.

P_4 is molariform; indeed, it is as large as the molars. It possesses four major cusps; the mesial two are set close together and connected by a protocristid while the distal two are more widely separated. A well developed stylid cusp is connected to the protoconid by a paracristid. A second accessory cusplet is present on this tooth. It is the tuberculum intermedium or what Remane (1960) called the tuberculum internum accessorius anterius. Hershkovitz prefers the term postmetaconulid (1971). The cusplet is located on the distal slope of the metaconid and undoubtedly develops from it embryologically (Plate 4). In the present sample it is well developed on P_4 in all but one specimen, in which it is completely lacking. Finally, it should be noted that these teeth display a small buccal cingulum which is limited to the protoconid.

Molars

Upper. Each upper molar possesses three major cusps, two bucally and onc lingually. Thus, thcy arc triangular in outline. The protocone is broad and connected to the paracone by a developmental ridge. A lingual cingulum can be seen circumscribing the protocone and is generally more pronounced along the distal portion of the cusp although, in our opinion, a hypocone is absent in the present sample. James (1960), however, believed a hypocone was "suggested" on the upper molars of *Hapalemur.* A complete buccal cingulum is present from which a mesostyle may be seen peeking up from between the paracone and metacone. These two structures are always present on M^{1-2} but variable on M^3 in the present sample. M^{1-2} are about equal in size, M^3 is always the smallest.

Lower. Morphologically the lower molars are quite similar to P_4. They present four major cusps and the same two accessory cusplets (mesiostylid and tuberculum intermedium). The protoconid and metaconid are close together and connected by the protocristid while the hypoconid and entoconid are more separated and are not connected by a transverse ridge. These teeth have been described as bilophodont by James (1960); however, the teeth examined in this investigation are not bilophodont and we believe this term is more appropriately reserved for describing the molars of Old World monkeys.

The tuberculum intermedium is present on M_1 in all specimens studied; on M_2, it is absent in one animal and on M_3, it is present in only one case. In all molars it is closely associated with the distal surface of the metaconid. The first two molars are subequal, the third somewhat smaller.

Odontometry

Measurements were not taken of these teeth since sample size was so small. The molar size formulae for the mesiodistal diameter are $M^1 \leqslant M^2 > M^3$ and $M_1 \leqslant M_2 > M_2$.

Genus *Cheirogaleus*

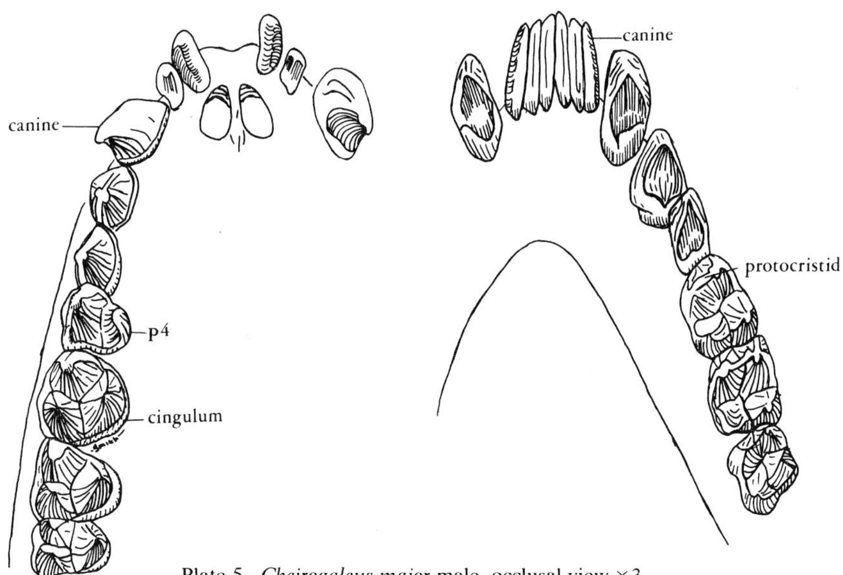

Plate 5. *Cheirogaleus major* male, occlusal view ×3.

Morphological Observations

	Sample	
	Male	Female
Cheirogaleus major (greater dwarf lemur)	1	

Incisors

Upper. I^{1-2} are set close together and I^1 is the largest. A diastema is present between the central incisors.

Lower. I_{1-2} are small, pin-like teeth. They are procumbent and display the greatest mesial angle of inclination at their bases.

Canines

Upper. The upper canine is a rather robust tooth with a crown tip above the occlusal plane. The crown is slightly recurved distally.

Lower. The lower canine is proclivous and only slightly larger than the incisors.

Premolars

Upper. P^2 is caniniform, being quite compressed buccolingually. There is a faint cingulum passing mesiodistally along the lingual surface of the tooth. The crown tip is higher than it is in P^{3-4}. P^3 has a single cusp which displays a cingulum pattern similar to that observed on P^2. The major difference between P^2 and P^3 is that the latter is somewhat wide buccolingually, although a protocone is not present. P^4 has a paracone and a protocone. In addition, this bicuspid tooth has a mesial, lingual and distal cingulum.

Lower. P_2 is large and somewhat procumbent. A small cingulum is present on the distolingual aspect of the tooth. P_{3-4} are similar morphologically, the salient difference between them being that P_3 is larger than P_4. All three premolars possess a single cusp, the protoconid.

Molars

Upper. M^{1-3} are tritubercular molars; there are no hypocones. The protocone is large and expanded mesiodistally. There is a lingual cingulum circumventing the protocone and it passes without interruption onto the mesial and distal surfaces of the molars. It is better developed on M^{1-2}. A faint buccal cingulum is present on M^{1-3}.

Lower. The protoconid and metaconid are close together and connected by a protocristid on M_{1-3}. A paracristid is present on all three molars. M_3 lacks the hypoconulid. A buccal cingulum is present on M_{1-3}, but it is somewhat better developed on M_{1-2}.

Odontometry

The single specimen was not measured. The molar size formulae for the animal are, $M_1 > M_2 > M_3$ and $M^1 = M^2 > M^3$.

Genus *Microcebus*

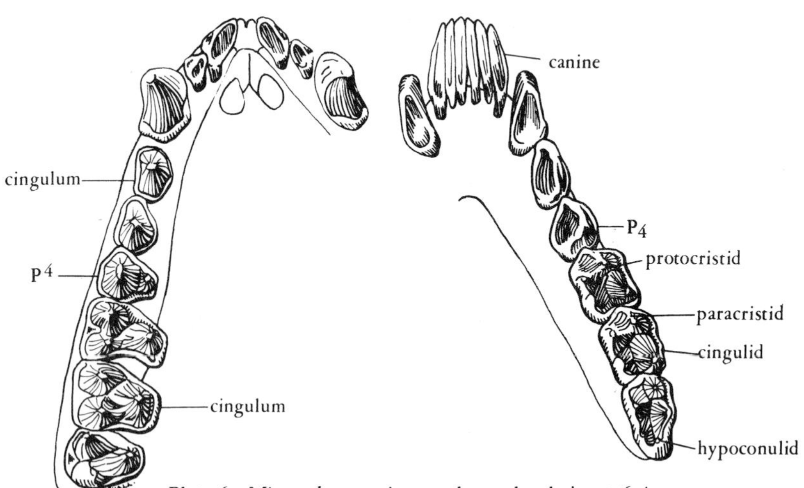

Plate 6. *Microcebus murinus* male, occlusal view ×6·4.

Morphological Observations

	Sample	
	Male	Female
Microcebus murinus (mouse lemur)	3	2

Incisors

Upper. I^1 is a small tooth with a slightly expanded crown and a mid-line diastema. I^{1-2} are set close together and I^2 is smaller than I^1.

Lower. I_{1-2} are styliform, elongated teeth of equal width and length. They are all set close together and form the dental comb.

Canines

Upper. The upper canine is a relatively large tooth with a crown tip projecting above the occlusal plane. A lingual cingulum is present which terminates as a distal stylid.

Lower. The lower canine is only slightly larger than the lower incisors and is procumbent. There is a suggestion of a distal marginal ridge.

Premolars

Upper. P^2 possesses a pointed paracone with no other cusps. A cingulum passes around the base of the tooth, and there are para- and distostyles present. P^3 is similar to P^2. P^4 presents a low protocone, thus making it a bicuspid tooth. The cingulum as well as the para- and distostyles are present.

Lower. P_2 has a single elongated protoconid and the tooth is somewhat procumbent and caniniform. The tooth possesses a weak stylar rim around its distal surface. P_3 is similar to P_2. As in P_{2-3}, a single cusp is present on P_4. The buccal cingulum is more pronounced than it is on P_3, as is the talonid portion of the tooth.

Molars

Upper. The upper molars are tribosphenic. The hypocone is probably absent although the slightly hypertrophied distolingual cingulum is thought by some to constitute a hypocone (James, 1960). The buccal cingulum is evident as well as the para- and metastyles. The metastyle is usually absent on M^3.

Lower. M_{1-2} have four cusps and M_3 has five. The protoconid is connected to the metaconid by a protocristid. The paracristid is better developed on M_{1-2}, less delineated on M_3. A buccal cingulum is present on all three molars. As noted above, M_3 has a hypoconulid.

Odontometry

Tooth size is virtually the same in both sexes and in both, the longest (mesiodistal) tooth is M_3 (female: 2·1 mm; male: 2·3 mm). The mandibular size formula is $M_1 = M_2 < M_3$, whereas the maxillary molars display a formula of $M^1 = M^2 > M^3$. Buccolingually M^2 is the widest molar in both sexes.

Genus *Phaner*

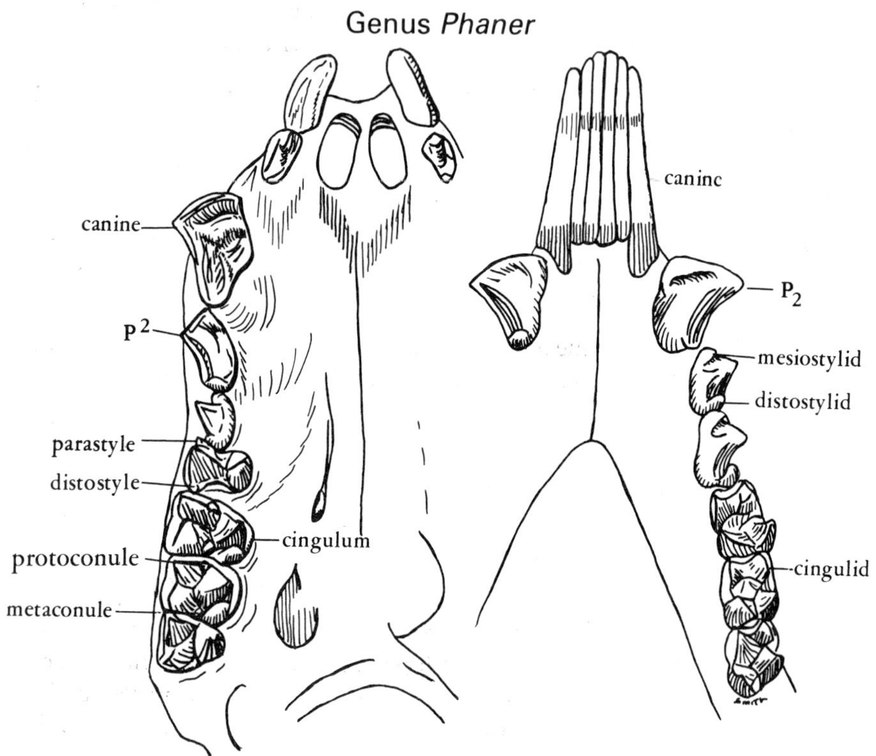

Plate 7. *Phaner furcifer* male, occlusal view ×4.

Morphological Observations

	Sample	
	Male	Female
Phaner furcifer (forked lemur)	2	

Incisors

Upper. I^1 is long and inclines mesially at a rather sharp angle, a condition not noticed in other lemurs. A diastema separates the two central incisors. I^2 also slopes mesially and is positioned immediately distal to I^1.

Lower. Both I_{1-2} are very elongated and narrow, displaying the extreme lemurine procumbancy. No surface markings are observable on the teeth.

Canines

Upper. The upper canine is large and dagger-like, displaying a slightly recurved tip. It is separated from I^2 by a small diastema. The crown projects well below the occlusal plane. A narrow mesiolingual groove passes from the base to near the crown tip.

Lower. The lower canine is in line with the incisors and only slightly larger. In lateral view, the dental comb curves upward near its end.

Premolars

Upper. P^2 is large and extremely caniniform. The crown is long and ends as a sharp point. A distostyle is obvious at the base of the tooth. Morphologically P^2 of *Phaner* is the most caniniform among extant lemurs. P^3 is small, presenting a single sharp paracone with a complete lingual cingulum. A narrow, incipient buccal cingulum is observable as well as para- and distostyles. P^4 is bicuspid; indeed, the protocone is well formed and a slight lingual cingulum skirts around its distal half. A complete cingulum is present buccally and again, para- and distostyles are obvious. Collectively, the three premolars are quite heteromorphic.

Lower. P_2 is large and caniniform as in other lemurs. The lanceolate-like crown rises well above the other teeth as it flares buccally. This alignment allows P_2 to glide in the space between the upper canine and P^2 during mastication, an ideal adaptation for piercing the chitinous covering of insects. P_3 has a single buccolingually compressed cusp giving the tooth a canine appearance. A lingual cingulum is present as well as a distinct distostylid. P_4 is also unicuspid and compressed. Both lingual and buccal cingula are present, as are well formed mesio- and distostylids. Note that in *Phaner*, P^4_4 is not molariform.

Molars

Upper. M^1 has four cusps although the hypocone is small. A narrow lingual cingulum passes around the protocone and connects to the hypocone. A complete buccal cingulum is present although only one style is present, the parastyle. Additionally, a miniscule protoconule and metaconule are present. M^2 is morphologically similar to M^1. M^3 lacks the hypocone, lingual cingulum and proto- and metaconules in the two specimens examined. A small parastyle and buccal cingulum are present, however.

Lower. M_1 possesses four cusps. The mesial two cusps are set close together and connected by a V-shaped protocristid. The distal two cusps are wider apart and there is no transverse crest. A cristid obliqua is present. A narrow buccal cingulum passes along the protoconid and in the two specimens studied, it gains the hypoconid. M_2 is morphologically similar to M_1, while M_3 is smaller, particularly buccolingually, and the buccal cingulum is limited to the protoconid. Hill (1953) reports a hypoconulid on M_3, but in the present sample ($n = 2$) it is absent.

Odontometry

The mesiodistal size gradient of the molars is $M_1^1 \geqslant M_2^2 > M_3^3$.

4

Family Indriidae

Present Distribution and Habitat

The members of this prosimian family live on the island of Madagascar. They are found throughout the island and at altitudes from sea level to 1,790 m (5,900 ft). Of the three living genera, *Propithecus* is represented in almost all the Madagascar forests, whereas *Indri* is restricted to the northern half of the eastern forest and *Avahi* is found in both the eastern and western forests (Petter, 1962).

The activity rhythm is dichototomous: *Propithecus* and *Indri* are purely diurnal, while *Avahi* is nocturnal. The Indriidae are essentially arboreal, although they may come to the ground on occasion. Their mode of locomotion is described as vertical clinging and leaping.

Dietary Habits

The diet of these animals is rather specialized, consisting mainly of bark, leaves, fruits and flowers. It is interesting to note that Petter (1962) stated that the Indriidae "which are more specialized from the locomotion standpoint (*vertical clinging and leaping*) (my italics) seem also to be more specialized in their diet" (p. 276).

General Dental Information

Permanent dentition: $I^2_1\ C^1_1\ P^2_2\ M^3_3$, or $I^2_2\ C^1_0\ P^2_2\ M^3_3$

Deciduous dentition: $i^2_2\ c^1_1\ m^3_3$, or $i^2_1\ c^1_1\ m^2_4$

Sequence of initial eruption of permanent teeth (Schwartz, 1974a);

Avahi laniger

I^2	M^1	I^3		P^4	M^2	P^3	C	M^3
	M_1	I_3	C	P_3	M_2	P_2		M_3

Propithecus verreauxi

M^1	I^2		M^2	I^3	P^4	P^3	M^3	C
M_1	I_3	C	M_2		P_3	P_2	M_3	

Propithecus diadema

M^1	I^2		I^3		M^2	P^4	P^3	M^3	C
M_1	I_3	C		M_2		P_3	P_2	M_3	

Indri indri

I^2	I^3	M^1	?
I_3	C	M_1	?

The difference between the two permanent dental formulae hinges on the interpretation of the most lateral tooth in the anterior portion of the mandible. If this tooth is a canine, the first formula is correct; if it is an incisor, the second formula is used. In the eruption sequence presented by Serra (1952a) he does not mention the appearance of a permanent canine, nor does Bennejeant (1936) mention this tooth in his description of the teeth of *Indri*. Indeed, Bennejeant states that the proclivous teeth of the mandible are incisors and the canines are absent. It should be noted that Bennejeant's statement is based upon the work of Friant (1935) who studied the teeth and jaws of infant prosimians (*Indri*). Recently Schwartz (1974a,b) studied 19 juvenile indriids. He concludes (1974b, p. 112) that the permanent lower lateral tooth of the indriid toothcomb is a canine since it meets the basic requirement of a lower canine, i.e. "that tooth which occludes in front of the upper canine".

The deciduous formula is generally agreed to be as presented here, although Hill (1953) offers $i^2_2\,c^1_1\,m^2_3$ for the Indriidae, while Schwartz (1974a,b) has suggested 1–1–4 for the lower number. Note the accepted presence of the lower deciduous canine in both formulae.

The upper incisors are set close together and of the three living genera, only in *Avahi* is the diastema between the central incisors as wide as it is in the lemurs. In both *Indri* and *Propithecus* this space is narrower; indeed, in *Propithecus* the central incisors frequently incline medially to the point of contact. The lower incisors are procumbent as in the Lemuridae. The difference between the lower anterior dentitions of the Lemuridae and Indriidae is simply a numerical problem. In both taxa the lower anterior teeth are procumbent; in the former taxon there are four incisors plus two canines forming the dental comb, while in the latter, there are either two incisors and two canines, or four incisors and no canines. The lateral tooth is larger than the central one and both possess longitudinal grooves on their buccal surfaces.

The upper and lower premolars are heteromorphic, that is, P^3_3 are quite different morphologically from P^4_4. The former teeth are compressed buccolingually and possess a single cusp, hence they are caniniform; the latter structures are broader buccolingually and they display a more complex crown configuration. In all Indriidae P^{3-4}_3 have either one or two cusps, while P_4 is normally unicuspid. Lingual cingula are usually present on P^{3-4} and lacking on P_{3-4}. Forbes (1894) described buccal cingula on P^{3-4} for *Avahi* and stated that these structures were not present in any other lemurs. Buccal cingula are present on these teeth in *Propithecus*, albeit quite rudimentary on P^3.

The upper molars are all quadritubercular with the third being slightly inset lingually and always the smallest of the series. The four cusps are well formed and pointed while being connected by a pattern of V-shaped crests. One crest is of particular interest, the postprotocrista which passes between the protocone and metacone in *Propithecus*, and several other prosimian genera. It is also found in certain Cebidae and Hominoidea. Functionally the postprotocrista occludes between the hypoconid of one tooth and the protoconid of the tooth immediately distal, so that it fills in the slight buccal embrasure between adjacent lower teeth and in this manner actively participates in the triturition of leaves, bark, etc. Buccal cingula are present on all three upper molars while para-, meso- and metastyles are well developed on M^{1-2}.

The lower molars display well formed cusps and connecting crests which when viewed from the buccal side form a W-pattern in *Avahi* and *Propithecus*. This crown pattern is less obvious in *Indri*. There are four major cusps on M_1 and M_2, and usually five on M_3, although James (1960) reports five on M_1 and M_3 in *Avahi*. A parastylid is present on M_1 and M_2 in *Avahi* and *Propithecus* and absent in *Indri*.

The masticatory habits of this family of prosimians are not well known. Their diet is known, but the methods employed in triturating food have not been thoroughly investigated to date. Undoubtedly, since the molar crowns possess crests and styles reminiscent of earlier fossil forms their function is that of puncturing and crushing the food. The cutting edges of these crests are such that during mastication they shear against one another in nearly vertical planes. For an excellent recent discussion of masticatory evolution in the primates see Mills (1973).

Genus *Propithecus*

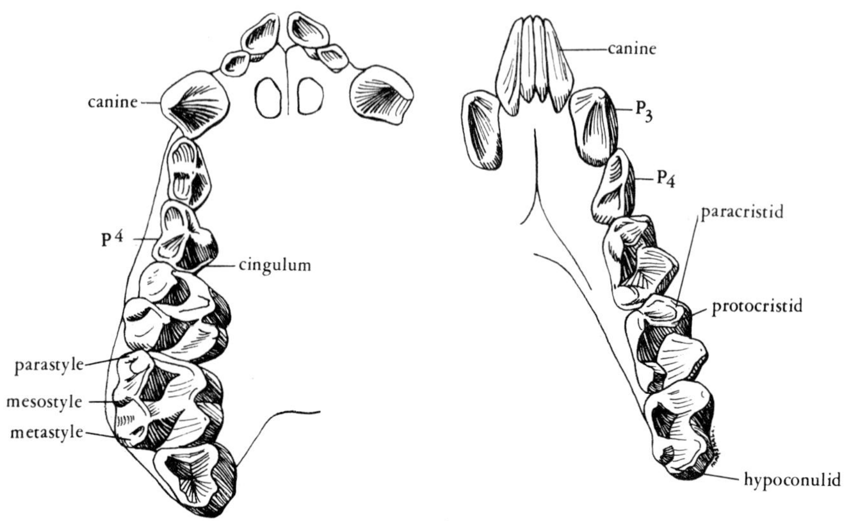

Plate 8. *Propithecus verreauxi* male, occlusal view ×1·5.

Morphological Observations

	Sample	
	Male	Female
Propithecus verreauxi (Verreaux's sifaka)	3	8

Incisors

Upper. The upper central incisors are broad and have slightly concave lingual surfaces which become exaggerated with wear. The crowns lean in a mesial oblique direction which reduces the width of the central diastema. The lateral incisors are small, positioned directly distal to the central incisors and often in contact with them. The lingual surface is concave.

Lower. These teeth are procumbent and rather robust. They are elongated and frequently present longitudinal grooves on their buccal surfaces.

Canines

Upper. These teeth are separated from the lateral incisors. They are elongated, presenting slightly recurved crowns ending in sharp points. Each canine presents a mesiolingual groove.

Lower. If the lateral most procumbent anterior tooth is a canine it is an elongated, robust tooth. It is the larger of the two and presents a slight thickening near the cervical portion. The four procumbent teeth are approximately the same length.

Premolars

Upper. P^3 has a single, large buccal cusp. The tooth is compressed buccolingually, resulting in a blade-like structure. Sharp crests pass mesially and distally from the tip of the cusp to terminate as small accessory cusplets. The buccal surface presents a vertical ridge which divides the surface into concave portions. There is also a well defined distolingual cingulum. P^4 is larger than P^3. It has a large buccal cusp as well as two small distolingual cusplets. A buccal cingulum is also present. The tooth is widened lingually, presenting an enlarged cingulum and a small nipple-like elevation in the region of the protocone. This trait is present in all *Propithecus* specimens examined. The two premolars are quite heteromorphic.

Lower. Both P_{3-4} possess large, single cusps. P_3 is blade-like, being markedly compressed buccolingually, and is thus caniniform. There is a slight distolingual cingulum terminating as a distal enamel spur. In occlusion, P_3 interdigitates between the upper canine and third premolar. Indeed, the distobuccal surface of P_3 and the mesiolingual surface of P^3 form an excellent slicing surface for triturating the bark, leaves, etc. which these prosimians eat. In addition to the single cusp, P_4 has a pronounced lingual crest passing obliquely and distally from the cusp tip to the base of the crown. This creates two concave surfaces on the lingual face of the tooth.

Molars

Upper. M^1 is four-cusped, the buccal two being larger and more pointed than the lingual two. The protocone is connected with the metacone by a slight postprotocrista. The buccal cingulum presents the para-, meso- and metastyles. Of these, the mesostyle is always the largest. A well developed marginal crest swings mesiobuccally from the protocone to terminate on the parastyle. Additional crests form the mesial and distal slopes of the paracone and metacone and end at the bases of the three buccal styles. Thus, a connecting series of V-shaped crests form a crescent configuration when this tooth is viewed from the buccal side. Morphologically M^2 is similar to M^1 except that it is slightly larger. M^3 is much smaller than M^{1-2} and is set lingually to these teeth. The tooth possesses four small cusps, of which the mesial two are connected by a low transverse crest. The buccal styles are absent.

Lower. M_1 possesses four cusps. The mesial two are set close together and are connected by a protocristid while the distal two are much wider apart. Both buccal cusps are mesial to their lingual counterparts. The protoconid is connected to a well formed parastylid by the paracristid and the cristid oblique interlocks the distal surface of the metaconid to the hypoconid. Thus, the buccal half of M_1 appears W-shaped when viewed from above. The mesial portion of the tooth is higher than the distal part. M_2 is similar to M_1 with two exceptions: the parastylid is smaller and the protoconid and metaconid are set wider apart resulting in a deeper trigonid basin. M_3 has five cusps and is quite different from the preceding two molars. The most striking contrast regards the much wider mesial portion of the tooth. The protoconid and metaconid are widely separated and lie opposite each other as do the hypoconid and entoconid. The hypoconulid is attached to the middle of the distal surface of the tooth.

Odontometry (Tables 23–26, Appendix)

T-tests were performed to determine the amount of sexual dimorphism present in the genus. There is no sexual dimorphism in the anterior teeth and only one case in the premolar region: P^3 ($P<0{\cdot}03$) is significantly longer in the males. The male molars are significantly larger than the female molars in the following manner: M_2 is significantly longer mesiodistally ($P<0{\cdot}009$) while the same dimension is less for M_3 ($P<0{\cdot}02$). The males are also larger in the buccolingual direction, M_1 ($P<0{\cdot}02$), M_2 ($P<0{\cdot}01$), and M_3 ($P<0{\cdot}03$). Interestingly, the difference in M_1 is the talonid while in M_{2-3} it resides in the trigonids. Two upper molars display a significant difference mesiodistally, M^2 ($P<0{\cdot}002$) and M^3 ($P<0{\cdot}001$), while all three are significantly wider, M^1 ($P<0{\cdot}007$), M^2 ($P<0{\cdot}003$) and M^3 ($P<0{\cdot}002$).

The mandibular molar formula is $M_1<M_2<M_3$, while for the maxillary molars it is $M^1<M^2>M^3$.

Genus *Indri*

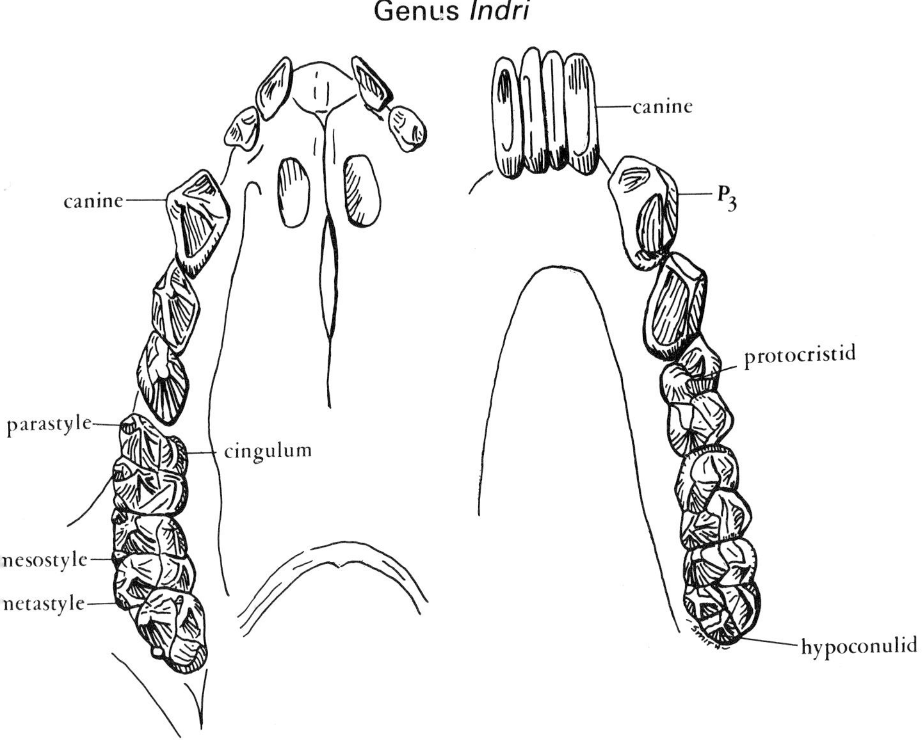

Plate 9. *Indri indri* male, occlusal view ×1·75.

Morphological Observations

	Sample	
	Male	Female
Indri indri (indri)	1	

Incisors

Upper. The upper central incisors are separated by a wide diastema. The crown is elongated mesiodistally and positioned just labial to the crown of I^2. I^2 is compressed in a similar manner to I^1 although the crown is more spatulate.

Lower. I_1 is a long, narrow tooth presenting a slight lingual bulge at its base. It is procumbent. The lateral tooth is larger than the central one and has a marginal ridge. Schwartz (1974b) believes this tooth is a canine rather than an incisor, and our study supports this conclusion.

Canine

Upper. An appreciable diastema separates the upper canine from I^2. The canine is robust, presenting a high, recurved crown. A distal stylid is concealed beneath the mesial projection of P^3.

Lower. Discussed above under lower incisors.

Premolars

Upper. P^3 presents a single, buccolingually compressed cusp surrounded by a feeble buccal and lingual cingulum. The result is a rather caniniform tooth. P^4 is similar to P^3 except that it has a wider buccolingual diameter, although a true protocone is lacking.

Lower. P_3 is large and compressed buccolingually. A distinct mesial flange is a characteristic of the tooth. The crown tip is well above the occlusal plane of the other mandibular teeth. A diastema is present between P_3 and the canine. P_4 is mesiodistally elongated and presents rather sharp crests descending from the crown tip. The crests are also noticeable on P_3, and on both they reinforce the blade-like nature of the teeth.

Molars

Upper. M^{1-2} has four cusps of approximately equal size. There is no postprotocrista. M^3 is the smallest of the three molars and the hypocone is lacking. Buccal cingula are present on the three molars as are the para-, meso- and metastyles. The mesostyle is largest on M^1.

Lower. There are four cusps on M_{1-2} and five on M_3. The protoconid and metaconid on M_1 are very close together and are connected by a protocristid. Distally, the protoconid and metaconid become more widely separated, until on M_3 they are as wide apart as the hypoconid and metaconid. A similar arrangement is observed in *Propithecus*. There are no buccal cingula.

Odontometry

No measurements were taken of the single specimen. The mesiodistal size gradient is $M_1 > M_2 > M_3$ and $M^1 > M^2 > M^3$.

Genus *Avahi*

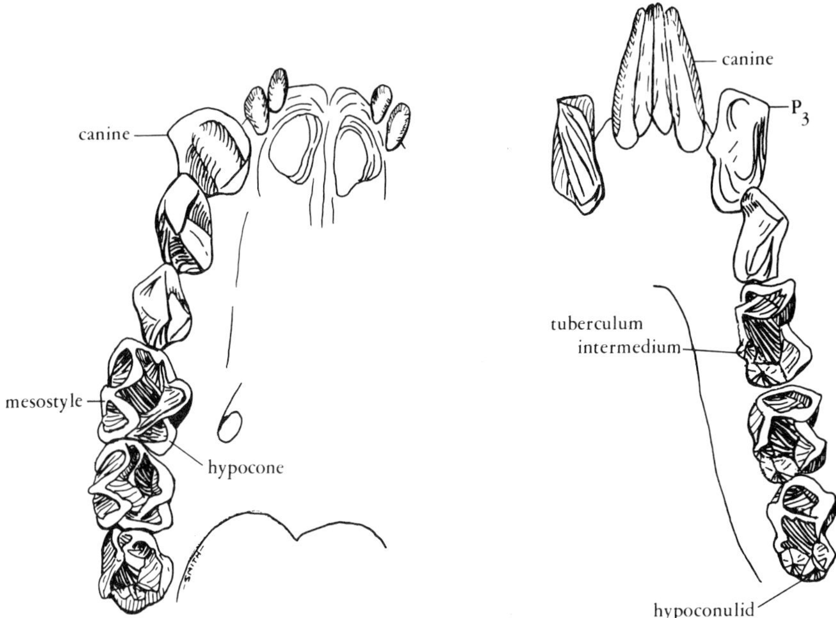

Plate 10. *Avahi laniger* male, occlusal view × 2.

Morphological Observations

	Sample	
	Male	Female
Avahi laniger (woolly lemur)	2	

Incisors

Upper. I^1 is a very small peg-shaped tooth. A wide diastema separates the central incisors. I^2 is only slightly larger than I^1 and both teeth are compressed.

Lower. I_1 is long, narrow and procumbent. A slender longitudinal crest extends the length of the lingual surface of the tooth. I_2 is absent.

Canines

Upper. The upper canine is compressed buccolingually, so that it is quite blade-like. A median longitudinal crest separates the lingual surface into two parts. The distal part ends as an enamel flange articulating with the lingual surface of P^3, thus there is no diastema between the canine and P^3.

Lower. The lower canine is larger than I_1 but both present longitudinal lingual crests and are equal in length. The dental comb is appreciably wider at the base than at its incisal border.

Premolars

Upper. P^3 is blade-like, being more compressed buccolingually than the canine. A vertical crest passes buccally from the tip of the paracone to the cervix where it meets the mesio- and buccal cingula. A similar vertical crest is present on the lingual surface of P^4. P^4 is morphologically similar to P^3 except that it is somewhat smaller and possesses a slight lingual cingulum. Both teeth have a single cusp and are more homomorphic than those in other indriids, particularly *Propithecus.*

Lower. P_{3-4} are unicuspid. P_3 is compressed buccolingually and is somewhat procumbent. The crown is blade-like and separated by a diastema from the canine. P_4 is also elongated mesiodistally. An oblique crest passes distally from the protoconid to the base of the crown.

Molars

Upper. M^{1-3} have four cusps. The buccal cusps are larger than the lingual ones, with the hypocone always the smallest. The para-, meso- and metastyles are well developed. Occlusobuccal crests interconnect the apices of the styles with those of the para- and metacones to form a dilambdodont outline. A thin postprotocrista connects the metacone and protocone to add to the cutting efficiency of the molars. M^{1-2} are very similar, whereas M^3 is smaller and lacks the metastyle.

Lower. M_{1-2} possess four cusps while M_3 adds a small hypoconulid. On M_1 the protoconid lies mesial to the metaconid and both are connected by a protocristid. In addition, a sharp paracristid connects the protoconid with the parastylid. A distinct cristid obliqua rounds out the contour of the buccal surface of M_1. M_{2-3} are similar to M_1 with one major exception: on M_2 the protoconid and metaconid are almost opposite each other, and on M_3 they are opposite. This results in a noticeable mesiodistal size gradient regarding the buccolingual diameter of these teeth. Thus, the trigonid increases in width from M_{1-3}. A well delineated tuberculum intermedium is attached to the metaconid on M_{1-3} in both of these specimens. Bennejeant (1936) reports that the tuberculum intermedium is found "very frequently" in *Avahi laniger.* In the present specimens it decreases in size from M_1 to M_3, supporting a field theory concept regarding its development.

Odontometry

The mesiodistal size gradient of the premolars and molars is: $P_3^3 > P_4^4$, and $M_1^1 > M_2^2 > M_3^3$.

5

Family Daubentonidae

Present Distribution and Habitat

The aye-aye lives on the island of Madagascar. Today it is found in the forests and bamboo groves of the eastern part of the island as well as in limited regions of the north-west sectors. It was, at one time, more widely distributed since fossil teeth (incisors) have been found in several different parts of the island. At the present time there is one genus and one species: *Daubentonia madagascariensis.*

Living mainly in rain forests, *Daubentonia* displays a quadrupedal gait moving along the thick branches with its body horizontal to the limb. On the ground they are also quadrupedal and they frequently elevate their tails as counterbalances. Napier and Napier (1967) believe the locomotor pattern of *Daubentonia* represents a possible modification of a vertical clinging form since they can climb vertical trunks much in the manner of *Lepilemur.* They are purely nocturnal animals.

Dietary Habits

Aye-ayes prefer insects and beetle larvae, although they also eat fruits and flowers. A recent and quite interesting suggestion by Cartmill (1972) is that *Daubentonia* occupies the ecological niche normally inhabited by the woodpecker. Their specialized incisors and elongated third digit serve as the woodpecker's bill and tongue.

General Dental Information

Permanent dentition: $I^1_1\ C^0_0\ P^1_0\ M^3_3$ or $I^0_0\ C^{(1)}_{(1)}\ P^1_0\ M^3_3$

Deciduous dentition: $i^2_2\ c^1_0\ m^2_2$ or $i^1_0\ c^{(1)}_{(1)}\ m^3_2$

Sequence of eruption of permanent teeth (Schwartz, 1974a):

Daubentonia madagascariensis

$$\frac{M^1 \quad M^2 \quad P^4 \quad M^3}{M_1 \quad M_2 \quad \quad \quad M_3}$$

Genus *Daubentonia*

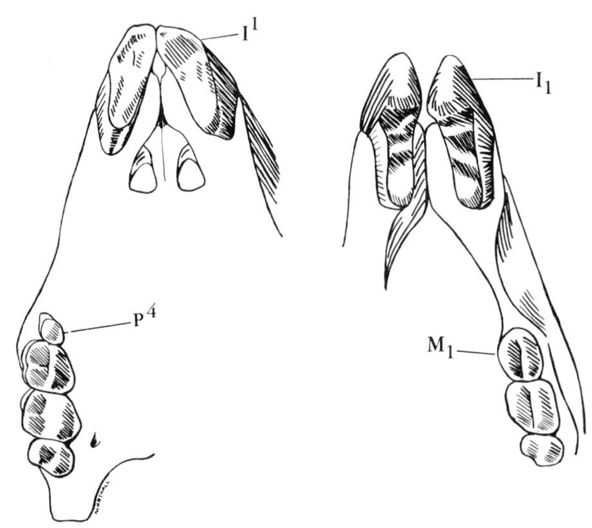

Plate 11. *Daubentonia madagascariensis* male, occlusal view ×2.

Morphological Observations

	Sample	
	Male	Female
Daubentonia madagascariensis (aye-aye)	1	

Schwartz (1974a) presents a lengthy discussion of the development and eruption of the teeth in the aye-aye. He concludes the following: (a) the large continuously growing anterior teeth are canines and are functionally associated with both the deciduous and permanent dentition (therefore, placed in brackets in the dental formulae) and (b) the permanent incisors were phylogenetically lost during the transition from an ancestral indriine stock by selection of continually growing teeth, the canines.

There was a single specimen available for study. The following comments are therefore made from this specimen and the literature. Of course no statement regarding dental variability was possible.

The specialized dental system is depicted in Plate 11. There is a single large incisor (canine) in each jaw, one P^4, and three upper and lower molars.

The anterior teeth of *Daubentonia* are frequently compared with those of rodents but Schwartz (1974a) believes this to be invalid. He presents convincing evidence that the anterior teeth in these two groups of animals are not homologous since in *Daubentonia* they develop within the maxilla and simply pass through the premaxilla, whereas in rodents they form within the premaxilla, therefore they are true incisors. Also, he cogently reminds the student that the aye-aye is not a rodent but a prosimian primate. It should be noted that another authority (Gregory, 1922) believed the lower front teeth were enlarged canines rather than incisors.

The unusual wear pattern of the anterior teeth results from the gnawing of hard wood which the animal must penetrate in its quest for the beetle larvae and grubs it so enjoys. Also the chisel-like nature of these teeth is accentuated through time since the distal surface is dentin which is softer and wears faster than the enamel covering the rest of the crown (James, 1960).

The single upper premolar (P^4) is extremely small and peg-like. It is separated by a wide diastema from I^1. There are no permanent lower premolars. The upper molars are small and rather nondescript. M^{1-2} present four rounded cusps, two buccally and two lingually being separated by an obvious mesiodistally coursing developmental groove. M^3 is somewhat oval when viewed occlusally and has three cusps. Since there are no lower premolars, the molars are separated by a wide diastema from the single lower incisor (canine). There are generally four cusps present, although, as in the upper molars, the occlusal surfaces are so badly worn it is difficult to make them out. Indeed, early wear of an extreme nature has been mentioned by several other investigators (Owen, 1840–1845; Gregory, 1922; James, 1960). A narrow mesiodistal groove is perceptible, separating the crown into two nearly equal halves.

Odontometry

Measurements of these teeth were not taken since there was only a single specimen. The mandibular molar formula is $M_1 < M_2 > M_3$ and the upper molars have an identical formula.

6

Family Lorisidae

Present Distribution and Habitat

The family is divided into two subfamilies, the Lorisinae and Galaginae. The former taxon includes the lorises and pottos; the latter, the bushbabies (galagos). Of the four genera constituting the Lorisinae, *Loris* and *Nycticebus* live in Asia and *Perodicticus* and *Arctocebus* live in Africa. The *Galago* is exclusively African.

All lorisids are arboreal. The two subfamilies are easily separated with respect to their locomotor behavior into the quadrupedal slow-moving Lorisinae and the vertical clinging and leaping Galaginae. The animals prefer tropical rain forests although they are occasionally found in montane forests, and apparently *Perodicticus* has been observed on the ground (Napier and Napier, 1967). Both subfamilies are nocturnal.

Dietary Habits

The diet of these prosimians is catholic. They have been observed eating insects, nuts, leaves, fruits, birds' eggs and in captivity even cereals (Sauer and Sauer, 1963). It is quite apparent from these observations that lorisids are omnivorous. This statement is supported by dental morphology, especially for *Perodicticus* according to Gregory (1922).

General Dental Information

Permanent dentition: $I^2_2\ C^1_1\ P^3_3\ M^3_3$

Deciduous dentition: $i^2_2\ c^1_1\ m^3_3$

Sequence of initial eruption of permanent teeth (Schwartz, 1974a):

Loris tardigradus

I^2	I^3		M^1		C	P^2	M^2	M^3	P^4	P^3
I_2	I_3	C	M_1	P_2	M_2			M_3	P_4	P_3

Nycticebus coucang

I^3	I^2		M^1	C		P^2	M^2	M^3	P^4	P^3
I_2	I_3	C		M_1	P_2		M_2	M_3	P_4	P_3

Perodicticus potto edwardsi

I^2				I^3		M^1	C	P^2		M^2	P^4		M^3	P^3
	I_2	I_3	C		M_1		P_2		M_2		M_3	P_4		P_3

Galago senegalensis moholi

		M^1		C		P^2	I^2	I^3	M^2	M^3	P^4	P^3
M_1	I_2	I_3	C		P_2		M_2			M_3	P_4	P_3

Galago demidovii

M^1	M^2	I^2	I^3		M^3	C	P^2	P^4	P^3
M_1	M_2	I_2	I_3	C	M_3	P_2		P_4	P_3

Galago crassicaudatus

M^1	I^2	I^3		M^2	C	P^2	P^4	M^3	P^3
M_1	I_2	I_3	C	M_2	P_2		P_4	M_3	P_3

Galago alleni

M^1	I^2	I^3		M^2		P^2	C	M^3	P^4	P^3
M_1	I_2	I_3	C	M_2	P_2		P_4	M_3	P_3	

The permanent and deciduous formulae are as in the Lemuridae rather than as in Indriidae. Schwartz (1974a) demonstrated the variability in the eruption sequences in the lorisids, and particularly in the galagines. Indeed, in the latter group he found subspecific differences in the pattern of eruption.

The upper incisors display few morphologic details. The central ones are widely separated and this space receives the dental comb when the teeth are occluded. Both I^1 and I^2 are small and I^2 is sometimes absent, particularly in *Nycticebus*. The lower incisors are procumbent as in the lemurs and indriids. They are merely narrow, styliform teeth possessing no noticeable details.

The upper canines are relatively large teeth projecting above the occlusal line. They are most robust in *Perodicticus* and *Nycticebus*. The lower canines are in line with the procumbent incisors.

The upper premolars are morphologically different from P^2 to P^4. The foremost is always caniniform and possesses a complete lingual cingulum which is also usually present on P^{3-4}. All three upper premolars in lorisids may display para- and distostyles although they are generally better developed in *Nycticebus*. P^3 is frequently smaller than P^2, the reverse of which is found in lemurs. A protocone is present on P^4 in all genera, and in addition, a metacone and hypocone adorn the crown of P^4 in *Galago*, making this a molariform tooth. Of the lower premolars, P_2 is always caniniform and it frequently presents a lingual cingulum terminating as a distal step. A diastema separates this tooth from the canine. P_{3-4} have lingual cusps and P_4 is quite molariform in most lorisids, particularly in *Loris* and *Galago*.

The upper molars are generally quadricuspid although the hypocone is quite small in *Perodicticus*. In all forms, the hypocone is either extremely small or absent on M^3. Para- and metastyles are frequently present on M^{1-2}, while M^3 has only a parastyle if it possesses either. Proto- and metaconules are occasionally observed on M^{1-2}, rarely on M^3. In general, the upper molars of the lorisids, with perhaps the exception of *Galago*, have retained the crescentic or V-shaped configuration of the principal cusps usually seen in Eocene prosimians. This is particularly true of *Nycticebus*. Indeed, Gregory (1922) believed *Nycticebus* to have the least specialized dentition among extant lorisids.

In the mandibular molars, M_{1-2} have four well developed cusps while M_3 possesses five. A protocristid is observed connecting the protoconid and metaconid in all genera, especially in *Galago* and *Loris*. In the present sample of lorisids, buccal cingula are found on the mandibular molars of *Loris*, *Nycticebus* and *Arctocebus*. The mandibular buccal cingulum represents a primitive trait in primates and can be traced back to *Purgatorius* of the Late Cretaceous (Van Valen and Sloan, 1965). Two excellent reviews of the status of the cingulum among primates can be found in Frisch (1965) and Kinzey (1973).

As stated previously, the lorisids are omnivorous although they all seem to prefer insects. All genera eat fruit, particularly *Nycticebus*. *Loris*, when fortunate enough to capture a bird, decapitates the animal for the brain and discards the rest (Hill, 1953). When procuring food lorises employ their hands and not their teeth, and once the food is in the mouth it is quickly moved to the premolar–molar area since the incisors are virtually useless for biting.

Hiiemae and Kay (1973) report that the masticatory function for *Galago* consists of three major strokes: (a) the preparatory stroke (up-stroke), (b) the power stroke (triturition), and (c) the recovery stroke (down-stroke). A similar functional mechanism is undoubtedly employed by all members of the family, although detailed studies have not been conducted to date.

The morphology, as well as the wear facets of the molars, indicate that chewing involves orthal retraction of the mandible resulting in crushing-puncturing strokes (see Gingerich, 1972 for a complete discussion of this type of mastication). Additionally, the reciprocal arrangement of molar crests are such that they constitute a shearing mechanism well adapted for mashing and mincing food.

Genus *Loris*

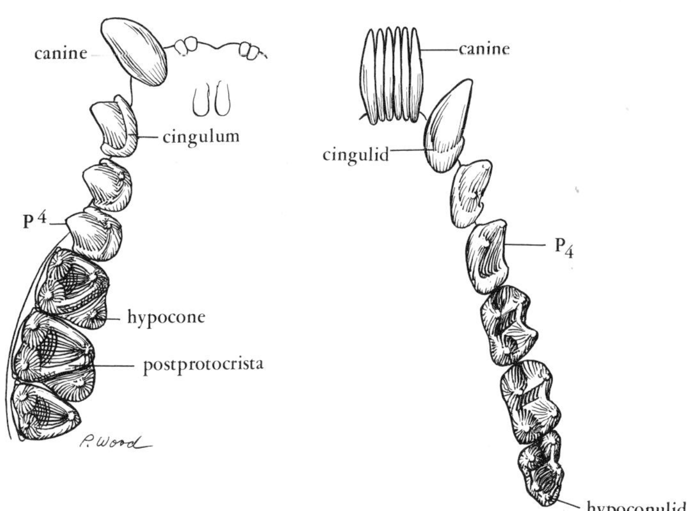

Plate 12. *Loris tardigradus* male, occlusal view ×3·1.

Morphological Observations

	Sample	
	Male	Female
Loris tardigradus (slender loris)	4	1

Incisors

Upper. I^{1-2} are small and peg-shaped. They are paced vertically in the premaxillary bone and the central incisors are separated by a diastema.

Lower. I_{1-2} are procumbent, although not as much as in lemurs. They are elongated and are in close contact with each other. All four are the same length.

Canines

Upper. The upper canines are rather robust and project above the occlusal line. There is a slight longitudinal groove on the buccal surface.

Lower. The lower canines are part of the dental comb and are only slightly larger than the incisors.

Premolars

Upper. P^2 has a large, pointed paracone that rises above the occlusal line of the other premolars. There is no lingual cusp, although the well formed lingual cingulum terminates as the parastyle and distostyle. P^3 is similar to P^2 except that the lingual cingulum displays an enlargement (protocone) opposite the paracone. P^4 is more molariform than P^{2-3}.

Lower. P_2 is the largest of the lower premolars and quite caniniform. It displays a complete lingual cingulum skirting the base which ends as an enamel spur. P_3 has a single, large buccal cusp and a short distal extension of enamel. In addition to the large protoconid, P_4 possesses an incipient metaconid the apex of which is attached to the distolingual surface of the protoconid. A well formed distal marginal ledge circumvents the base of the tooth. P_4 is molariform.

Molars

Upper. M^{1-2} present four cusps: paracone, metacone, protocone and hypocone. The para- and metastyles are visible along the buccal margins of the teeth and in addition, the protoconule is noticeable as a slight elevation on the preprotocrista. The metaconule has been reported on M^1 (Hill, 1953), although it is not present in our sample. Neither of these conules is present in our sample on M_3. A well defined distal cingulum passes from the region of the hypocone to the metastyle on M^{1-2}. M^3 has three cusps; the hypocone is lacking, albeit there is a distolingual cingulum. All three molars possess postprotocristae.

Lower. M_{1-2} possess four well formed cusps, with the mesial two close together and connected by the protocristid. Of the two cusps, the protoconid lies mesial to the metaconid. A slight mesial marginal ridge runs uninterrupted between the two cusps and this ridge is connected to the protoconid by a short paracristid. The distal moiety is somewhat wider and lower than the mesial one and the hypoconid lies mesial to the entoconid. The functional result of this tooth pattern is that the talonid basin can amply accommodate the protocone when the teeth are in centric occlusion. M_3 is morphologically similar to M_{1-2}, except that it possesses the hypoconulid.

All three have a cristid obliqua passing from the hypoconid to near the base of the protoconid. A buccal cingulum is present on M_{1-2} passing along the entire surface of the protoconid and onto the mesial portion of the hypoconid, while being limited to the protoconid on M_3.

Odontometry

Because of the small sample size no measurements were taken.

Genus *Nycticebus*

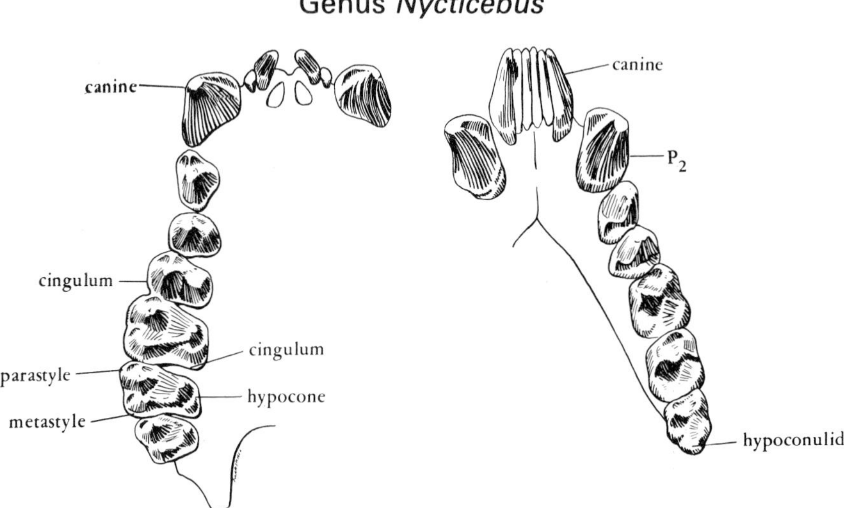

Plate 13. *Nycticebus coucang* male, occlusal view ×2·4.

Morphological Observations

	Sample	
	Male	Female
Nycticebus coucang (slow loris)	18	12

Incisors

Upper. I^1 is much larger than I^2. There is a narrow mesial marginal ridge passing along the lingual surface of the tooth. The two central incisors are separated from each other and are inclined mesially. I^2 is very small and in contact with I^1. It may be shed early in life or be entirely absent (Hill, 1953; James, 1960). In the present sample, I^2 is absent in 27·7% of males and 41·6% of females. Indeed, Schwartz (1974a) suggests that there are "two-incisor"-*Nycticebus* and "four-incisor"-*Nycticebus* groups at the subspecific level. This difference coupled with several other features prompted this student to recommend that the subspecies *Nycticebus coucang borreanus* and *Nycticebus coucang javanicus* be elevated to the species level.

Lower. I_{1-2} are procumbent and styliform. I_1 is slightly smaller than I_2. The dental comb is not quite as procumbent as in lemurs.

Canines

Upper. The upper canine is a large, broadly-based tooth presenting a recurved crown. A consistent but shallow groove is noticeable along the mesiolingual surface.

Lower. The lower canine is part of the dental comb. It is somewhat larger than the incisors and presents a distal marginal ridge which swings across the lingual surface to ascend along the mesial border as the mesial marginal ridge.

Premolars

Upper. P^2 possesses a single large cusp, the paracone. A parastyle and distostyle are present and in addition, a narrow lingual cingulum passes between the two styles. A slight external cingulum is also present. A longitudinal lingual ridge passes from the crown tip to the cingulum. The general appearance of P^2 is caniniform. P^3 is similar to P^2 except that it may be bicuspid. Gregory (1922) noted this internal cusp (protocone) on P^3 and therefore likened the tooth to those found in Eocene lemurs. In the present study, the protocone is present in 61·1% of males and 66·6% of females. The para- and distostyle complex is present as well as buccal and lingual cingula. P^4 is always biscuspid in our sample. The protocone is well formed but never as large as the paracone. The para- and distostyles are quite noticeable; moreover the buccal and lingual cingula are more pronounced than in P^{2-3}. Occasionally the lingual cingulum is slightly hypertrophied in the area where the hypocone forms; however, I hesitated to designate it by this epithet. The tooth is molariform.

Lower. P_2 is a large, caniniform tooth possessing a single high cusp. There is a well developed mesial marginal ridge which passes into the lingual cingulum to terminate distally as an enamel ledge. P_3 has a single, large cusp displaying a lingual cingulum. P_4 is similar to P_3, except that the distal enamel ledge is better developed.

Molars

Upper. M^{1-2} possess four cusps whereas M^3 usually has three cusps. In this sample the hypocone is present on M^3 in only 22·2% of males and is altogether absent in females. The occlusal pattern is reminiscent of the V-shaped configuration seen in Eocene lemurs. A postprotocrista is present on all three molars. The protoconule and metaconule are present although they wear down early, making them difficult to identify. Para- and metastyles are present on M^{1-2}, while generally only the parastyle is present on M^3. Buccal and lingual cingula are present.

Lower. M_{1-2} have four cusps; M_3 has five cusps. The protoconid and metaconid are set close together and connected by a protocristid. The former cusp is always mesial to the latter one and both are slightly higher than the distal cusps. The hypoconid and entoconid are spaced, and the cristid obliqua passes between the hypoconid and metaconid. A very slight buccal cingulum is perceptible on the protoconid of each lower molar. M_3 hypoconulid is present.

Odontometry (Tables 27–30, Appendix)

$I^1 > I^2$ in both dimensions. The lower incisors and canines are not measured. P_2 is the largest lower premolar, whereas P^4 is the largest upper premolar. In the upper jaw the equation is $M^1 > M^2 > M^3$, but the lower molars are $M_1 < M_2 < M_3$, although the mean differences are miniscule between each of the lower molars.

T-tests reveal only one instance of sexual dimorphism. The mesiodistal diameters of the upper canines are significantly larger ($P < 0{\cdot}03$) in males.

Genus *Perodicticus*

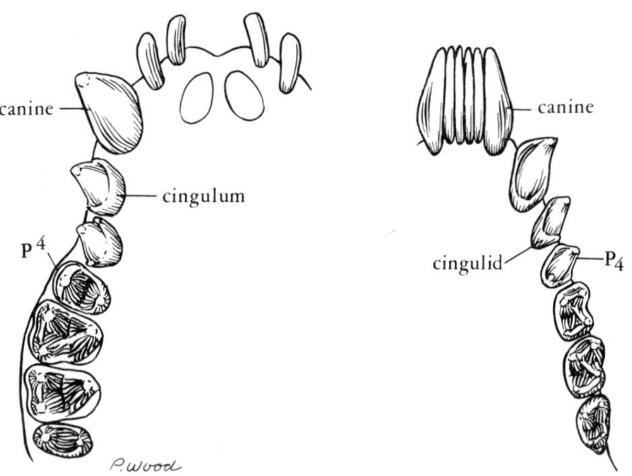

Plate 14. *Perodicticus potto* female, occlusal view ×2·15.

Morphological Observations

	Sample	
	Male	Female
Perodicticus potto (potto)	11	9

Incisors

Upper. I^{1-2} are about equal in size and have a very slight suggestion of lingual, cervical and distal marginal ridges. This development is barely perceivable, and disappears quickly with wear. The central incisors are separated in the mid-line and I^{1-2} do not normally contact each other.

Lower. I_{1-2} are long, narrow, identical in length and procumbent.

Canines

Upper. The upper canines are large, elongated teeth having a single pointed crown and mesiolingual groove. They are implanted vertically.

Lower. The lower canines are in line with the proclivous incisors although they are much more robust. A slight but noticeable mesial marginal ridge is present.

Premolars

Upper. P^2 is elongated and caniniform. The crown tip is well above the occlusal line of the other premolars or molars. Contrary to reports by Hill (1953) and James (1960), this tooth possesses a narrow lingual cingulum which in the present sample, is always visible on the unworn tooth. P^3 presents a paracone adorned fore and aft by small, pointed para- and distostyles. It also has a well delineated lingual cingulum. There is no suggestion of a protocone. P^4 is similar to P^3 with the notable exception of the presence of the protocone. The cusp is present on all P^4 in this study. Also, the buccal cingulum is more apparent on P^4 than on P^{2-3}. The para- and distostyles are present.

Lower. P_2 is separated by a diastema from the canine. It is large and caniniform, presenting a slightly recurved crown. A lingual cingulum passes from the mesial marginal ridge distally to end as a slight heel. P_{3-4} are similar to P_2, but smaller. All three have pointed crown tips. Indeed, all three lower premolars are quite caniniform in *Perodicticus*, and of the three teeth, P_4 is more caniniform in this genus that it is in any other genera of extant lorisids.

Molars

Upper. M^{1-2} are morphologically and functionally tribosphenic teeth. The lingual cingulum is present and its distolingual portion forms a small ledge which is particularly well developed on M^2. Concerning these two teeth Gregory (1922, p. 178) wrote "the first and second upper molars are tritubercular, with rounded cusps lacking external cingula and with very feeble hypocones". However, in our opinion the hypocone is absent and there are slight external cingula from which arise the para- and metastyles. A postprotocrista is present on both M^{1-2}. M^3 is small and presents only two cusps, the paracone and protocone, although a metacone is present in one male specimen. A very small parastyle is observable as well as a slight buccal cingulum.

Lower. M_{1-2} have four cusps while M_3 normally has five cusps. In the present sample, one male and one female lacked the hypoconulid. The protoconids and metaconids are connected by protocristids on all molars while short paracristids are present mesial to the protoconids. The cristid obliqua is present on all lower molars.

Odontometry (Tables 31–34, Appendix)

The upper central incisor is larger than the lateral one whereas the lower incisors are subequal, although no direct measurements were taken. The upper canine has approximately the same mesiodistal diameter in both sexes but displays a significant sex difference buccolingually ($P < 0{\cdot}04$). This is the only significant sex difference found in the present study, suggesting that tooth size differences between the sexes of *Perodicticus* are practically nil.

The lengths of the mandibular molars express a similar relationship in both sexes, $M_1 < M_2 > M_3$. Likewise, M^2 is usually the longest upper molar.

Genus *Galago*

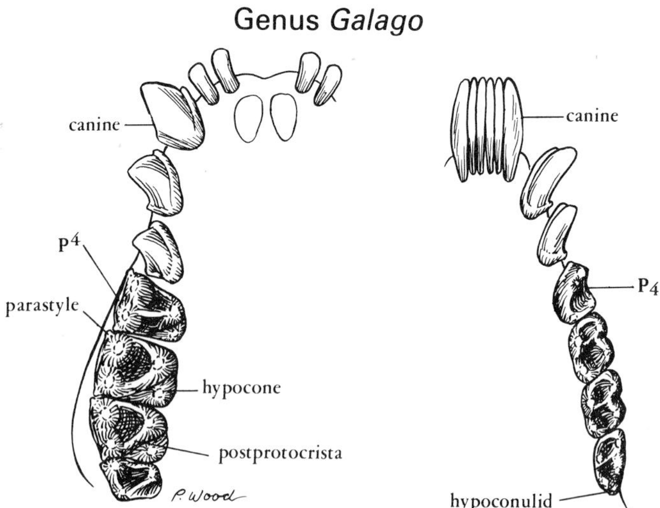

Plate 15. *Galago crassicaudatus* male, occlusal view ×2.

Morphological Observations

	Sample Male	Female
Galago crassicaudatus (greater bushbaby)	17	14
Galago senegalensis (lesser bushbaby)	22	17

Incisors

Upper. The upper central incisors are widely separated in the mid-line. The lingual surfaces of I^{1-2} have weakly developed marginal ridges with no additional demarcation.

Lower. I_{1-2} are long, narrow and emblazon the lemurine pattern.

Canines

Upper. The upper canine is well developed and implanted vertically. Mesial and distal marginal ridges are present as well as a slight lingual cingulum. Distally, there is an indication of an enamel spur.

Lower. The lower canine is in line with the incisors. It is larger than these teeth and has a distal marginal ridge.

Premolars

Upper. P^2 has a single cusp and is compressed buccolingually, giving the impression of a canine. Buccal and lingual cingula are present as well as sharp para- and distostyles. P^3 is morphologically more variable than P^2 in that it tends to develop a lingual cusp. The incidence is low but nevertheless warrants consideration. In the present sample P^3 is bicuspid in 11·8% of male and 7·07% of female *Galago crassicaudatus* and in 27·7% of male but 0% of female *Galago senegalensis.* The para- and distostyles and buccal and lingual cingula are present. P^4 is molariform in that a metacone and hypocone are present, resulting in a four-cusped premolar with both para- and metastyles. Structurally, the three premolars are often quite different in the same individual and in all animals observed; the morphologic transition between P^3 and P^4 is conspicuous. The only other extant prosimians displaying such an abrupt change in premolar morphology are *Hapalemur* and *Arctocebus* (Plates 4 and 16).

Lower. P_2 is caniniform and the crown rises well above the occlusal surface of the other teeth. A complete cingulum sweeps around the lingual aspect of the tooth to terminate as a distostylid. P_3 is slightly procumbent and except for size, resembles P_2. Both the lingual cingulum and distostylid are present. P_4 is molariform, possessing four cusps as well as several crests. The protoconid and metaconid are close together; indeed, the metaconid is merely an enlargment, with an independent apex, on the distolingual surface of the protoconid. A sharp paracristid passes mesially from the protoconid to terminate as a small stylid. The metaconid is differentially expressed by the two species studied. In *Galago crassicaudatus* it is absent in 52·9% of the males and 64·2% of the females, while in *Galago senegalensis* it is absent in 22·7% of males and 23·5% of females. Thus, there is a rather significant difference in the development of this trait between the two species.

Molars

Upper. The molars are essentially tribosphenic even though a hypocone is present on M^{1-2}. The hypocone arises from the cingulum and lies on a lower plane than the protocone; it is normally larger on M^2 and absent on M^3. A postprotocrista is present in all three molars, while the para- and metastyles are always better developed on M^{1-2}. If a style is present on M^3 it is the parastyle.

Lower. M_{1-2} possess four cusps. The mesial two cusps are close together and connected by a protocristid, while the distal two cusps are wider apart and not connected. The cristid obliqua is present but it is better developed on M_{1-2} than on M_3. It should be noted however, that the general development of this crest in *Galago* is not as pronounced as it is in other lorisids. A buccal cingulum is not present in the specimens studied. M_3 has a hypoconulid.

Odontometry (Tables 35–42, Appendix)

Of the two species studied, *Galago crassicaudatus* is the larger animal and it also has slightly larger teeth. There is little sexual dimorphism revealed by the t-test in either species. although there are a few more sex differences in the teeth of *Galago senegalensis*. In this species the mesiodistal diameter of the upper canine ($P<0·03$), and P^2 ($P<0·02$), is significant as well as the mesiodistal diameter of P_2 ($P<0·01$). At the same time, the buccolingual diameter of the upper canine ($P<0·01$) and P_3 ($P<0·03$) are significant. Likewise in *Galago crassicaudatus* both dimensions of the upper canine are significantly different between the sexes (M–D = $P<0·04$; B–L = $P<0·05$). In this species, it is M^1 ($P<0·02$) rather than P^2_2 that is distinctive.

It is apparent from this brief survey of lorisid odontometrics that there is little sexual dimorphism in tooth size exhibited by the members of the taxon. The upper canine is the only tooth consistently different between the sexes of this primate genus.

Genus *Arctocebus*

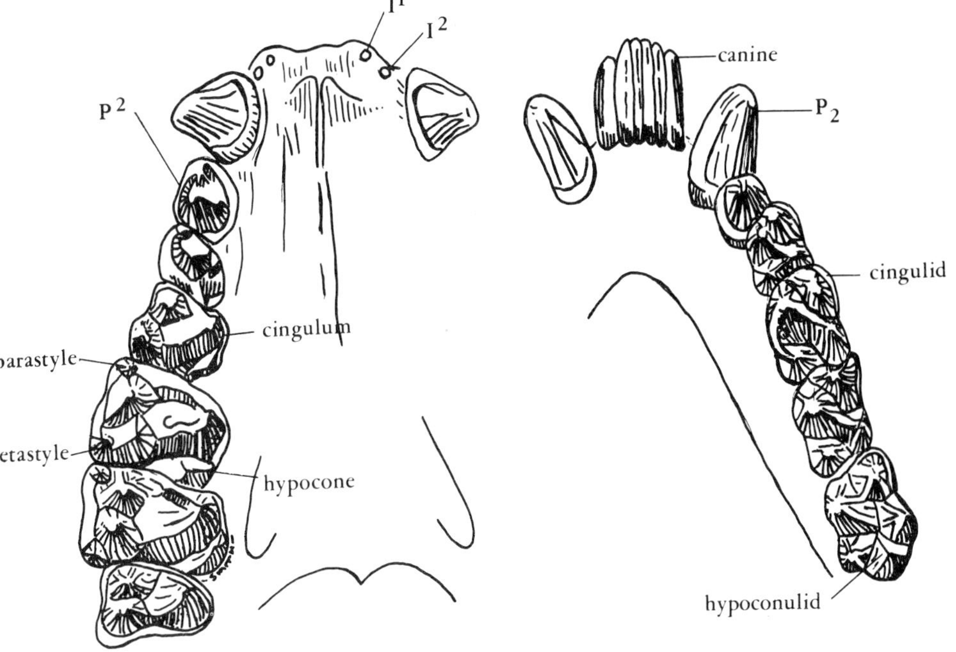

Plate 16. *Arctocebus calabarensis* male, occlusal view ×4.

Morphological Observations

	Sample Male	Sample Female
Arctocebus calabarensis (golden potto)	1	

Incisors

Upper. The upper incisors are small, peg-shaped teeth with I^1 slightly larger than I^2. The central incisors are separated by a mid-line diastema while I^{1-2} are set close together and just mesiolingual to the canine.

Lower. The lower incisors are narrow, pin-shaped structures and are of course procumbent. Both are of equal length.

Canines

Upper. The upper canine is a robust tooth well elevated above the occlusal surface of the other teeth. It presents a buccal and lingual cingulum.

This is the only lorisid with a buccal cingulum on the canine. There is a diastema between the canine and P^2.

Lower. The lower canines are procumbent and only slightly wider than the incisors. There is a narrow marginal ridge circumventing the lingual surface of the tooth.

Premolars

Upper. P^2 has a single, large paracone surrounded by a complete cingulum. The tip of the paracone rises above the occlusal surface of the other premolars. In passing distally, the buccal and lingual cingula incline toward the neck of the tooth. In the specimen we examined the para- and distostyles are absent, nor are they reported by Hill (1953) or James (1960). P^3 is smaller than P^2, which is typical of all living lorisids. A paracone is surrounded by a cingulum on which a barely perceptible protocone can be observed. The distolingual portion of the cingulum is slightly enlarged but in our opinion there is no hypocone, although Hill (1953) reports two lingual cusps for this tooth. P^4 is molariform, presenting a well formed paracone, protocone and metacone. The hypocone is hypertrophied to the extent of having an independent apex. A buccal cingulum is present as well as para- and distostyles.

Lower. P_2 is caniniform. The crown is long and slightly recurved distally. It rises above the occlusal plane of the other teeth. P_2 presents a well formed lingual cingulum, and a narrow buccal cingulum is also discernible. As in the upper premolars, P_3 is small and its single cusp appears to be wedged in between P_{2-4}. A narrow but complete cingulum surrounds the solitary cusp. P_4 has four cusps. The protoconid and metaconid are very close, the latter being little more than an enlargement on the distolingual slope of the former. It is however, a discrete cusp. The hypoconid and entoconid are present forming the small, narrow talonid. The cristid obliqua is observable on the talonid. Buccal and lingual cingula are present.

Molars

Upper. M^{1-2} are very similar in size and morphology. Each presents three sharp cusps, the protocone, metacone and paracone. The postprotocrista is well defined, passing between the protocone and metacone. The hypocone is present as a small fourth cusp. An obvious cingulum passes around the base of M^{1-2}, and buccally it forms a well developed stylar shelf on which reside the para- and metastyles. M^3 is much smaller than M^{1-2}, while morphologically only three cusps are present, the paracone, metacone and protocone. The hypocone is absent in the present specimen, although Hill (1953) and

James (1960) report small or rudimentary hypocones for the genus. A small stylar shelf is present, albeit para- and metastyles are absent.

Lower. There are four cusps on M_{1-2} and five on M_3. All cusps are sharp and the mesial two are slightly higher than the distal two. The former are positioned closer together than the latter and connected by a V-shaped protocristid. The cristid obliqua is observable as a sharp crest passing between the hypoconid and metaconid, while a weak paracristid terminates on a mesiostylid. A well formed buccal cingulum is present on all three molars. On M_{1-3} the cingulum is also perceptible on the mesial, lingual and distal surfaces, although its development on the lingual surface is very weak. It is not present on the distal side of M_3.

Odontometry

No measurements were taken since only a single specimen was studied. For this specimen the molar size formula is $M_1 = M_2 < M_3$ and $M^1 = M^2 > M^3$.

7

Family Tarsiidae

Present Distribution and Habitat

The living tarsiers are found only in south-east Asia, although their earliest antecedents have been discovered in Eocene formations in both North America and Europe. They occupy many islands of the Malay archipelago, from Sumatra in the west to the southern Philippines in the north, and east as far as the Celebes. They have not been found on the mainland of Asia.

Tarsiers live in low tropical rain forests and are rarely seen much above sea level. They frequent coastal areas and seem to prefer regions of secondary growth. Their locomotion is classified as vertical clinging and leaping (Napier and Walker, 1967). They have been observed leaping as much as 1·8 m (6 ft) in a manner not unlike that employed by frogs. Tarsiers are nocturnal and to some degree crepuscular.

Dietary Habits

Tarsiers are essentially insectivorous, eating beetles, grasshoppers, spiders, etc. They are, however, not averse to eating lizards and small crustaceans. Tarsiers seize their prey with both hands and convey it to their mouths for eating. They prefer to capture live prey and eat it immediately, although food may be stored in the mouth for some time before mastication begins.

General Dental Information

Permanent dentition: $I_1^2\ C_1^1\ P_3^3\ M_3^3$

Deciduous dentition: $i_0^0\ c_1^1\ m_2^2$

Sequence of initial eruption of permanent teeth (Schwartz, 1974a):

Tarsius spectrum

Upper: P^2 M^1 I^3 M^2 I^2 C P^4 M^3 P^3

Lower: P_2 I_3 M_1 M_2 C M_3 P_4 P_3

Tarsius syrichta

Upper: P^2 I^3 M^1 M^2 I^2 P^4 C P^3 M^3

Lower: P_2 I_3 M_1 M_2 P_4 C P_3 M_3

Tarsius bancanus

Upper: P^2 M^1 M^2 I^3 I^2 C P^4 M^3 P^3

Lower: I_3 P_2 M_1 M_2 P_4 C M_3 P_3

The permanent dentition of *Tarsius* differs from that of other prosimians in the following manner. Tarsiers possess three premolars and one lower incisor, whereas other prosimians either have two premolars and one lower incisor (indriids), or three premolars and two lower incisors (lorisids and lemurs). Of course, *Daubentonia* and *Lepilemur* display additional numerical contrasts.

The eruption pattern in *Tarsius* differs from that in most other prosimians in the late odontiasis of M^3_3, although the rather late appearance of the upper canine is also observed in *Propithecus* and *Lemur.*

The upper central incisors are vertically implanted, conical teeth set close to the mid-line. There is a complete cingulum encircling the bases which is more pronounced on the lingual side of the tooth. I^2 is much smaller than I^1 and it too presents a complete cingulum. There is little space between it and the upper canine. The single lower incisor is practically vertical in the mandible, in marked contrast to the procumbent nature of I_1 in lemurs.

The upper canine has a sharp crown tip while the base of the crown is conical, being surrounded by a cingulum. The lower canine is slightly hypertrophied and vertical, rather than procumbent. A cingulum skirts three sides of the tooth, being absent only from a small portion of the mesial surface.

The upper premolars are heteromorphic. P^2 is small and possesses a single sharp cusp and a complete cingulum. Except for size, it resembles the upper canine. P^3 is expanded buccolingually, although in our opinion it lacks a protocone. It does have a cingulum. P^4 has a large paracone and a somewhat smaller protocone; it lacks a true metacone. It does possess a cingulum as well as incipient para- and distostyles. Thus P^4 is molariform. The lower premolars are not heteromorphic; rather, they appear very similar from front to back (homomorphic). They each possess a single cusp surrounded by a cingulum.

The upper molars have three cusps, the paracone, protocone and metacone. The lingual cingulum is slightly thickened in the region of the hypocone but the cusp is wanting. A postprotocrista is present as well as buccal cingula on all molars. Small para- and metastyles are present; mesostyles are absent. The lower molars are tribosphenic, possessing a true trigonid with paraconid, protoconid and metaconid, and a talonid with entoconid and hypoconid. Note that the paraconid is absent, except in *Tarsius*, in most recent primates, having generally disappeared during the Eocene. Hershkovitz (1971, p. 127), however, claims that the paraconid may exist in extant platyrrhines and catarrhines as an individual trait, "usually in posterior premolars and more frequently in the deciduous than in the permanent series". The lower molars retain the buccal cingulum.

The chewing habits of *Tarsius* are not well known. It is recognized that they use their hands for capturing and holding the prey while they rip pieces from it with the mouth (Davis, 1962). Apparently the animals eat very slowly; for example, Davis reports that it took approximately 25 minutes to dispose of a four-inch katydid, while Hill (1955, p. 226) describes periods of 30 seconds to one minute "between acts of mastication". The morphology and wear facets of the teeth strongly suggest the orthal retraction motion previously described for the Lorisidae which results in crushing-puncturing strokes. Hill (1955) mentions deliberate side-to-side translations of the mandible. Mills (1955, 1963) also analyzed the positions of wear facets on the molars of *Tarsius* and concluded that chewing in these prosimians involves translation combined with rotation. He further suggests that this type of jaw movement is more primitive than that displayed by the higher primates. Finally it should be noted that the incisors are undoubtedly much more useful in masticatory maneuvers in *Tarsius* than they are in most other prosimians for it is well known that tarsiers kill their prey by biting with their anterior teeth (incisors and canines) (Wharton, 1950).

Genus *Tarsius*

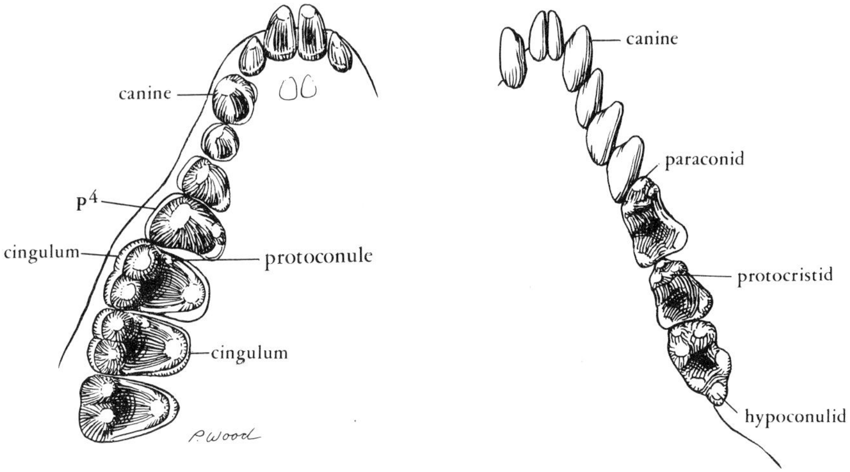

Plate 17. *Tarsius spectrum* female, occlusal view ×3·7.

Morphological Observations

	Sample	
	Male	Female
Tarsius spectrum (spectral tarsier)	1	5
Tarsius bancanus (Horsfield's tarsier)	1	1
Tarsius syrichta (Philippine tarsier)	7	9

Incisors

Upper. I^{1-2} are vertically implanted and I^1 is considerably more elongated than I^2. Both teeth are completely surrounded by a cingulum. A narrow diastema resides between I^2 and the canine.

Lower. I_1, the only lower incisor is not procumbent as in other prosimians, but is small and conical, placed in close opposition between the larger lower canines. Each possesses a basal cingulum.

Canines

Upper. The upper canine is small compared with canines in other prosimians. It is caniniform, presenting a sharp crown tip that is barely higher than the premolars. The tooth is girdled by a cingulum which develops into a mesial flange or style.

Lower. The lower canine is not procumbent but is nearly vertical in the jaw. It is more robust than the incisors or premolars and extends well above the occlusal surface of these teeth. It has a basal cingulum.

Premolars

Upper. P^2 is conical and resembles the canine. A sharp crest passes down the lingual surface of the tooth from the crown tip to the lingual cingulum. This crest divides the tooth into two nearly equal parts. P^3 is similar to P^2 except that it is larger and the lingual cingulum is more pronounced, particularly in the distolingual area. P^4 tends to be molariform. It possesses two cusps, paracone and protocone, which are always present in the present study. The protocone is small but an independent apex is identifiable. A weak buccal cingulum is present.

Lower. P_{2-4} are rather plain, uniform teeth possessing a single cusp encircled by a cingulum. P_2 is very small in contrast to that of lemurs.

Molars

Upper. The upper molars are tribosphenic. They have three well developed cusps, the paracone, protocone and metacone, and when unworn they are high and sharp. The hypocone is absent in our opinion, or at most, it is nothing more than a small bulge on the distolingual portion of the cingulum. A diminutive protoconule is present on all three molars in all specimens studied, whereas a metaconule is indicated on M^{1-2} in approximately 25% of our *Tarsius syrichta* and *Tarsius spectrum* specimens. It is normally absent on M^3. In *Tarsius bancanus*, it is present on M^{1-2} in the female and absent in the male. A buccal cingulum is present on all three molars.

Lower. The molar molars are tribosphenic. The trigonid with protoconid, paraconid and metaconid is substantially elevated above the talonid which bears the hypoconid and entoconid on M_{1-2} and the hypoconulid on M_3. Also present is the cristid obliqua and a V-shaped protocristid. A narrow but distinct buccal cingulum is observable on all three molars.

Odontometry (Tables 43–46, Appendix)

I^1 is larger than I^2; indeed it is about twice the size of I^2. There is a progressive mesiodistal size increase in the premolars, $P^2_2 < P^3_3 < P^4_4$. The size relations of the maxillary molars is $M^1 \leqslant M^2 > M^3$, while the mandibular formula is $M_1 \leqslant M_2 < M_3$.

Sexual dimorphism in tooth size was studied in *Tarsius* by pooling the different species. A single difference is noted: the buccolingual diameter of the trigonid of M_3 is significantly larger in the male ($P < 0{\cdot}02$). Otherwise, tooth dimensions are similar for the sexes.

8

Family Callithricidae

Present Distribution and Habitat

The family Callithricidae (marmosets and tamarins) is separated into two subfamilies, the Callithricinae and the Callimiconinae. The former group includes the genera *Callithrix*, *Cebuella*, *Saguinus* (*Saguinus*), *Saguinus* (*Oedipomidas*), *Saguinus* (*Marikina*) and *Leontideus*, while the latter taxon contains a single genus, *Callimico*. The Callithricidae live in the tropical and montane rain forests of Middle and South America. *Saguinus* (*Oedipomidas*) is found as far north as Panama while the other genera are limited to South America.

The Callithricidae are arboreal primates living the majority of their lives in the trees. Their locomotion is generally described as squirrel-like for they progress by quick, jerky movements and are said to use their claws for climbing. Their locomotion is described as quadrupedal by Napier and Napier (1967), while Erikson (1963) places them in his springer category. Both subfamilies are diurnal.

Dietary Habits

The Callithricidae are essentially insectivorous, although they eat fruit, seeds and other vegetable matter. They also eat birds' eggs as well as birds when they can capture them. Indeed, Napier and Napier (1967) believe *Saguinus* (*Oedipomidas*) to be omnivorous. In captivity marmosets eat a wide variety of food as do most captive primates.

General Dental Information

Permanent dentition: $I^2_2\ C^1_1\ P^3_3\ M^2_2$, Callithricinae

Permanent dentition: $I^2_2\ C^1_1\ P^3_3\ M^3_3$, Callimiconinae

Deciduous dentition: $i^2_2\ c^1_1\ m^3_3$

Sequence of eruption of permament teeth (Serra, 1952a):

Callithrix penicillata

$$\frac{M^1 \quad I^1 \quad M^2 \quad P^4 \quad I^2 \quad P^3 \quad P^2 \quad C}{M_1 \quad I_1 \quad M_2 \quad P_4 \quad I_2 \quad P_3 \quad P_2 \quad C}$$

The eruption sequence for *Cebuella pygmaea* is not known

Saguinus (*Oedipomidas*)

$$\frac{M^1 \quad I^1 \quad M^2 \quad . \quad . \quad . \quad . \quad . \quad C}{M_1 \quad I_1 \quad M_2 \quad . \quad . \quad . \quad . \quad . \quad C}$$

(the premolar eruption sequence was not presented)

Saguinus (*Leontocebus*)

$$\frac{M^1 \quad I^1 \quad I^2 \quad M^2 \quad P^4 \quad P^2 \quad P^3 \quad C}{M_1 \quad I_1 \quad I_2 \quad M_2 \quad P_4 \quad P_2 \quad P_3 \quad C}$$

The eruption sequence for *Callimico goeldii* is not known

The number of permanent molars differs between the two subfamilies. The aberrant *Callimico* possesses three rather than two permanent molars, thus differing numerically from all other Callithricidae.

The upper central incisor is spatulate while I^2 is narrow and tends to have a pointed crown tip; consequently, the incisors are heteromorphic. A lingual cingulum is present on both I^{1-2}, as well as mesial and distal marginal ridges. The lower incisors are used for separating the family into two major groups, the long-tusked tamarins and the short-tusked marmosets. *Callimico* fits into the long-tusked group. The appellation refers to the relation between the incisors and canines: the short-tusked animals have elongated incisors; indeed, the incisors are practically as long as the canines. The difference between the two varieties is quite obvious (Fig. VII).

The upper canines are rather well constructed, elongated teeth. In all forms, the canines project below the occlusal surface of the other teeth. Even in *Cebuella*, the smallest of living platyrrhini, the canine protrudes well beyond the tips of the other teeth. The lower canine is a relatively robust tooth in all species. It presents a lingual cingulum which terminates as a distal stylid, particularly in *Leontideus* and *Callimico*.

The upper premolars are bicuspid. The paracone is always the larger of the two cusps. Buccal and lingual cingula are present although they are better developed on P^{3-4}. Also, the para-, meso- and metastyle complex is

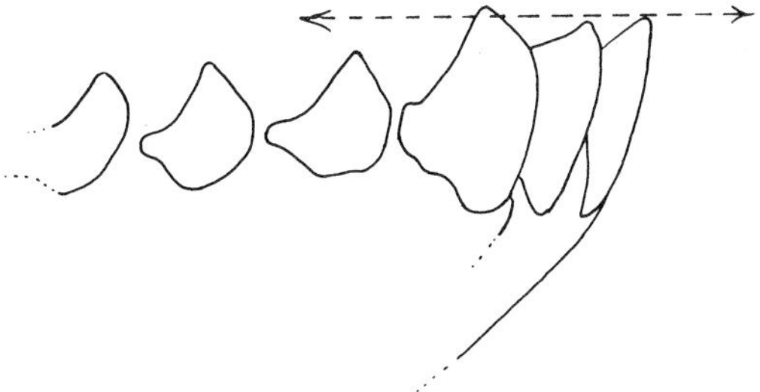

A. *Cebuella pygmaea*, short—tusked marmoset

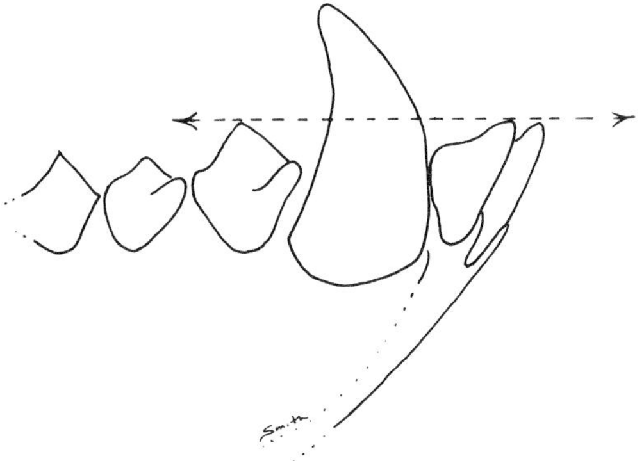

B. *Leontideus rosalia*, long—tusked tamarin

Fig. VII. Canine–incisor relations in Callithricidae.

present. P_2 is somewhat caniniform; it presents a single buccal cusp, and may display a weak cingulum around the mesiolingual neck of the tooth, especially in *Saguinus* and *Leontideus*. A buccal cingulum is occasionally found on the tooth. P_3 is generally bicuspid, although the protoconid is very small and, if worn, is difficult to identify in *Cebuella* and *Callithrix*. P_3 may possess a buccal cingulum. P_4 is biscuspid with a small distal platform. The protoconid and metaconid are connected by a protocristid. A buccal cingulum may be present.

The upper molars are essentially three-cusped: paracone, protocone and metacone. The hypocone is wanting in marmosets with the exception of *Saguinus* and *Callithrix*, where its presence is variable (Kinzey, 1973; also present study). Buccal and lingual cingula as well as a variety of stylar formations display an inconstant appearance on the molars. M^2 deviates from this pattern: it is much reduced and lacks the styles, except in *Callimico*, where it is only slightly smaller than M^1. In this taxon, M^3 is present but petit. The lower molars have four cusps; the hypoconulid is absent. The protoconid and metaconid are set close together and connected by a protocristid. Distally, the hypoconid and entoconid are farther apart and there is no transverse crest. M_2 is reduced except in *Callimico*, in which M_{1-2} are of about equal dimensions. M_3 is the smallest of the series but still possesses four cusps. The expression of buccal and lingual cingula is variable on all lower molars.

There is little known about the actual masticatory habits of marmosets. The morphology of the molars, however, would seem to suggest a pattern not unlike that observed in prosimians: essentially a vertical movement complicated by some mesiodistal and buccolingual components during the chewing cycles. Puncture-crushing cycles probably predominate (vertical movement) in these preponderantly insect-eating creatures. This is further supported by the anatomy of the temporomandibular joint which, according to Hill (1957), "is mainly of the hinge type", permitting little protraction, retraction or lateral movement.

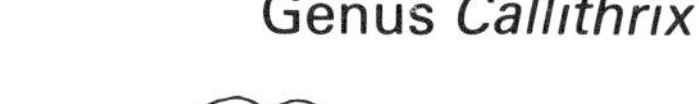

Genus *Callithrix*

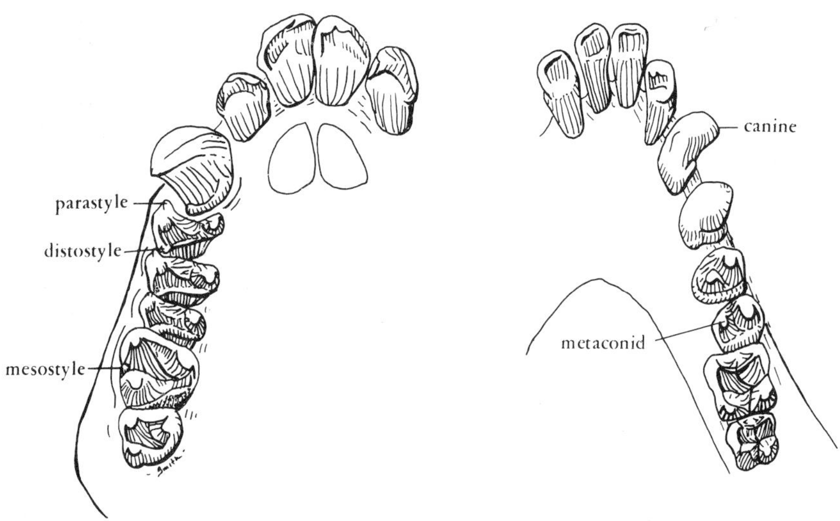

Plate 18. *Callithrix penicillata* male, occlusal view ×4·15.

Morphological Observations

	Sample Male	Sample Female
Callithrix penicillata (blackbeard marmoset)	3	2

Incisors

Upper. The upper central incisors are high, spatulate teeth with a lingual cingulum. They protrude slightly forward. The lateral incisors are narrow with pointed crown tips and each displays narrow lingual cingula. I^{1-2} are heteromorphic.

Lower. The lower central and lateral incisors are about equal in size. They are relatively elongated when compared with the canines, and therefore, *Callithrix* is considered a short-tusked marmoset (Fig. VII). I_2 has a more pointed crown than I_1 and is slightly thicker.

Canines

Upper. The upper canines are relatively large in both sexes. They extend an appreciable distance beyond the occlusal surface of the other teeth. A longitudinal groove indents the mesial surface of each tooth.

Lower. The lower canines are slightly larger than the incisors and resemble I_2 morphologically. They have a distal stylid which helps to increase their mesiodistal diameter.

Premolars

Upper. P^{2-4} have two cusps, the paracone and protocone, with the former being much larger and higher than the latter. A buccal cingulum is present in all specimens as well as para- and distostyles on all three premolars. Kinzey (1973) reports a continuous buccal cingulum on P^4 in 6% of 31 specimens of *Callithrix argentata* and a mesostyle in 7%. Para- and distostyles are present on P^{2-4} in all specimens studied by Kinzey. He also reports the presence of lingual cingula on P^{3-4}, none on P^2. A similar condition obtains for *Callithrix penicillata.*

Lower. P_2 is caniniform, possessing a single compressed cusp. A small distostylid is present as well as an incipient lingual cingulum. P_3 has one cusp although the tooth is wider buccolingually than P_2. A narrow mesiolingual cingulum is present. P_4 is bicuspid, presenting a large protoconid and a much smaller metaconid. The two cusps are set close together and connected by a thin protocristid. A small talonid platform is present. Kinzey (1973) reports protostylids and distostylids as variable structures on all three premolars in *Callithrix argentata.* In the present study, only distostylids are present.

Molars

Upper. M^1 has three cusps; the hypocone is lacking. Three cusps are apparently the rule in *Callithrix* (Hill, 1957; James, 1960; Kinzey, 1973). It should be mentioned that Kinzey found a small hypocone on M^1 in one of 31 cases examined. A very small remnant of the lingual cingulum is present on the distolingual surface of the protocone. A slight buccal cingulum is present plus a mesostyle. M^2 is small albeit consisting of three cusps. The postprotostyle is absent but an extremely narrow buccal cingulum is observable. M^3 is absent.

Lower. M_{1-2} possess four cusps. The protoconid and metaconid are connected by a protocristid, whereas there is no transverse nexus between the hypoconid and entoconid. Kinzey (1973) reports buccal cingula as variable entities on M_{1-2} in *Callithrix argentata*; they are absent in the five animals we examined. M_3 is absent.

Odontometry

No measurements were taken because of the small sample size; however, it was observed that in all animals the following mesiodistal tooth size formulae obtained: $P^2 > P^3 < P^4$, $P_2 > P_3 > P_4$ and $M^1_1 > M^2_2$. There is no obvious sexual dimorphism in tooth size.

Genus *Cebuella*

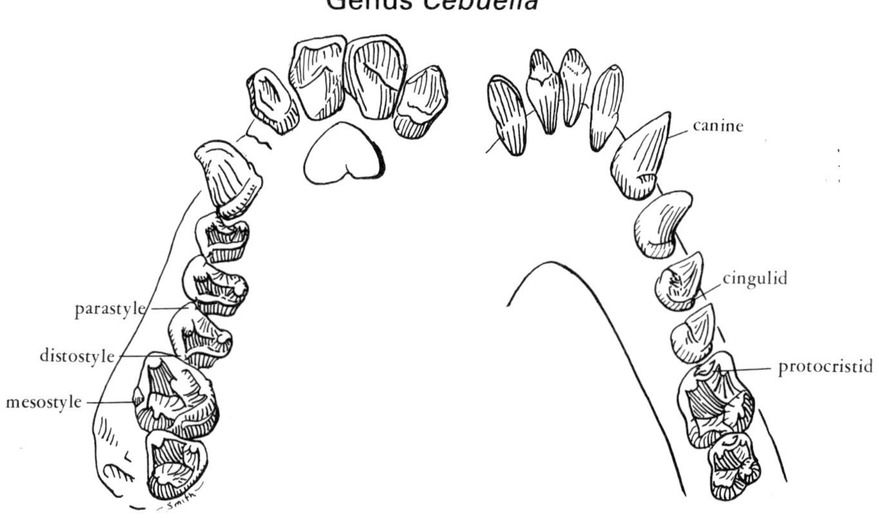

Plate 19. *Cebuella pygmaea* male, occlusal view ×5·6.

Morphological Observations

	Sample Male	Sample Female
Cebuella pygmaea (pygmy marmoset)	2	1

Incisors

Upper. I^1 is larger than I^2 and somewhat spatulate. A narrow lingual cingulum is present in the unworn state. I^2 is more conical and pointed than I^1. There is a diastema between I^2 and the canine.

Lower. I_{1-2} are subequal in size. They are as high as the canines and there is little differentiation between I_{1-2}.

Canines

Upper. The upper canine is a relatively robust tooth projecting well below the occlusal line of the other teeth. A narrow lingual cingulum festoons the base of the tooth, terminating as a small distostyle.

Lower. The lower canine is only slightly larger than the lower incisors; therefore, *Cebuella* is classified as a short-tusked marmoset. A small distostylid is present.

Premolars

Upper. P^2 has a single cusp, the paracone. A complete primary lingual cingulum is present. P^3 is bicuspid, as is P^4. All three premolars display para- and distostyles.

Lower. P_{2-3} are small single-cusped teeth. P_4 is bicuspid, the metaconid being a slight elevation connected to the protoconid by the protocristid. Buccal cingula are not present in the three animals studied.

Molars

Upper. M^1 has three cusps. The hypocone is absent although a very weak postprotostyle is present on M^1. A small mesostyle intrudes between the paracone and metacone on M^1 in the three specimens. Para- and metastyles are present on M^1 in one of the three animals studied. M^2 is similar to M^1 except that it is smaller and lacks stylar development. M^3 is absent.

Odontometry

Both the upper and lower premolars increase in size from mesial to distal; thus, $P^2_2 < P^3_3 < P^4_4$, whereas the molar mesiodistal size gradient is $M^1_1 > M^2_2$. The premolars and molars are practically the same size in the three animals studied and we doubt if sexual dimorphism exists in the species.

Genus *Saguinus*

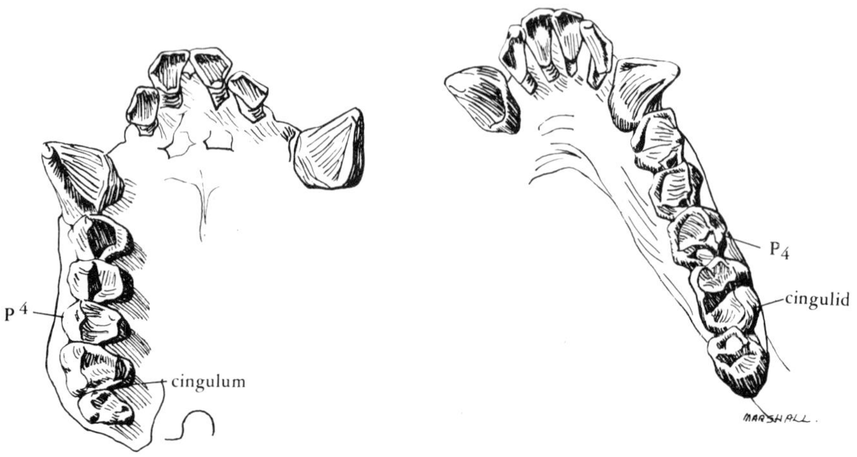

Plate 20. *Saguinus (Oedipomidas) geoffroyi* male, occlusal view ×2·9.

Morphological Observations

	Sample Male	Sample Female
Saguinus (Oedipomidas) geoffroyi (Geoffroy's tamarin)	12	15

Incisors

Upper. I^1 is a broad, spatulate tooth with a well developed lingual cingulum which becomes the mesial and distal marginal ridges. The crown of I^2 is not as broad as in I^1; indeed, it is more often pointed, resulting in heteromorphic incisors.

Lower. I_{1-2} are subequal in size and approximately the same height. A small enamel flange is present on the distal margin of I_2.

Canines

Upper. The upper canine is separated from I^2 by a diastema but is in contact with P^2. The tooth is elongated beyond the occlusal plane. A complete lingual cingulum is present as well as a deep longitudinal groove on the lingual surface of the tooth.

Lower. The lower canine is a long, robust tooth. It is larger than the incisors and extends well above them, thus placing *Saguinus* in the long-tusked tamarin group. A lingual cingulum terminates as a distostylid.

Premolars

Upper. P^2 is bicuspid, although the protocone is very small. Para- and distostyles are always present, a condition Kinzey (1973) also reports for *Saguinus mystax* and *Saguinus oedipus.* P^3 is similar to P^2 except larger. P^4 is the largest of the series. The protocone is well developed and a postprotostyle occurs in over 85% of the present sample. Kinzey (1973) found a protostyle in 100% of his sample of *Saguinus mystax* and *Saguinus oedipus.* Para- and distostyles are present on P^{3-4}.

Lower. P_2 is caniniform. The protoconid is large and has a well formed lingual cingulum. Kinzey (1973) reports a lingual cingulum on the lower premolars of a single individual (*Saguinus mystax*). P_2 also displays consistent mesio- and distostylids in our sample as well as those reported by Kinzey (1973). P_3 has two cusps, the protoconid and metaconid. Stylids are variable, although present in over 80% of our sample. P_3 is bicuspid as is P_4, however the development of stylids is less frequent (approximately 65%).

Molars

Upper. M^1 has three cusps. In the present sample, a small hypocone is present in one male. Kinzey (1973) reports a hypocone incidence of 16% of 32 *Saguinus mystax*, and failed to find it in *Saguinus oedipus.* Obviously, the hypocone is a relatively rare entity in *Saguinus.* The protostyle and post-protostyle are always present in M^1 in *Saguinus geoffroyi.* An identical frequency is noted by Kinzey for *Saguinus mystax* and *Saguinus oedipus.* Para- and metastyles also enjoy a high incidence in our sample as well as in Kinzey's (between 94 and 100%), whereas the mesostyle is much less frequent in our material (25% compared with 100% in *Saguinus mystax* and 41% in *Saguinus oedipus*). M^2 is much smaller than M^1, although three cusps are generally present. A hypocone is absent in the present sample. The postprotostyle has a greater frequency than the protostyle: 100% compared with 66%. A similar reduction is noted by Kinzey (1973). No sex differences are present in *Saguinus geoffroyi.* M^3 is absent.

Lower. M_{1-2} have four cusps, and no hypoconulid. The mesial two cusps are connected by a protocristid while there is no transverse nexus between the distal cusps. A buccal cingulum is a variable entity on the protoconid. In *Saguinus geoffroyi* it is present on M_1 in approximately 90% of the animals sampled, while being slightly less common on M_2. Kinzey's (1973) findings are not appreciably different from those reported here.

Odontometry (Tables 47–50, Appendix)

Our findings on sexual dimorphism are interesting. In general, the teeth of *Saguinus geoffroyi* are approximately the same size in both sexes. Where statistically significant size differences exist, the female has larger teeth. T-tests reveal only three important differences: the mesiodistal dimensions of M^1 ($P<0{\cdot}02$) and M_1 ($P<0{\cdot}02$), and the buccolingual diameter of M^2 ($P<0{\cdot}003$). Additional studies of larger samples should be performed to see if our findings hold up.

Genus *Leontideus*

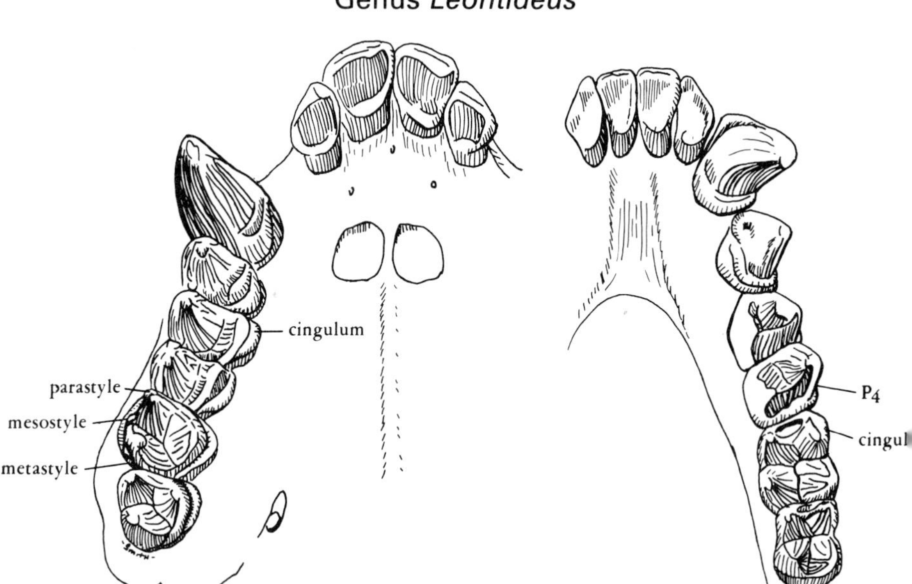

Plate 21. *Leontideus rosalia* female, occlusal view ×3·5

Morphological Observations

	Sample	
	Male	Female
Leontideus rosalia (golden lion tamarin)		2

Incisors

Upper. I^1 is broad and spatulate. The lingual surface is completely outlined by well developed marginal ridges. I^2 has a more pointed incisal border than I^1; thus, the incisors are heteromorphic. Well demarcated marginal ridges are present on I^2.

Lower. I_{1-2} are of about equal dimensions, I_1 being only slightly narrower than I_2. Both teeth are approximately the same height. Narrow marginal ridges are present on both incisors.

Canines

Upper. The upper canine is long and trenchant, displaying a narrow cingulum along its lingual base. A small longitudinal groove incises the mesial surface of the tooth.

Lower. The lower canine is much larger than the incisors and projects well above them. A marked lingual cingulum skirts the base of the canine to form a distostylid. *Leontideus* belongs to the long-tusked tamarins.

Premolars

Upper. P^2 is caniniform. A complete primary lingual cingulum increases the buccolingual diameter of the tooth, although a true protocone is wanting. P^3 is bicuspid as is P^4. A small but distinct postprotostyle is present on P^{3-4}. P^{2-4} have para- and distostyles.

Lower. P_2 is somewhat caniniform with a single cusp rising sharply above the other premolars. An obvious primary lingual cingulum is present. P_3 is also unicuspid and has the appearance of having been squeezed between P_2 and P_4. P_4 is biscuspid and presents a small talonid shelf. Buccal cingula are not present in the two animals studied.

Molars

Upper. M^{1-2} have three cusps plus a well formed lingual cingulum which is more complete on M^1, although in neither is a hypocone present. M^1 has para-, meso- and metastyles, while M^2 has only the parastyle. M^3 is absent.

Lower. M_{1-2} have four cusps; both lack the hypoconulid. The mesial two are close together and only slightly elevated above the two distal cusps. A buccal cingulum is limited to the protoconid on both molars.

Odontometry

No measurements were made of the two specimens.

Genus *Callimico*

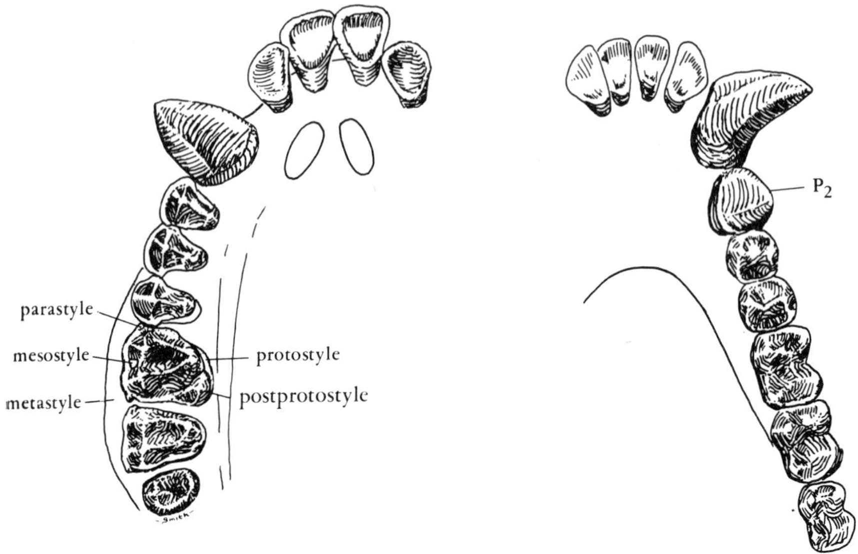

Plate 22. *Callimico goeldii* male, occlusal view ×3·5.

Morphological Observations

	Sample	
	Male	Female
Callimico goeldii (Goeldi's marmoset)	3	

Incisors

Upper. I^1 is spatulate and the incisal border is broad and flat. A complete marginal ridge delineates the lingual surface of the crown. The ridge becomes more pronounced along the cervical region of the tooth. I^2 is smaller than I^1 and the incisal border is pointed, resulting in heteromorphic incisors. Lingual marginal ridges are quite noticeable on I^2.

Lower. I_{1-2} are subequal. On I_1 the incisal border is flat, while on I_2 it inclines distally.

Canines

Upper. The upper canine is a relatively robust, conical tooth with a sharp tip. A diastema separates the canine from I_2. A slight cingulum skirts the lingual neck to pass onto the distostyle at the base of the tooth. A vertical groove on the mesial surface extends from near the crown tip to the neck.

Lower. The lower canine is large with a sharply tapering crown. It presents a groove mesiolingually and a well formed distostylid. There is no diastema between it and I_2. The size discrepancy between the canines and incisors places *Callimico* with the long-tusked tamarins.

Premolars

Upper. P^{2-4} are bicuspid and increase in size from front to back. The paracone is much larger and higher than the protocone, particularly on P^2. Para- and distostyles are present on P^{2-4} as well as a postprotostyle on P^4.

Lower. P_2 has a single cusp and the tooth is caniniform. A complete primary lingual cingulum is present. P_{3-4} are biscuspid and both have incipiently developed talonid basins, especially P_4. No cingula are visible on P_{3-4}.

Molars

Upper. M^{1-2} have three cusps, and the hypocone is absent in the three specimens studied. M^1 has a well developed lingual cingulum which, according to Hill (1959), may enlarge into a diminutive hypocone. Para- and metastyles are present on M^{1-2}, while a small mesostyle is just visible on M^1. M^2 has a lingual cingulum but it is small compared with the one on M^1. M^3 is present but very small. Of the three specimens studied, two had M^3 with three cusps while one had an M^3 with two cusps, the metacone being absent. Hill (1959) reports M^3 as bicuspid. Also, it lacks the extra adornments present on M^{1-2}. It should be mentioned that the presence of M^3_3, more vertically implanted lower incisors, and certain crown features of the lower molars have led some students to regard *Callimico* as a primitive cebid (Simpson, 1945; Cabrera, 1957), or demanding a separate family, Callimiconidae (Dollman, 1937). The most recent assessment of the animal's taxonomic status is that of Hill (1959), who places *Callimico* in the Callithricidae. Napier and Napier (1967) subscribed to this interpretation, and so do we.

Lower. M_{1-3} have four cusps; the hypoconulid is absent. The mesial two cusps are connected by a V-shaped protocristid and the entire mesial area of the tooth is slightly higher than the distal portion. There is no transverse crest between the hypoconid and entoconid. Buccal cingula are absent.

Odontometry

Since *Callimico* is such a rare species the teeth were measured in the three male animals. The means are recorded below in millimeters. These are all small teeth, the largest being M^1. The molar formula is $M_1^1 > M_2^2 > M_3^3$.

Measurement	I^1	I^2	C	P^2	P^3	P^4	M^1	M^2	M^3
M–D	2·1	1·7	2·5	2·1	2·0	2·2	2·8	2·6	1·4
B–L	2·0	1·5	2·4	2·6	3·1	3·4	4·0	3·5	2·0
	I_1	I_2	C	P_2	P_3	P_4	M_1	M_2	M_3
M–D	1·4	1·2	2·1	2·5	1·8	1·7	3·0	2·3	1·6
B–L	1·5	1·3	2·3	2·0	2·0	2·1	2·2	2·1	1·8

9

Family Cebidae

Present Distribution and Habitat

The family Cebidae includes five or six subfamilies depending on whether Callicebus is considered a subfamily (Callicebinae) by itself (Pocock, 1925a; Hill, 1960), or placed in the subfamily Aotinae (Simpson, 1945; Napier and Napier, 1967). The latter classification is followed here. The Cebidae are located from Mexico to Argentina and from the east coast of South America to the Peruvian Andes. They range from sea level to approximately 2,000 m (6,500 ft) in altitude.

The Cebidae live their lives in the trees and are well adapted for this existence. There are no forms comparable to some of the terrestrial Old World monkeys, although numerous parallels can be drawn between several of the arboreal species of the two areas. Their mode of locomotion ranges from brachiation to quadrupedalism (Erikson, 1963; Napier and Napier, 1967). The family possesses one unique characteristic: five genera (*Cebus*, *Lagothrix*, *Ateles*, *Alouatta* and *Brachyteles*) possess prehensile tails. The tails are used not only for suspending the body but for conveying food to the mouth, and in this function there is no parallel among the Old World monkeys. The little night monkey (*Aotus*) is nocturnal; all other genera are diurnal.

Dietary Habits

The cebid diet is mostly vegetarian, i.e. fruits, leaves and shoots. According to Hill (1960) the smaller species (*Saimiri*, *Callicebus* and *Aotus*) also have a predilection for insects, particularly spiders, beetles and butterflies, as well as small lizards and birds' eggs. Among the cebids, *Alouatta* and *Brachyteles* stand out as highly specialized leaf-eaters. Indeed, Hill (1960) has likened the econiche occupied by *Alouatta* to that occupied by the langurs of Asia and the indrisine lemurs of Madagascar.

General Dental Information

Permanent dentition: $I^2_2\ C^1_1\ P^3_3\ M^3_3$

Deciduous dentition: $i^2_2\ c^1_1\ m^3_3$

Sequence of eruption of permanent teeth (Serra, 1952a):

Aotus

M^1	M^2	M^3	I^1	I^2	P^4	P^3	P^2	C
M_1	M_2	M_3	I_1	I_2	P_4	P_3	P_2	C

Callicebus

M^1	I^1	I^2	M^2	P^4	P^2	P^3	C	M^3
M_1	I_1	I_2	M_2	P_4	P_2	P_3	C	M_3

Ateles

M^1	I^1	I^2	M^2	P^2	P^4	P^3	C	M^3
M_1	I_1	I_2	M_2	P_2	P_4	P_3	C	M_3

Brachyteles arachnoides

M^1	I^1	I^2	M^2	P^4	P^3	P^2	M^3	C
M_1	I_1	I_2	M_2	P_4	P_3	P_2	M_3	C

Lagothrix

M^1	I^1	I^2	M^2	P^2	P^4	P^3	M^3	C
M_1	I_1	I_2	M_2	P_2	P_4	P_3	M_3	C

Pithecia

M^1	I^1	M^2	I^2	M^3	P^4	P^2	P^3	C
M_1	I_1	M_2	I_2	M_3	P_4	P_2	P_3	C

The eruption sequence for *Chiropotes* is not known

Cacajao

M^1	I^1	M^2	I^2	M^3	P^4	P^2	P^3	C
M_1	I_1	M_2	I_2	M_3	P_4	P_2	P_3	C

Cebus

M^1	I^1	I^2	M^2	P^2	P^4	P^3	C	M^3
M_1	I_1	I_2	M_2	P_2	P_4	P_3	C	M_3

Saimiri sciureus (Long and Cooper, 1968)

M^1	M^2	I^1	I^2	P^3	P^4	P^2	M^3	C
M_1	M_2	I_1	I_2	M_3	P_4	P_2	P_3	C

Alouatta

M^1	I^1	I^2	M^2	P^2	P^4	P^3	M^3	C
M_1	I_1	I_2	M_2	P_2	P_4	P_3	M_3	C

All cebids have three permanent molars although many genera display markedly reduced third molars. M^3_3 reduction is most noticeable in *Aotus*, *Saimiri*, *Pithecia*, *Ateles* and *Cebus*.

The upper incisors vary in shape and size among the genera. In some, e.g. *Alouatta*, they are homomorphic, whereas in *Ateles* and the other Atelinae they are heteromorphic. Lingual marginal ridges are usually present and occasionally lingual tubercles are observed. In addition, both I^{1-2} may display vertical grooves or ridges on their lingual surfaces. The mandibular incisors also vary a great deal among the genera. In general I_1 is rectangular while I_2 is more caniniform.

The upper canines display different degrees of sexual dimorphism among the genera. They are elongated and always project beyond the occlusal plane. The lower canines are generally robust teeth with large bases. They usually recurve and present sharp crown tips. Occasionally, lingual cingula are noticed and these usually terminate as distostylids.

The upper premolars are usually bicuspid, especially in *Cebus*. P^2, on the other hand, may lack any development of the protocone in *Alouatta*. Orlosky (1973) found the protocone absent in 33 to 45% of *Alouatta* on P^2. The paracone is always the larger of the two cusps. Buccal cingula vary on P^{2-4} both within and among species, and their development is related to the presence or absence of the para- and distostyles. The latter adornments may be absent on P^{2-4} as in *Ateles geoffroyi*, or have an incidence of 78 to 100% as in *Alouatta seniculus* (Kinzey, 1973). Hypocones may be found on P^{3-4} and their frequency increases from P^3 to P^4 (Orlosky, 1973). The lower premolars tend to add cusps from P_{2-4}. Thus, P_2 is frequently unicuspid, P_3 is normally bicuspid and P_4 may have three cusps. Of course this morphologic gradient can be expected to vary among the different taxa. As noted earlier when discussing *Tarsius*, the paraconid is a rare commodity among extant primates, but they have been reported as vestigial structures in New World monkeys by Butler (1939) and Hershkovitz (1971). In his excellent study of the teeth of living and fossil cebids, Orlosky (1973) notes the appearance of the paraconid on P_2 in *Alouatta*, *Ateles*, *Lagothrix*, *Cebus*, *Saimiri* and *Cacajao*.

The upper molars normally possess four cusps, although M_3 frequently lacks a hypocone resulting in a triangular occlusal outline. In *Callicebus* the upper molars possess four cusps; however, the hypocone is small and the occlusal outline is somewhat reminiscent of tritubercular molars. Styles and cingula are variously developed on all three molars as well as an intricate system of cristae. Of the latter, the postprotocrista is frequently seen passing between the protocone and metacone. The lower molars usually have four cusps; however, a hypoconulid may be present in some genera, e.g. *Ateles*. The mesial two cusps are connected by a protocristid whereas the distal two lack a transverse connection. Also, a series of additional cristids are observable interconnecting the different cusps. Mesio- and distostylids are differentially developed among cebids and reach their maximum expression in *Saimiri sciureus* (Kinzey, 1973).

It is well known that food is mechanically broken into smaller and smaller pieces as it is sheared across the cusps and crests forming the crown surfaces of the premolars and molars. Primate molar tooth design has maximized the occlusal surface for triturating a wide assortment of foods. The teeth of cebid monkeys exemplify this mechanical–functional–food relationship in a variety of ways. James (1960, p. 132) is quite aware of the importance of molar design among cebids for he writes:

> The narrow molars of the movable mandible acting against the fixed wide molars of the maxillae are an excellent example of the arrangement for trituration of leaves, the chief food of howlers.

The careful and thorough cinefluorgraphic studies of jaw movement in *Tupaia*, *Galago*, *Saimiri* and *Ateles* (Hiiemae and Kay, 1973; Kay and Hiiemae, 1974) permit an analysis of jaw dynamics and tooth use which was previously impossible to make. The path of the mandible during mastication is outlined as well as the details of normal feeding behavior. These four genera chew their food in remarkably similar ways. Thus, different animals given the same food will eat it in the same way irrespective of molar morphology and the same animal expresses more behavioral variation in eating different foods than transpires between animals of different species. Such findings emphasize the significance of the consistency of food and suggests to these students:

> The importance of aggregate changes in diet as the selection pressure for dietetic adaptations (Kay and Hiiemae, 1974, p. 255).

In contrast to prosimians, *Ateles* and *Saimiri* and probably all higher primates use their incisors for biting and nipping food. If the morsel is hard it is quickly passed to the posterior teeth. However, the increased importance of the incisors as food-reducing elements possibly occurred *pari passu* with the evolution of spatulate incisors and the fusion of the mandibular symphysis (Hiiemae and Kay, 1973). Whatever the sequence proves to be, these dental and anatomical characters are common to all higher primates.

The work of Zingeser (1966, 1968a, 1973a) on the function of the masticatory apparatus in various cebid monkeys again reflects the integrated nature of the system. Zingeser suggests that underbite as exemplified by *Alouatta*, and also the Colobinae, is an adaptation to a predominately leaf-eating diet. Interestingly, he finds this herbaceous occlusal adaptation more frequent (100%) in island populations of *Alouatta* (Zingeser, 1970). The underbite occlusion also has a high incidence in *Brachyteles* and *Lagothrix* and is seen occasionally (1·5%) in *Cebus* (Colyer, 1936). According to this author, *Aotus*, *Saimiri*, *Cacajao*, *Pithecia* and *Callicebus* have either an edge-to-edge incisor relationship, or the maxillary incisors bite mesial to the lower ones. In his odontologic study of *Brachyteles*, Zingeser

Genus *Aotus*

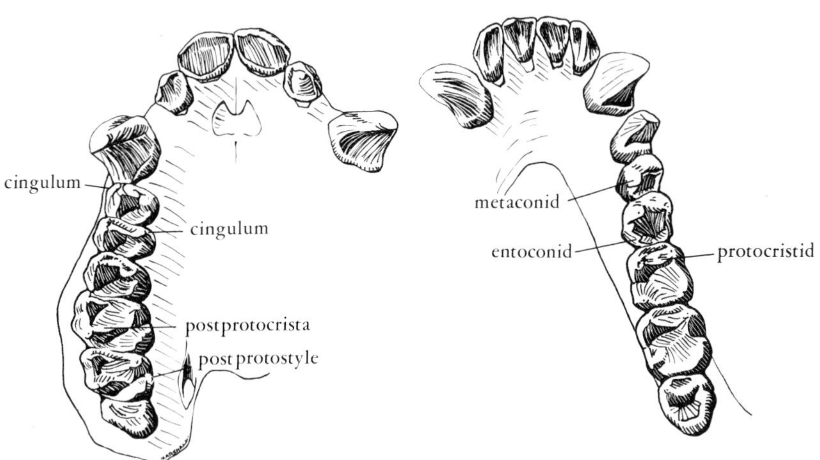

Plate 23. *Aotus trivirgatus* male, occlusal view ×2·2.

Morphological Observations

	Sample	
	Male	Female
Aotus trivirgatus (night monkey)	9	15

(1973b) clearly demonstrates the leaf-eating adaptations of the dentition of this endangered species.

As far as I know the eruption pattern of the permanent molars in *Aotus* is unique among living primates. The eruption of the three permanent molars before any other permanent tooth is primitive and probably represents the original mammalian sequence. The insectivore *Tupaia*, it should be noted, erupts the three permanent molars before adding other permanent teeth. The explanation for the progressively later eruption of M^2_2, M^3_3 in higher primates is not completely understood although Gregory (1922) offers an interesting explanation for the later appearance of these teeth. He believes this pattern is probably secondary resulting from infantile jaws not being able to accommodate the larger teeth. In other words, they tend to form later, *ergo*, they tend to erupt later. In addition, Gregory (1922) writes that *Aotus* as well as *Callicebus* have skull and dental characteristics making them the most primitive and "tarsioid" of all living cebids.

Incisors

Upper. The maxillary incisors are heteromorphic, I^1 being large and broad while I^2 is narrow and pointed. Marginal ridges are well developed in both incisors.

Lower. The lower incisors are both relatively flat buccolingually. The incisal border of I_2 is somewhat more inclined distally than it is on I_1. Narrow marginal ridges outline the lingual surfaces of both incisors.

Canines

Upper. The upper canine is moderately large and projects well beyond the occlusal plane. There is an obvious lingual cingulum skirting the cervical portion of the tooth.

Lower. The lower canine is small and projects only slightly above the other teeth. The crown is pointed and presents a small distostylid.

Premolars

Upper. P^{2-4} are bicuspid, presenting a large paracone and a smaller protocone. The protocone is mesiolingual to the paracone. Lingual cingula are not present on P^2 in the present sample but do appear on P^{3-4}. Kinzey's (1973) findings are not dissimilar to these for *Aotus.* Buccal cingula are much more common and appear as continuous bands or remnants thereof on all three premolars. Para- and distostyles are present on P^{2-4} approximately 75% of the time.

Lower. P_2 is caniniform, usually with a single large protoconid that projects above the other premolars. Occasionally (20–25%) a small metaconid is identifiable attached to the protocristid. P_3 is always bicuspid; however, on rare occasions small hypoconids and entoconids are observable, whereas P_4 displays these two additional cusps much more often. Thus, there is a molariform gradient from P_{2-4}.

Molars

Upper. M^{1-2} possess four cusps and are quandrangular in occlusal outline, while M^3 is usually bicuspid. The variable cusps on M^3 are the metacone and hypocone. Postprotocristae are present on M^{1-2} but are rarely observed on M^3. Buccal cingula are not common on the molars of *Aotus* but when they occur they are normally found on both M^{1-2} (Orlosky, 1973). Lingual cingula, on the other hand, have a greater frequency, particularly the postprotostyle. This trait is present in over 30% of the present sample on M^{1-2}, slightly less on M^3 (15%). Kinzey (1973) found much higher frequencies in his sample of *Aotus*, thus M^1 (100%), M^2 (88%) and M^3 (74%).

Lower. M_{1-3} displays four cusps. There is no hypoconulid on M_3. Protocristids are present on all three molars although they are more pronounced on M_{1-2}. Buccal cingula are not present in these specimens but again Kinzey (1973) reports a low incidence (5–18%) of the distobuccal cingulum.

Odontometry (Tables 51–54, Appendix)

Sexual dimorphism in tooth size is located in the canines and molars and in all cases the males are larger than the females. Except for the mesiodistal diameter of M_3 ($P < 0{\cdot}02$) the sexual differences reside in the upper canines and molars and pertain to the buccolingual dimensions. Thus, canine width is significant ($P < 0{\cdot}01$) as are the widths of M^2 ($P < 0{\cdot}03$) and M^3 ($P < 0{\cdot}01$).

Genus *Callicebus*

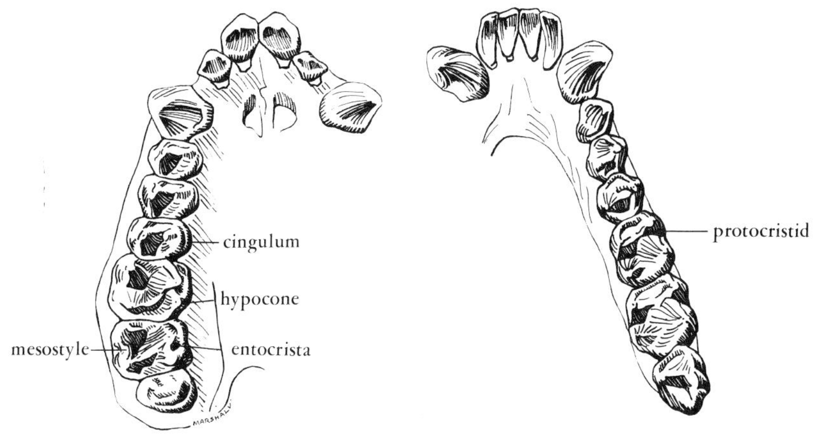

Plate 24. *Callicebus moloch* male, occlusal view ×2·1.

Morphological Observations

	Sample	
	Male	Female
Callicebus moloch (dusky titi)	3	1

The eruption pattern for *Callicebus* is similar to that presented for *Ateles* and *Cebus* with the exception that P^4_4 erupts before P^2_2 (Serra, 1952a).

Incisors

Upper. I^{1-2} are heteromorphic. I^1 is broad and presents a relatively horizontal incisal border while I^2 is narrow and pointed. Marginal ridges are present on I^{1-2} and the lingual surfaces are somewhat concave. A lingual cingulum is present in both teeth.

Lower. I_{1-2} are subequal. The labial surfaces of both teeth are convex while the lingual surfaces are concave. Narrow marginal ridges are present, with the distal one better developed.

Canines

Upper. The upper canine presents a stout base from which it tapers to end above the occlusal plane. The distolingual surface is slightly concave and a noticeable lingual cingulum skirts around the cervical portion of the tooth. No mesial groove is present.

Lower. The lower canine curves mesially and then distally to extend above the occlusal plane. A lingual cingulum is present as well as a distal heel.

Premolars

Upper. P^{2-4} are biscuspid and, on each, the paracone is larger than the protocone. As in *Aotus*, the protocone is mesiolingual to the paracone and connected by a preprotocrista. Cingular remnants are present on the lingual surfaces of P^{2-4} in the present sample; they are also reported by Kinzey (1973), who found the postprotostyle to increase in frequency mesiodistally: P^2 (53%), P^3 (79%) and P^4 (100%). He also identified para- and distostyles in his sample of 40 *Callicebus torquatus*, but they are less common than the lingual cingulum. The parastyle is present on P^2 (13%), P^3 (13%) and P^4 (8%) while the distostyle is only slightly more prevelant, P^2 (18%), P^3 (23%) and P^4 (23%). Thus, the incidence of buccal and lingual cingula is reversed in *Aotus* and *Callicebus*.

Lower. P_2 is caniniform and its single large protoconid is usually higher than on the other premolars. P_{3-4} are bicuspid although the metaconid is always much smaller than the protoconid. A protocristid connects the two cusps on P_{3-4}. Kinzey (1973) found distostylids on P_2 (24%), P_3 (19%) and P_4 (8%). Orlosky (1973) found entoconids (91%) and hypoconids (77%) on P_4 in *Callicebus moloch.*

Molars

Upper. M^{1-2} possess four cusps and are rather square in appearance. The hypocones of M^{1-2} are relatively large and connected to the distal slopes of the protocones by the entocrista, a fact which led Gregory (1922, p. 220) to consider them "apparently pseudohypocones like those of the Notharctidae". This is in contrast to the situation in *Aotus* where the hypocone is connected with the cingulum. In our opinion, as stated earlier (p. 20), such a morphological distinction has little taxonomic value. The hypocone is very small on M^3 and may be entirely absent. A lingual cingulum is usually well developed and may extend around the protocone to become continuous with the hypocone. The cingulum is better developed on M^{1-2} where Kinzey (1973) found it present in 100% of his *Callicebus torquatus* sample. On M^3 the postprotostylar portion displays an identical frequency; however, the protostyle is present in only 25% of the sample. Buccal cingula are not as common on the molars as are lingual cingula. The mesostyle is all that remains of the buccal cingulum and Kinzey (1973) identifies it on M^1 (95%) and M^2 (53%) in 40 *Callicebus torquatus.* It is present in 70% of our small sample on both M^{1-2}.

Lower. M_{1-2} possess four cusps while the hypoconulid is present on M_3. The trigonid basin is narrow relative to the wide talonid basin. A protocristid, particularly well developed on M_1, separates the two regions. On all three molars, the metaconid lies distolingual to the protoconid. A cristid obliqua is well defined on M_{1-2}, lacking as a definitive entity on M_3. A distostylid is present on M_1 (56%), M_2 (83%) and M_3 (10%) in 40 *Callicebus torquatus* (Kinzey, 1973). It is also present more often on M_2 in our material.

Odontometry

No measurements were taken of our small sample. The data presented by Orlosky (1973), revealed no sexual dimorphism in tooth size in *Callicebus moloch.* An interesting study by Kinzey (1972) attributes the lack of sexual dimorphism in canine size to the low level of aggression in *Callicebus moloch.* Additional studies are needed to help clarify the relationship between behavior and canine tooth size in the primates.

The mesiodistal molar relations are $M^1 > M^2 > M^3$ and $M_1 < M_2 > M_3$.

Genus *Ateles*

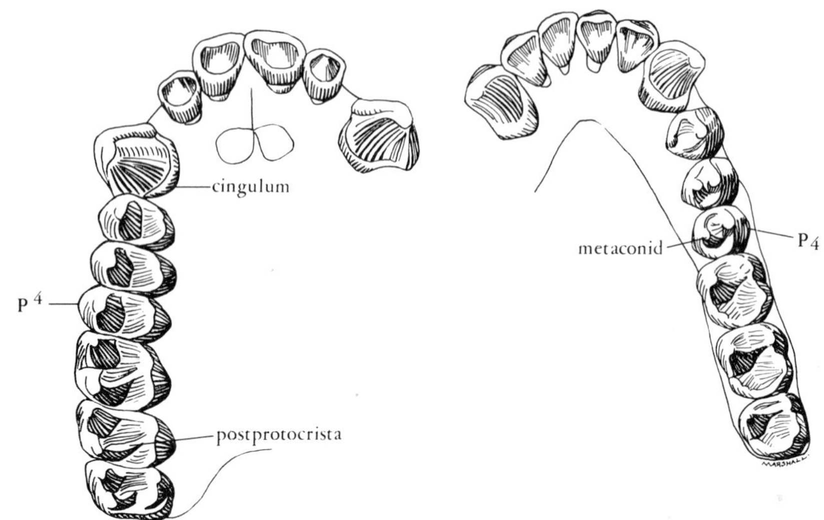

Plate 25. *Ateles geoffroyi* female, occlusal view 1·9.

Morphological Observations

	Sample	
	Male	Female
Ateles geoffroyi (blackhanded spider monkey)	15	12
Ateles belzebuth (longhaired spider monkey)	11	8

According to Serra (1952a), among New World monkeys the eruption of M_3^3 after the canine is found only in *Ateles, Cebus* and *Callicebus.* The late eruption of M_3^3 is regarded as a more specialized condition characteristic of the catarrhine primates, with the exception of some of the langurs where it erupts before the premolars and canines (Schultz, 1935).

Incisors

Upper. I^{1-2} are heteromorphic teeth. I^1 is broad and roughly square whereas I^2 is more triangular. Both display marginal ridges around the lingual surface and the cervical portion is particularly well developed. A vertical lingual ridge is frequently present on the lingual surface of I^1 as well as an occasional lingual tubercle. These structures are not present on I^2.

Lower. In contrast to the upper incisors, the lower ones are homomorphic. Both are flat with only a suggestion of marginal ridge development.

Canines

Upper. The upper canine is long and has a stout base. A deep mesial groove passes from the cervical region to near the tip of the tooth. A lingual cingulum is present. The sexual dimorphism of these traits is significant: in females the lingual cingulum is more prominent and the mesial groove opens onto the lingual surface, while in males the cingulum is slight and the groove is restricted from the lingual surface (Orlosky, 1973).

Lower. The lower canine is robust, rising above the occlusal plane as it splays buccally. A lingual cingulum is obvious as well as a small distal shelf. The morphology is similar in both sexes.

Premolars

Upper. P^{2-4} are bicuspid with the paracone always the largest. A small hypocone is almost always present on P^{3-4} in *Ateles belzebuth*, but is wanting in *Ateles geoffroyi*. Moreover, P^{3-4} in *Ateles belzebuth* have a significantly greater frequency of para- and distostyles ($P < 0{\cdot}05$) than in *Ateles geoffroyi* (Orlosky, 1973). Kinzey (1973) found a complete absence of buccal and lingual cingulum remnants in both *Ateles geoffroyi* and *Ateles paniscus*. It would be interesting to have data on *Ateles fusciceps* regarding cingular derivatives since this is the remaining species constituting the genus *Ateles*.

Lower. P_2 is almost always unicuspid whereas P_{3-4} are bicuspid and in *Ateles belzebuth* P_4 is frequently tricuspid. When P_4 is tricuspid it is the hypoconid that is represented. A protocristid is present on P_{2-4} connecting protoconid and metaconid, or when the metaconid is lacking (P_2), it passes distolingually to merge with the marginal ridge. Orlosky (1973) reports that paraconids and buccal cingula are limited to P_2 and that both are more frequent in *Ateles belzebuth*. Kinzey (1973) records no derivatives from the buccal cingulum on P_{2-4} in either *Ateles paniscus* or *Ateles geoffroyi*.

It is apparent from these observations that the premolars of *Ateles belzebuth* are much more complicated than those of the other species of *Ateles*. More demographic and ecologic (detailed field studies) information is needed to help answer what appears to be microevolutionary problems at the species, or even the population, level regarding the presence and absence of many of these dental traits.

Molars

Upper. M^{1-2} possess four cusps, while M^3 displays anything from two to four. The variable cusps are the hypocone and metacone. Hypocones may be lacking in 50% of *Ateles geoffroyi* of the present study while Kinzey (1973) reports their absence in 62% of *Ateles geoffroyi* and 55% of *Ateles paniscus*. A postprotocrista is always present on M^{1-2}, variable on M^3. Mesostyles are rare, but are found on M^{1-2} in the present study, particularly in *Ateles belzebuth* (30%). Kinzey (1973) records mesostyles only on M^1 in *Ateles paniscus* (20%) and *Ateles geoffroyi* (4%). Small postprotostyles are present in all species on M^{2-3}. Their frequency varies from 20% in *Ateles belzebuth* to 45% in *Ateles geoffroyi.*

Lower. The lower molars are usually characterized as having four cusps, but in both species small hypoconulids can be present on all three molars. They are more common in *Ateles belzebuth* (50%) than in *Ateles geoffroyi* (15%). The occlusal outline is more quadrate than oblong, especially on M_{2-3}. A well formed protocristid connects the protoconid and metaconid, thereby delineating the distal border of a rather narrow trigonid fossa. The talonid fossa on the other hand is a broad and shallow reentrant basin for the protocone when the teeth are in occlusion. Buccal and lingual cristids connect their respective cusps forming rather sharp edges in the unworn state.

Odontometry (Tables 55–58, Appendix)

The minor display of sexual dimorphism in *Ateles* is sporadic and may involve either dimension and both upper and lower teeth. In *Ateles geoffroyi* the female is often larger than the male and this includes the buccolingual diameter of the upper canine, although it is not significant. Orlosky (1973) examined interspecies differences between *Ateles belzebuth* and *Ateles geoffroyi* and found significant differences in the anterior teeth but not in the molars. In most instances *Ateles belzebuth* has larger teeth than *Ateles geoffroyi.*

Genus *Brachyteles*

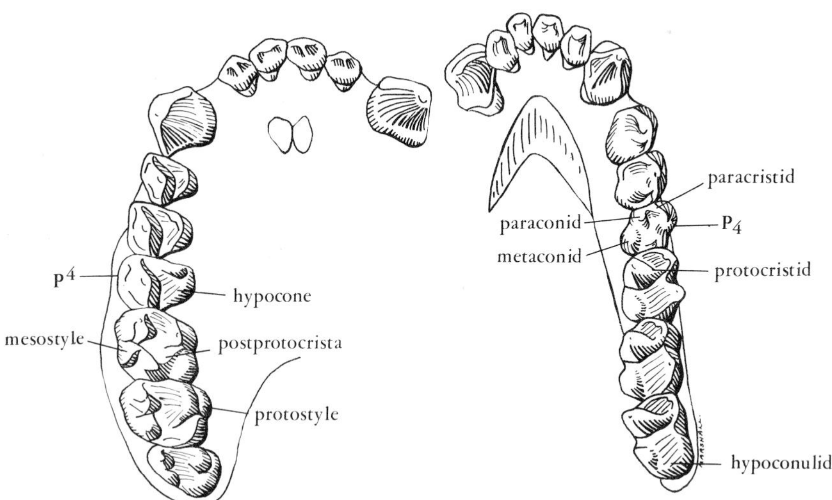

Plate 26. *Brachyteles arachnoides* male, occlusal view ×1·3.

Morphological Observations

	Sample	
	Male	Female
Brachyteles arachnoides (woolly spider monkey)	1	

The eruption pattern of *Brachyteles* differs from *Ateles* in having the canine erupt after M^3_3, and P^4_4 erupt before the other premolars. The available evidence on odontiasis in New World primates suggest a great deal of sequential polymorphism for the premolars.

Incisors

Upper. I^{1-2} are heteromorphic in that I^1 is a broad, spatulate tooth and I^2 presents a more pointed contour. Marginal ridges are present on both teeth, and in particular, the lingual cingulum is well developed.

Lower. I_{1-2} are also heteromorphic. I_1 is narrow with a flat incisal border while I_2 is more lanceolate and larger than I_1. Both incisors are thickened buccolingually. *Brachyteles* has a high incidence of mandibular underbite, though not as high as *Alouatta* (Colyer, 1936; Serra, 1951; Zingeser, 1973b). Underbite occurs less frequently in *Lagothrix*. When *Brachyteles* depart from the epharmosis pattern they usually have an edge-to-edge bite.

Canines

Upper. The upper canines are not extremely large but they do present stout bases. They have a lingual cingulum as well as a mesiolingual groove, which is bordered buccally by a substantial vertical ridge. The mesial border is convex when viewed buccally, which led Remane (1960) to include *Brachyteles* in the same group with *Ateles* and *Alouatta* regarding canine morphology.

Lower. The lower canine is barely higher than adjacent teeth. A slight diastema separates it from P_2. A well formed lingual cingulum is visible on the canine and its distal portion (ledge) forms a part of the honing mechanism described by Zingeser (1968b, 1973b).

Premolars

Upper. P^{2-3} are bicuspid whereas P^4 tends to be tricuspid. In all premolars the paracone is always appreciably larger and higher than the protocone. The third cusp of P^4 is the hypocone; Zingeser (1973b) calls it a pseudohypocone since according to him it does not arise from the cingulum. All three premolars possess well formed buccal cingula and their derivates, the para- and distostyles. The crown morphology of P^{2-4} is very similar between *Brachyteles* and *Alouatta*. For further details see the comprehensive study of Zingeser (1973b).

Lower. P_2 is a caniniform tooth possessing but a single large protoconid. A narrow, ill-defined protocristid passes distolingually from the protoconid to merge with the lingual cingulum. The mesiobuccal surface of P_2, although functionally part of the honing mechanism, shows little if any elongation. P_{3-4} possess a well developed protoconid and a small metaconid; in addition, P_4 displays an obvious paraconid separated by a lingual notch from the metaconid. A paracristid is seen passing mesiolingually from the protoconid toward the paraconid. There are no buccal cingula in the specimen studied, nor are any reported by Zingeser (1973b).

Molars

Upper. M^{1-3} possess four cusps; on each molar the hypocone is the smallest and is separated from the protocone by a linguo-occlusal groove. A postprotocrista is present but not as distinct as in *Ateles* or *Lagothrix*. A metaconule is rare in *Brachyteles*; when present, it is an intricate part of the postprotocrista (Zingeser, 1973b). It is not present in our specimen. Protostyles are rather common (Zingeser, 1973b) and they are present in our specimen on M^{2-3}. Ectocingular derivatives are much more prevalent. According to Zingeser (1973b), M^1 possesses para-, meso- and metastyles; M^2 has parastyles and occasionally metastyles, while M^3 frequently lacks any stylar formation.

Lower. The morphological details of the lower as well as the upper molars are presented by several authors. Remane (1960) was so impressed with the artiodactyl nature of the buccal crests plus the bunodont cusps in both upper and lower molars that he coined the term "selenobuntodontie" to describe them. Zingeser (1973b) noted the similarity between the lower molars, particularly M_1, and the bilophodont molars of the colobine monkeys of the Old World. These molars present many interesting details of morphology and function. M_{1-3} have four cusps; the lingual two are particularly high and conical. A small hypoconulid may adorn the distal surface of M_3. A paraconid, when present, is limited to M_1. M_1 also presents a distinct hypoflexid on the buccal surface which gives it a bilophodont appearance. Finally, the protocristid is more pronounced on M_1 than on M_{2-3}.

Odontometry (Tables 59–60, Appendix)

These data are taken by permission from the work of Zingeser (1973b). Sexes are not separated; the total sample is 21 skulls. The mesiodistal molar size formula is $M^1_1 > M^2_2 > M^3_3$.

Genus *Lagothrix*

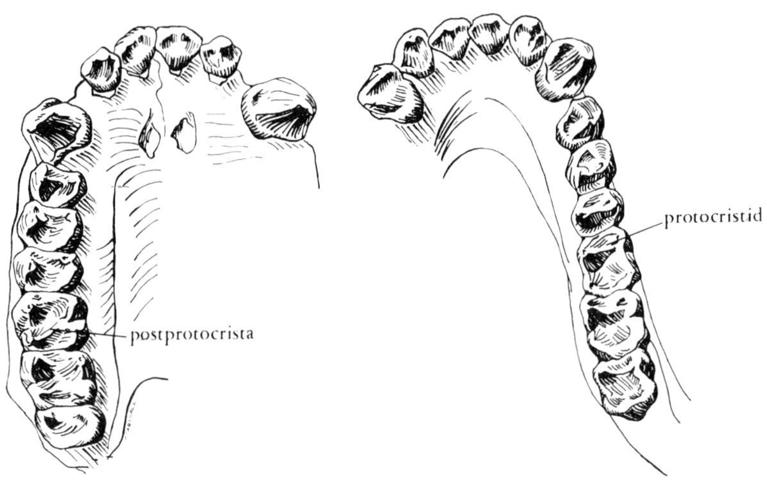

Plate 27. *Lagothrix lagothricha* female, occlusal view ×1·5.

Morphological Observations

	Sample	
	Male	Female
Lagothrix lagothricha (Humboldt's woolly monkey)	3	3

The eruption pattern of *Lagothrix* differs from *Brachyteles* in the appearance of the premolars, while both of these species differ from *Ateles* in the (M^3_3, C^1_1) sequence. Thus the Atelinae express quite different patterns of odontiasis, indicating a need for more comprehensive studies of eruption among these taxa.

Incisors

Upper. The upper incisors are heteromorphic. Marginal ridges outline the lingual surfaces of both I^{1-2}, whereas a lingual tubercle is present only on I^1. A sizable diastema intervenes between I^2 and the canine.

Lower. The lower incisors are similar in morphology and both display vertical lingual grooves. They are also approximately equal in size.

Canines

Upper. The upper canine is a large, robust tooth projecting well beyond the occlusal plane. Both a lingual cingulum and mesial groove are present and both are better developed in females, a condition similar to that seen in *Ateles.*

Lower. The lower canine is stout, rising above the occlusal plane, particularly in males. A noticeable lingual cingulum becomes a distal ledge.

Premolars

Upper. P^{2-4} are bicuspid, with the paracone larger than the protocone. A hypocone is suggested on P^4. There are no stylar formations on P^{2-4} in the present sample. Kinzey (1973) reports an absence of styles in *Lagothrix lagothricha*, with the exception that P^2 has mesostyles in 7% of 29 specimens.

Lower. P_2 is usually unicuspid; if present, the metaconid is quite small. A well formed primary lingual cingulum is present. P_{3-4} are essentially bicuspid but both display expanded talonids upon which small entoconids and hypoconids may appear, particularly on P_4. Thus, cusp number increases mesiodistally on the lower premolars. Mesio- and distostylids are absent in our small sample; however, Kinzey (1973) identified them on P_2 in 4% and 7% respectively, of 29 animals.

Molars

Upper. M^{1-3} possess four cusps and as in *Ateles*, M^3 may lose the hypocone. Kinzey (1973) reports a M^3 hypocone absenteeism of 25% for *Lagothrix lagothricha.* A postprotocrista is present on M^{1-2}, variable on M^3. Mesostyles are infrequent, but did appear on M^1 in 28% of Kinzey's 1973 sample. One M^1 in the present sample possesses a mesostyle. The lingual cingulum remnants are extremely rare; there are none in the present study, while Kinzey (1973) records a 3% incidence of the protostyle on M^2 in 29 *Lagothrix lagothrichda.*

Lower. M_{1-3} have four cusps. There is no hypoconulid on M_3 in the present sample. A protocristid connects the protoconid and metaconid and at the same time defines the distal border of the narrow trigonid basin. The talonid fossa is wide and shallow; indeed, as in the other teeth, the lower molars of *Lagothrix* are morphologically similar to those of *Ateles*.There are no remnants of a buccal cingulum on these molars or in the specimens studied by Kinzey (1973).

Odontometry

The only significant sexual dimorphism was the mesiodistal dimension of the upper canine ($P<0{\cdot}01$) and the buccolingual diameter of the lower canine

($P < 0{\cdot}05$). In both cases, males were larger than females. As in *Ateles*, females frequently have larger teeth than males.

Genus *Pithecia*

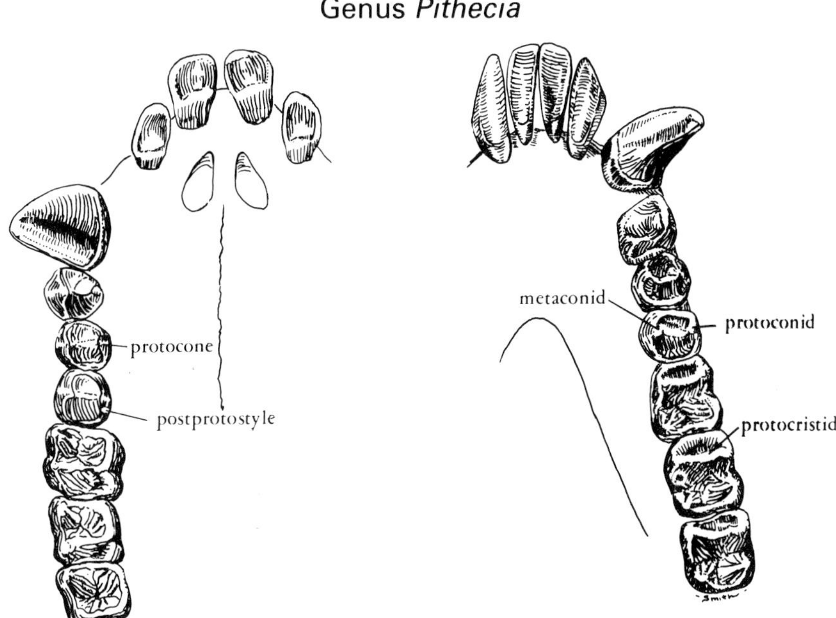

Plate 28. *Pithecia pithecia* male, occlusal view ×2·5.

Morphological Observations

	Sample	
	Male	Female
Pithecia pithecia (paleheaded saki)	2	1

The eruption sequence ($I^1_1 M^2_2 I^2_2 M^3_3$) is unique to the genera of the Pitheciinae as far as we can determine. Because of this eruption sequence Serra (1952a) believes it justifies placing them between the lemurs and tarsiers on the one hand, and the remainder of the platyrrhine and catarrhine on the other. Unfortunately, nothing is known about eruption sequence polymorphism in this group.

Incisors

Upper. I^1 is larger than I^2 and both present well defined lingual fossae. Marginal ridges outline the lingual surfaces of both teeth and an obvious lingual tubercle resides on I^1. Orlosky (1973) describes a lingual tubercle on 60% of 11 specimens of *Pithecia monachus*. Apparently, lingual tubercles are infrequent or lacking on I^2.

Lower. Both incisors are elongated, with I_2 slightly larger than I_1. I_2 presents a sloping distal marginal surface on which is a perceptible ridge. Both upper and lower incisors are somewhat procumbent, a condition not seen among other extant cebids (Orlosky, 1973).

Canines

Upper. The upper canine is robust and tilts buccally, being separated from I^2 by a wide diastema. A vertical groove is present on the mesial surface running from the cervical region to near the crown tip.

Lower. The lower canine is large and inclines buccally. It presents three surfaces and three borders, of which the distolingual border is rather sharp. A lingual cingulum increases in thickness distally to terminate as a distostylid.

Premolars

Upper. P^2 has two cusps, a large, pointed paracone and a much smaller protocone. The protocone is variably developed in *Pithecia* since Orlosky (1973) found it absent in 31% of eight *Pithecia monachus*. Moreover, P^2 is much more caniniform than the other premolars. A narrow primary lingual cingulum is present. P^{3-4} are bicuspid; indeed, the protocone is nearly as large as the paracone. The occlusal surfaces of P^{3-4} present numerous delicate crenulations within the boundaries of the marginal ridges. Such occlusal markings are rare in other cebid subfamilies. A small postprotostyle is observable on P^4 and Kinzey (1973) reports the presence of these styles on both P^{3-4} in 14% and 25% respectively of 29 animals.

Lower. P_2 presents a mesiodistally elongated protoconid with a moderate degree of mesiobuccal flattening. The general appearance of P_2 resembles the sectorial P_3 of Old World monkeys and we suggest that the functional relationship between P_2 and the upper canine in *Pithecia* is similar to that between P_3 and the upper canine in Old World monkeys, i.e. a honing mechanism (see Zingeser, 1969). Metaconids may be present on P_2 but they are very small and attach to the distolingual surface of the protoconid. Orlosky (1973) found them in 24% of 27 specimens of *Pithecia monachus*, and found paraconids in 70% of these same animals. P_{3-4} are always bicuspid and frequently present hypoconids and entoconids. The protoconid and metaconid are connected by a well formed protocristid. Small mesio- and distostylids are present on P_{2-4} and the occlusal surfaces of P_{3-4} show small crenulations.

Molars

Upper. M^{1-3} possess four cusps and are quite quadrangular in occlusal view. A postprotocrista swings between the protocone and metacone although it is not as obvious in *Pithecia* as in some other cebids. The occlusal surface is crenulated on all three molars. Mesostyles are not present in the specimens examined; however, Kinzey (1973) found them present on M^1 in 14% of 29 *Pithecia pithecia*, but absent in an equal sample of *Pithecia monachus*. Protostyles and postprotostyles occur between 70 and 100% of the time (Orlosky, 1973; Kinzey, 1973).

Lower. M_{1-3} have four cusps. There is no hypoconulid. A protocristid is present and a narrow trigonid basin lies just mesial to it. The occlusal surface of M_{1-3} is crenulated; indeed, of all Pithecinae, *Pithecia* seems to possess the most complicated crenulation pattern on both upper and lower premolars and molars. A mesiostylid is present in our specimens on M_{1-3} and Kinzey (1973) reports a mesiodistal frequency gradient, i.e. M_1 (26%), M_2 (41%) and M_3 (63%).

Odontometry

The three specimens were not measured; however, mesiodistal size gradients are $M^1 < M^2 > M^3$, while $M_1 < M_2 > M_3$. Orlosky (1973) reports no significant sex differences for any tooth dimension of *Pithecia monachus*.

Genus *Chiropotes*

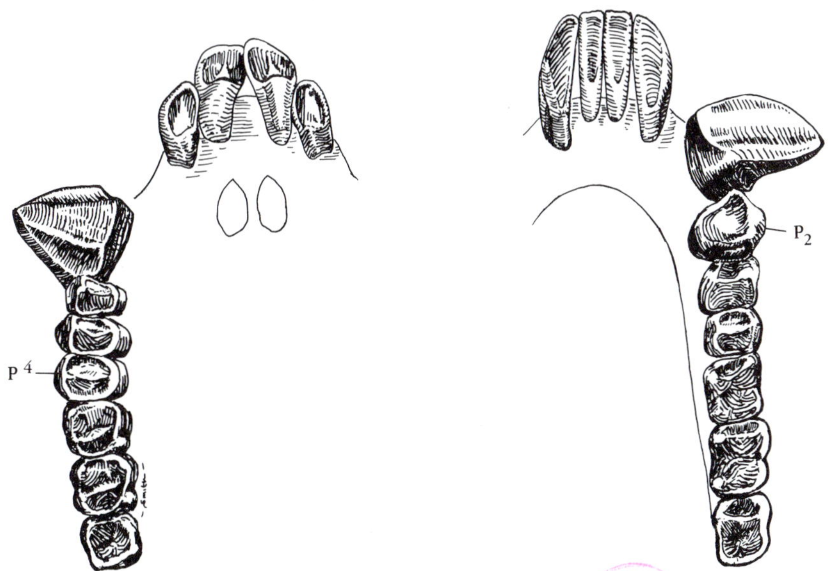

Plate 29. *Chiropotes satanas* male, occlusal view ×2.

Morphological Observations

	Sample	
	Male	Female
Chiropotes satanas (black saki)	1	1

In his study of New World monkey eruption patterns, Serra (1952a) does not mention *Chiropotes* although I should imagine it to be similar to *Pithecia* and *Cacajao* sequences which are presented.

Incisors

Upper. I^1 is larger and more spatulate than I^2. Both incisors have complete marginal ridges circumscribing lingual fossae. A lingual tubercle is present on I^1, absent on I^2. These incisors are more heteromorphic than in *Pithecia.*

Lower. Both I_1 and I_2 are elongated and slightly procumbent. I_2 is larger than I_1 and the distal border tapers mesially.

Canines

Upper. The upper canine is a large, robust tooth in both sexes. It projects well below the occlusal line and is directed buccally. A well defined mesial groove is present, running from the cervix to near the crown tip. A narrow lingual cingulum is present.

Lower. The lower canine is triangular and separated by a wide diastema from I_2. The crown splays buccally as it projects well above the occlusal line.

Premolars

Upper. P^{2-4} are bicuspid although the protocone is much smaller on P^2. Mesial and distal marginal ridges are present on P^{2-4} as well as a transverse crest between the protocone and paracone. Crenulations adorn the occlusal surfaces of P^{3-4}. Postprotostyles are not present in these two animals but Kinzey (1973) reports an incidence of 14% on P^4 in his sample of 27 *Chiropotes satanas.* He also records mesostyles on P^{3-4} (8 and 4% respectively).

Lower. P_2 is elongated mesiodistally and has a small metaconid in both specimens. Orlosky (1973) reports the presence of metaconids in 40% of *Chiropotes satanas.* Morphologically then, P_2 again reminds one of the sectorial P_3 of catarrhine monkeys. P_{3-4} are bicuspid and in addition, hypoconids and entoconids are often developed distally. Occlusal crenulations are also present on P_{3-4}. Although mesio- and distostylids are not observable in the present sample ($n = 2$), Kinzey (1973) reports them in his study of *Chiropotes satanas.*

Molars

Upper. M^{1-3} have four cusps and are quadrangular in occlusal outline. Multiple crenulations embellish their occlusal surfaces after the manner observed in *Pithecia.* Protostyles and postprotostyles are present in our two specimens on M^{1-3} and Kinzey (1973) records a variable frequency (60–90%) for these structures on M^{1-3}. While we did not observe metastyles on M^{1-3}, Kinzey reports them on M^{1-3} in 4% of 27 animals.

Lower. M_{1-3} have four cusps; the hypoconulid is absent. A protocristid connects the protoconid with the metaconid while a transverse nexus is lacking between the distal two cusps. As in *Pithecia* and to a lesser extent in *Cacajao,* crenulations are present on the occlusal surfaces of M_{1-3}. Diminutive mesio- and distostylids appear on M_{1-3} in both specimens. Kinzey (1973) states their presence as between 16 and 70% with appreciably higher frequencies for the mesiostylid.

Odontometry

The mesiodistal size gradients for the molars are, $M^1_1 = M^2_2 > M^3_3$.

Genus *Cacajao*

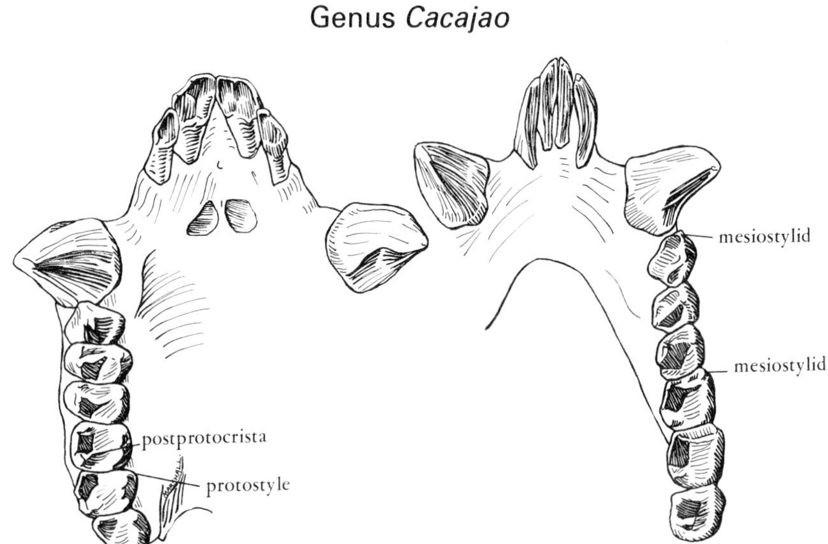

Plate 30. *Cacajao calvus* male, occlusal view ×1·65.

Morphological Observations

	Sample	
	Male	Female
Cacajao calvus (bald uakari)	3	3

The eruption sequence of *Cacajao* is identical to that of *Pithecia* according to Serra (1952a).

Incisors

Upper. I^1 is much larger than I^2 and both present marginal ridges circumscribing their lingual surfaces. A lingual tubercle is present on I^1, absent on I^2. A vertical lingual groove resides within the lingual surface of I^1. I^2 lacks any development of a lingual groove.

Lower. Both I_{1-2} are elongated, narrow teeth, and both are slightly procumbent as in other Pithecinae. Marginal ridges are absent.

Canines

Upper. The upper canine is large and projects well below the occlusal plane. A faint lingual cingulum is present as well as a vertical mesial groove. According to Orlosky (1973), *Cacajao* does not display sexual dimorphism in these traits as do *Cebus* and *Saimiri.*

Lower. The lower canine is large and widely separated from I_2. The canine inclines buccally as it passes beyond the occlusal plane. As the lingual cingulum passes distally a distinct ledge becomes apparent.

Premolars

Upper. P^2 is bicuspid but the protocone is small. A narrow crista connects the two cusps. P^{3-4} are bicuspid and the two cusps are of nearly equal proportions. As in P^2, a crista bridges the two cusps and thereby divides the occlusal surface into two portions. A buccal cingulum is not represented in our sample, but Kinzey (1973) reports a 56% frequency of mesostyles on P^2 in 17 *Cacajao rubicundus.* Derivatives of the lingual cingulum are absent from our material as well as Kinzey's (1973).

Lower. P_2 is somewhat elongated mesiodistally as in *Pithecia.* A small metaconid may reside on P_2 and if present, it appears on the distolingual aspect of the protoconid. P_{3-4} are bicuspid and additionally, small hypoconids and entoconids are occasionally present on the expanded talonids. A protocristid is present on P_{3-4} and when unworn their occlusal surface displays small crenulations. Kinzey (1973) found mesiostylids on P_2

(31%) and P_3 (13%), and distostylids on P_2 (75%) and P_3 (4%) of *Cacajao rubicundus.* We found both stylids on P_{2-3} in our survey.

Molars

Upper. M^{1-3} have four cusps. As in many other cebids, M^3 may lack the hypocone. A postprotocrista is present on all three molars, although with wear it quickly disappears. Occlusal surface crenulations are visible on M^{1-3} but they are less developed than in *Pithecia.* Remnants of the buccal cingulum are not present in our specimens nor does Kinzey (1973) report them for *Cacajao rubicundus.* However, protostyles are present in our material on M^{1-2} and in *Cacajao rubicundus* Kinzey (1973) records their incidence on M^1 (100%), M^2 (87%) and M^3 (50%). He also notes a postprotostyle on M^1 (6%) in these specimens.

Lower. M_{1-3} possess four cusps. There is no hypoconulid on M_3. A narrow protocristid connects the protoconid with the metaconid on M_{1-3}, but it is less clearly delineated on M_3. The trigonid fossa is narrow and more circumscribed than the talonid basin. Small crenulations are present on the occlusal surfaces in M_{1-3}. A mesiostylid is present in our sample on M_1, whereas Kinzey (1973) reports a frequency of 6% for this structure on M_{1-3}.

Odontometry

Tooth size sexual dimorphism is small in this genus and it mainly centers around the upper and lower canines and P_2.

Genus *Cebus*

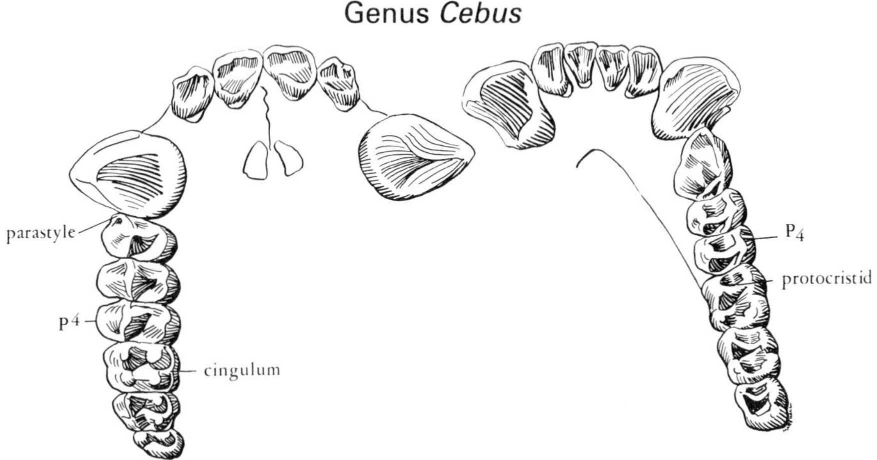

Plate 31. *Cebus apella* male, occlusal view ×1·3.

Morphological Observations

	Sample	
	Male	Female
Cebus apella (blackcapped capuchin)	33	16

The late appearance of M_3^3 should be noted in *Cebus*. It is also the last tooth to erupt in *Ateles* and *Callicebus*. In all other New World primates it erupts prior to the canine.

Incisors

Upper. I^1 is a broad, spatulate tooth while I^2 has a more pointed crown. Thus, the maxillary incisors are quite heteromorphic. The lingual surfaces of both incisors are outlined by distinct marginal ridges. The cervical one is particularly well developed.

Lower. I_{1-2} are implanted vertically and their crowns are relatively flat. Lingual marginal ridges are present and these are better developed on I_2. Incidentally, I_2 is larger than I_1.

Canines

Upper. The upper canine is a large, robust tooth projecting well below the occlusal plane. A lingual cingulum is present; however, it is much better developed in females and according to Orlosky (1973) its development is significantly different between the sexes. A mesial crest runs from the cervix to near the crown tip.

Lower. The lower canine is a large, well constructed tooth. The robust crown is sharp and projects well above the other teeth. A lingual cingulum terminates as a pronounced distostylid.

Premolars

Upper. P^{2-4} are bicuspid and on all three, the paracone is larger than the protocone although the protocone increases in size from P^{2-4}. Para- and distostyles are present in these specimens on all three premolars, although parastyles are more common than distostyles and both are more frequent on P^2. In his study of the Ceboidea cingulum, Kinzey (1973) notes that the parastyle on P^2 is present most often (52%) in *Cebus capucinus*, while being present in 15% of *Cebus apella* and completely absent in *Cebus albifrons*. Regarding remnants of the lingual cingulum in *Cebus apella*, only the postprotostyle is present and only on P^{3-4} (P^3: 22%; P^4: 44%). Kinzey's (1973) figures are slightly less: 19 and 26% respectively. As with buccal cingula, *Cebus capucinus* possesses the greatest number of postprotostyles (P^2: 9%; P^3: 26%; P^4: 44%). Such marked interspecific variability in animals inhabiting similar environments and eating similar foods may be the result of random genetic drift operating on small populations (Swindler and Orlosky, 1972).

Lower. P_{2-4} are bicuspid although the metaconid on P_2 is small and attached to the distolingual surface of the protoconid. Also, P_2 is incipiently sectorial in that it is elongated mesiodistally, and thereby forms a honing platform for the distolingual surface of the upper canine. P_{3-4} normally display hypoconids and entoconids around their narrow talonid basins; these are more frequent on P_4. A protocristid is always observable connecting the protoconid to the metaconid. A mesiostylid is present on P_2 in 55% of our sample while Kinzey (1973) reports it in 46% of the 27 *Cebus apella* he studied. What we believe to be the mesiostylid, Orlosky (1973) designates the paraconid. He found it in over 60% of *Cebus apella*, and also noted a lack of sexual dimorphism. Mesiostylids are much less common on P_{3-4} (less than 5%) in both our sample and that of Kinzey. Distostylids are present in decreasing frequency from P_{2-4} in both this study (P_2: 58%; P_3: 30%; P_4: 10%) and in Kinzey (1973) (P_2: 50%; P_3: 31%; P_4: 8%).

Molars

Upper. M^{1-2} possess four cusps while M^3 frequently lacks the hypocone. Kinzey (1973) records M^3 hypocones in 82% of 29 *Cebus capucinus* and 23% of 29 *Cebus albifrons*. Indeed, M^{1-2} tend to become separated into mesial and distal moieties giving the appearance of bilophodont molars. Gregory (1922, p. 225) says the upper molars of *Cebus* "may be a progressive adaptation to insectivorous and frugivorous habits". Whatever the ultimate relationship between molars and diet may prove to be, Hershkovitz (1971) clearly demonstrates, by analogy, the mechanics of crista capture necessary to produce a bilophodont molar from the triangular platyrrhine pattern. The point to be made at this time is that according to Hershkovitz (1971) several of the progessive stages of crista capture and transformation are identifiable in *Cebus* alone. The current study unequivocally supports this thesis: e.g. in over 50% of the present sample the postprotocrista and entocrista merge to form a transverse crest between the metacone and hypocone. A postprotocrista is present in 40% of our sample. Both lingual and buccal styles are variable; the most common is the postprotostyle (M^1, 68%) while mesostyles occur less than 10% of the time on M^1, never on M^{2-3}.

Lower. M_{1-3} have four cusps, the hypoconulid being absent as in other Cebinae. A protocristid connects the protoconid and metaconid forming the distal wall of a narrow trigonid basin while the talonid fossa is somewhat more expanded. The molar morphology, as with the upper molars, occasionally parallels that seen in Old World monkeys, a situation noted more than 50 years ago by Gregory (1922). Orlosky (1973) mentions that the entoconid tends to disappear on M_3, 37% absent in his study. Buccal cingulum remnants vary from 0 to 49% in this study as well as in that of Kinzey (1973).

Odontometry (Tables 61–64, Appendix)

The t-test revealed that in the great majority of tooth dimensions, male *Cebus apella* were significantly larger than their female counterparts. This is particularly so for the mesiodistal diameters of the upper canine ($P<0{\cdot}01$), the lower canine ($P<0{\cdot}01$) and P_2 ($P<0{\cdot}01$). The latter dimension undoubtedly reflects the greater sectorial nature of this tooth in the male. Also, the buccolingual dimensions of both canines are significant ($P<0{\cdot}01$). It is obvious that both canines are large, powerful teeth in the male.

Genus *Saimiri*

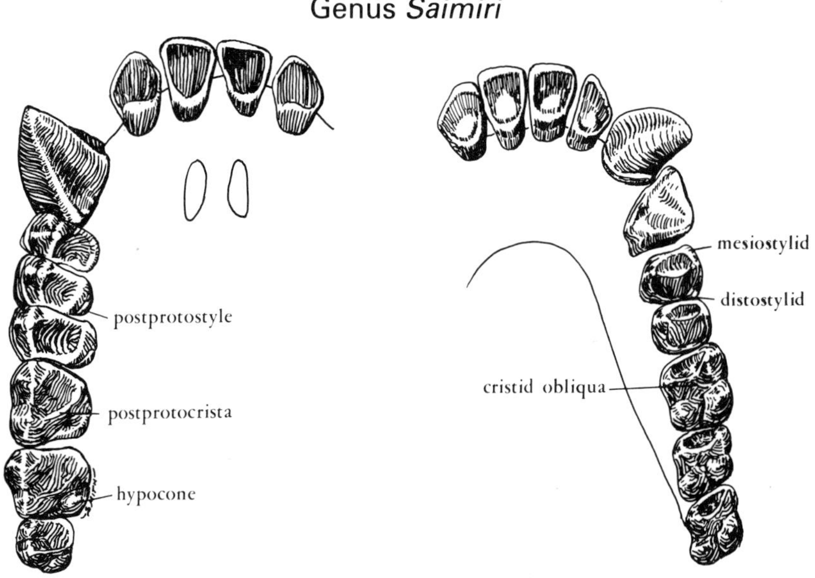

Plate 32. *Saimiri sciureus* male, occlusal view ×2·75.

Morphological Observations

	Sample Male	Sample Female
Saimiri sciureus (squirrel monkey)	20	6
Saimiri oerstedii (redbacked squirrel monkey)	11	6

To my knowledge the study of Long and Cooper (1968) is the first investigation of odontiasis in a New World primate using live animals of known age. The advantages of using live animals instead of skull material for determining the sequence of eruption and the ages at which they occur are obvious. Unfortunately there are very few such studies available for any primate species. The early eruption of M^2_2 and the late eruption of the canine are phylogenetically primitive characteristics. Of all extant New World primates only *Saimiri* and *Aotus* erupt M^1_1 and M^2_2 before I^1_1 appears.

Incisors

Upper. I^1 has a broad crown displaying a flat incisal border, while the crown of I^2 is more pointed and slopes distally. Both possess marginal ridges delimiting a shallow lingual fossa. I^{1-2} are heteromorphic.

Lower. I_1 has a flat incisal border and on I_2 this border slopes distally. Marginal ridges are present on both teeth but they are usually better developed on I_2.

Canines

Upper. The upper canine is robust, presenting a trenchant, sharply pointed apex projecting well below the occlusal plane. There is a longitudinal groove on the mesial surface passing from near the tip of the cervix. Males normally lack a lingual cingulum, whereas females have quite pronounced ones, a nonmetric sexual difference first noted by Orlosky (1973).

Lower. The lower canine is large and slightly recurved. There is no diastema fore or aft of the tooth. The lower canine also manifests nonmetric sexual dimorphism (Orlosky, 1973). Males possess salient mesiolingual grooves and lack continuous lingual cingula, whereas females lack the grooves and display lingual cingula. In Orlosky's (1973) study of 68 females and 69 males, a single male deviated from this pattern.

Premolars

Upper. P^2 is caniniform, normally possessing but a single high paracone. A protocone is present in less than 10% of our sample. P^{3-4} are bicuspid. P^2 lacks any development of a lingual cingulum in this study, nor is it reported by Kinzey (1973) or Orlosky (1973). Lingual cingula, however, abound on P^{3-4}. They are present in over 95% of the current sample, particularly the variety Kinzey (1973) calls the postprotostyle. Both Kinzey (1973) and Orlosky (1973) record high frequencies for the lingual cingulum. Remnants of buccal cingula are less common and are variably expressed on the three premolars. Of the three types, the parastyle occurs 60–65% of the time, the distostyle occurs 20–25% of the time, and the mesostyle is not recorded.

Lower. P_2 is unicuspid and elongated mesiodistally. Rarely is a metaconid observable (less than 15% of the time). P_2, therefore, is an incipient sectorial tooth very similar in morphology to that seen in *Cebus apella.* P_{3-4} are bicuspid, or may occasionally emblazon hypoconids and entoconids. The extra cusps are more common on P_4. Sharp protocristids connect the protoconids and metaconids on both P_{3-4}. Mesio- and distostylids are common on all three premolars; indeed, Kinzey's (1973) findings suggest that they have their highest frequencies among extant cebids in *Saimiri sciureus* (over 50%).

Molars

Upper. M^{1-2} have four cusps, while M^3 usually has but three. A series of cristae connect the various cusps and of these, the postprotocrista is the most pronounced. The trigon basin is clearly delineated by these cristae. The hypocones are small on M^{1-2} and sometimes absent (35%) on M^3. Kinzey (1973) found much higher hypocone absenteeism (86%) for his sample of 32 *Saimiri sciureus.* A lingual cingulum is present on M^{1-2} in every individual examined, whereas on M^3 it is present in 85% of the cases. Small para- and metastyles reside on all three molars and their incidence decreases from M^{1-3}.

Lower. M_{1-3} have four cusps. The hypoconulid is wanting and on occasion the entoconid is absent on M_3, thereby producing a three-cusped tooth (Orlosky, 1973). The protoconid is always mesial to the metaconid on M_{1-2}, while these cusps are opposite each other on M_3 and on all three molars they are connected by sharp protocristids. A short but distinct paracristid passes mesially to merge with the mesial marginal ridge. Also noticeable is a sharp cristid obliqua on the occlusal surface of each molar. Lingual cingula are wanting, but there are buccal cingula on all three molars. In our sample they are present on 94% of M_{1-2}, and somewhat less (75%) on M_3. These frequencies are similar to those found by Orlosky (1973) and Kinzey (1973).

Odontometry (Tables 65–72, Appendix)

Whenever sexual differences obtain in the squirrel monkey, the males are larger. The significant differences are few but when they occur they are in the dimensions of the maxillary and mandibular canines and P_2. Thus, the mesiodistal dimension of the upper canine is significant ($P < 0{\cdot}004$), and that of the lower canine is ($P < 0{\cdot}01$). Both dimensions are significantly different in P_2 (M–D, $P < 0{\cdot}001$ and B–L, $P < 0{\cdot}001$). It is also noteworthy that in males the mesiodistal diameter of the upper canine surpasses that of all other teeth.

Genus *Alouatta*

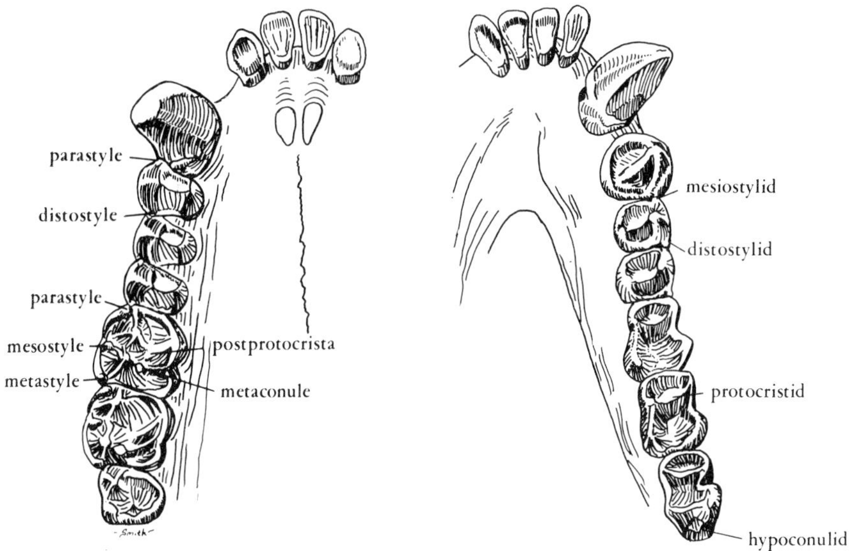

Plate 33. *Alouatta villosa* male, occlusal view ×1.

Morphological Observations

	Sample	
	Male	Female
Alouatta seniculus (red howler)	18	18
Alouatta villosa (mantled howler)	15	42
Alouatta belzebul (redhanded howler)	11	10

Schultz (1960) says M_3^3 erupts after the canine, thus agreeing with *Cebus* and *Ateles*. Such discrepancies suggest sequence polymorphism regarding the eruption of these two teeth and only adequate longitudinal data will settle the question.

Incisors

Upper. I^{1-2} are homomorphic. In I^1 the crown is broad and the incisal border tapers distally. I^2 is similar, differing only in being slightly narrower. The lingual surfaces of both teeth display marginal ridges of which the cervical one is more pronounced. Orlosky (1973) describes three types of lingual surfaces for the upper incisors; (a) smooth, (b) vertical groove and (c) vertical ridge. The three varieties are variably expressed on I^{1-2} as well as among the various species, although he did not find any significant sexual or species differences. The smooth lingual surface is the most frequent (12–43%) in both sexes and in all species.

Lower. I_{1-2} are morpholgically similar albeit I_2 tends to be somewhat more caniniform than I_1. Marginal ridges are present as well as the three types of lingual surfaces discussed for the maxillary incisors. Incisor occlusion is interesting in *Alouatta.* Instead of the edge-to-edge bite which is the normal condition in most monkeys—both Old and New World species—the howler monkey has a high incidence of mandibular incisor protrusion, i.e. underbite (epharmosis). Colyer (1936) notes that underbites range in frequency from 9·5% in *Alouatta seniculus* to 50% in *Alouatta villosa* (*palliata*), while Schultz (1960) finds only two cases of edge-to-edge bite in 493 skulls of *Alouatta villosa* (*palliata*) and Serra (1951) reports an incidence of 43·5% for the genus. Most recently, Zingeser (1968a) recorded that all specimens of *Alouatta caraya* demonstrated epharmosis. All of these monkeys came from islands in the Rio Parana in northern Argentina and Zingeser (1970, p. 184) believed that island populations have been favored in "the unhindered development of the herbaceous underbite trait". Supporting evidence for a dietary hypothesis from Old World leaf-eating monkeys will be considered later. The percentages of underbite in the present sample are 92·8% for *Alouatta villosa,* 47·1% for *Alouatta seniculus,* and 75% for *Alouatta belzebul.* There is no doubt of the magnitude of underbite variability in *Alouatta.* Zingeser (1970) presented one explanation; however, detailed studies both in the field and the laboratory are necessary before the complete story will be known.

Canines

Upper. The upper canines are large, lanceolate-like teeth in both sexes. A mesial groove is present from near the crown tip to the cervix. A lingual cingulum is feebly developed in both sexes, whereas an incipient buccal cingulum is variable in expression.

Lower. The lower canine is large and rises well above the other teeth. A lingual cingulum is present and as it passes distally it forms a distinct ledge.

Premolars

Upper. The premolars are essentially bicuspid, although P^2 lacks a developed protocone in 38% of the present sample. There are no species or sex differences in the expression of the protocone. Even in those cases lacking a protocone the primary lingual cingulum is large, thus simulating a bicuspid tooth. The protocone is always smaller and lower than the paracone even on P^4. In the present sample para- and distostyles are always observable on P^{3-4} but are variable on P^2. Thus, on P^2 both of the styles are present in over 65% of the animals. Kinzey's (1973) data do not conflict with these findings.

Lower. P_2 is usually unicuspid and somewhat caniniform while P_{3-4} are bicuspid, or on occasion tricuspid, particularly P_4 (Orlosky, 1973). Orlosky (1973) also detected a significant difference in the expression of paraconids (mesiostylids) and metaconids on P_2 among *Alouatta villosa*, *Alouatta belzebul* and *Alouatta seniculus*. *Alouatta villosa* always had a lower frequency of mesiostylids and a higher incidence of metaconids whereas the latter two species were similar. Entoconid development is variable on P_4 particularly in *Alouatta belzebul* where it is expressed in about half of the specimens studied. Buccal cingula may be present on all three premolars but they are rare on P_{2-3}, occurring more often (46%) on P_4.

Molars

Upper. The upper molars of *Alouatta* are perhaps the most interesting of all the Cebidae with the possible exception of *Brachyteles*. They possess four cusps and a complex system of cristae, conules and styles. The development of styles, particularly the mesostyle and the V-shaped nature of the para- and metacones led Gregory (1922) to see a secondary resemblance to the molar pattern of *Notharctus crassus*. Certainly, there are parallel adaptations to the leaf-eating Indriidae, and Todd (1918) compared the crescentic cusps of *Alouatta* to those of the leaf-eating koala bear. A small lingual cingulum, the postprotostyle is present approximately 36% of the time on both M^{1-2}, and less frequently (15%) on M^3. Buccal cingulum remnants are much more common on the molars. In most species both para- and metastyles are always present on M^{1-2}, while the parastyle is only slightly less common on M^3. The metastyle is absent from M^3.

The molars of *Alouatta* are famous for their mesostyles, which have for years been considered an integral part of the molars' occlusal anatomy (Todd, 1918; Gregory, 1922; Zingeser, 1968a). Recently Kinzey (1973, p. 111) proposed a name change from mesostyle to eoconule or mesoloph since the well developed cuspule between the metacone and paracone is "clearly derived from the ectoloph (eocrista)". This alleged derivation is not supported by the present material. In our sample the mesostyle on some molars appears at the confluence of the postmetacrista and the premetacrista but it is always continuous with either the mesial or distal cingulum. On other molars, and often in the same molar series, the mesostyle appears to be derived directly from the distal cingulum and the two structures are again continuous, and whatever connections the mesostyle has with the eocrista appear to be secondary. Kinzey (1973, p. 112) recognized the latter development but considered it "pinched-off and secondarily attached to the cingulum" and proposed the appellation, "pseudomesostyle" to account for its appearance. In our opinion, it is rather hazardous to make such morphological decisions without the proper odontogenetic and phylogenetic information. Finally, the variety of configurations seen on the erupted teeth suggest a complex answer and we can do no better than quote Simpson (1955, p. 435), who regarded a similar difference between a hypocone and pseudohypocone as a "distinction without a difference". Mesostyles are present 100% of the time on M^{1-2}, but are quite rare on M^3.

The occlusal surface also displays a postprotocrista and a metaconule. The postprotocrista is always observable on M^{1-2} but occurs only about one-half as frequently on M^3 which also occasionally lacks the metacone or hypocone, or both. The metaconule is discussed by Friant (1942) and Serra (1952b), and we document it here particularly on M^{1-2} and to a much smaller extent on M^3. The metaconule manifests on the postprotocrista as a slight enlargement near the base of the metacone (Plate 33), and its free border is sharp and in occlusion occupies the re-entrant space between the postmetacristid of one tooth and the paracristid of the next lower molar. The metaconule may be correlated with the transverse widening of the talonid and the shearing action of the paracristid so obvious in *Alouatta*, a functional mechanism thought by Serra (1952b) to be associated with the leaf-eating habits of the howler monkey.

Lower. M_{1-2} possess four large cusps. The protoconid and metaconid are united by a sharp protocristid and the trigonid area is somewhat higher than the talonid. In both teeth the protoconid is mesial to the metaconid and on M_1, the cusps are closer together than on M_2. The talonid basin is wider than the trigonid and in occlusion it receives the protocone of the maxillary molar. Both the paracristid and cristid obliqua are well defined and sharp, and undoubtedly play important triturating roles during mastication. M_3 is somewhat different morphologically in that (a) the trigonid cusps lie opposite each other, (b) the trigonid is as wide as the talonid, (c) the trigonid basin is expanded mesiodistally, (d) the cristid obliqua is not as oblique and (e) both the hypoconulid and tuberculum intermedium may be present. Indeed, the hypoconulid is present on M_3 in 67·1% of *Alouatta villosa*, 88% of *Alouatta seniculus* and 75·2% of *Alouatta belzebul*. These observations are contrary to those of Gregory (1922), but are in agreement with those of Le Gros Clark (1971), Zingeser (1968a) and Orlosky (1973). The tuberculum intermedium is much less frequent in the three species, being present in 23·8% of *Alouatta belzebul*, 13·7% of *Alouatta seniculus*, and only 1% of *Alouatta villosa*.

Finally it should be mentioned that small paraconids have been noted on the lower molars by Le Gros Clark (1971). These are absent in this study.

Odontometry (Tables 73–80, Appendix)

Sexual dimorphism is characteristic for these three howler species: in all cases, males are larger than females. T-tests demonstrate the high degree of sexual differences in the majority of the teeth: indeed, of 85 variables nine are significant at $P<0{\cdot}05$, while the remainder are significant at $P<0{\cdot}01$ in *Alouatta seniculus. Alouatta villosa* expressed somewhat fewer sexual differences in the same variables. The greatest mesiodistal and buccolingual diameter disparity occurred in the upper and lower canines in all three species. In other words, male *Alouatta* have large, robust canines. It should be noted that P_2 was highly significant in both species ($P<0{\cdot}01$), undoubtedly being correlated with the canine size. These pronounced sexual differences in tooth size are in part an allometric correlation with the large body-size differences between the sexes of these species (Zingeser, 1967).

The lower molars increase in the mesiodistal dimension from M_{1-3}. This phenomenon, unusual among New World primates, was probably first observed by Pocock (1925a), who wrote, "*Alouatta* stands alone among the Platyrrhini in having the last molar longer than the first".

10

Family Cercopithecidae

Present Distribution and Habitat

The family Cercopithecidae is divided into two subfamilies, the Cercopithecinae and Colobinae. The division is based on anatomy: the Cercopithecinae possess cheek pouches and nonsacculated stomachs, while the Colobinae lack cheek pouches and display elaborate sacculated stomachs. Both subfamilies are widely distributed over the Old World between latitudes 35° N and 30° S. A notable exception is the Japanese macaque (*Macaca fuscata*) which resides as far north as 41° on the island of Honshu. The Cercopithecinae enjoy a more extensive distribution than the Colobinae, ranging from North Africa in the west to Japan in the east; moreover, the Cercopithecinae are both terrestrial and arboreal while the Colobinae are predominately arboreal. The Colobinae are represented in Africa by a single genus, *Colobus*, while all other members of the subfamily live in India, China and Malaysia.

Members of both subfamilies range from sea level to the snow line; indeed, some langurs live in excess of 3,400 m (11,000 ft) in the Himalayas and one species of guenon, *Cercopithecus mitis*, is found up to 3,000 m (10,000 ft) in Africa (Napier and Napier, 1967).

The predominant mode of locomotion of all Old World monkeys is quadrupedal, whether in the trees or on the ground. Napier (1961) proposes the term "semi-brachiation" to describe the arm-swinging habits of *Colobus*, *Presbytis* and *Nasalis*. Both subfamilies are diurnal.

Dietary Habits

Old World monkeys are usually considered vegetarians since they eat mostly fruit, leaves, bark, nuts, grass, flowers and buds. The Colobinae eat mostly leaves; indeed, some students consider them to be almost exclusively phyllophagous; e.g. Booth (1956) did not find a particle of animal matter in the stomachs of 41 *Colobus polykomos* and Kern (1964) estimates the diet of *Nasalis* to consist of 95% leaves and 5% fruits and flowers. In most cercopithecids however, animal protein is obtained in varying amounts by eating birds' eggs, small birds and insects. It is well known that baboons supplement their basic diet of grass, roots and tubers with an occasional young gazelle (Washburn and DeVore, 1961). The chacma baboon (*Papio ursinus*) also has a predilection for marine shellfish, as do the crab-eating *Macaca fascicularis* (Hall, 1963).

Diets are frequently altered in captivity and Old World monkeys are famous for their catholic tastes in the laboratory: they eat onions, apples, monkey chow, lifesavers, sugar cubes, sweet cookies and of course bananas. It is well known that the Japanese macaque eats sweet potatoes, and the dietary plasticity of this animal has been well demonstrated (Itani, 1958).

General Dental Information

Permanent dentition: $I^2_2\ C^1_1\ P^2_2\ M^3_3$

Deciduous dentition: $i^2_2\ c^1_1\ m^2_2$

Sequence of eruption of permanent teeth (Schultz, 1935):

Cercopithecus

	M^1	I^1		I^2		M^2	P^3	P^4			C		M^3
M_1		I_1	I_2		M_2			P_4	P_3	C		M_3	

The eruption sequence for *Erythrocebus* is not known

The eruption sequence for *Cercocebus* is not known

Macaca mulatta

	M^1	I^1	I^2	M^2	P^3	P^4	C	M^3
M_1		I_1	I_2	M_2	P_3	P_4	C	M_3

(Hurme and van Wagenen, 1961)

Papio

	M^1		I^1		I^2		M^2		P^3	P^4		C		M^3
M_1		I_1		I_2		M_2		P_3	P_4		C		M_3	

The eruption sequence for *Mandrillus* is not known

The eruption sequence for *Theropthecus* is not known

Colobus

		M^1	I^1		M^2	I^2	P^3		P^4			C	M^3
	M_1		I_1	M_2	I_2		P_3	P_4		C	M_3		

Nasalis

		M^1	I^1		I^2	M^2		P^3	P^4	C		M^3
	M_1		I_1	I_2	M_2		C	P_3	P_4		M_3	

Pygathrix

		M^1		M^2		I^1		I^2		M^3			P^4	P^3	C
	M_1		M_2		I_1		I_2		M_3		P_4	P_3	C		

The eruption sequence for *Presbytis* is not known

The cercopithecid formula differs from the platyrrhine formula by having one less permanent premolar in each jaw (P^2_2), leaving P^{3-4}_{3-4}. There is also one less deciduous molar in each jaw.

The upper incisors are generally broad, spatulate teeth, particularly I^1. I^2 is much narrower and frequently peg-shaped or pointed. Because of the morphological contrast between I^{1-2}, Remane (1960) and Swindler and Sirianni (1969) use the term "heteromorphic" to describe the upper incisors in the majority of Old World monkeys. Also a median lingual sulcus is present on I^1 as well as frequent lingual tubercles, whereas on I^2 the sulcus is more often replaced with a median enamel ridge. I_1 is usually larger than I_2. Both I_{1-2} possess labial surfaces which are convex mesiodistally, while their lingual surfaces are flat or only slightly convex mesiodistally. A small lingual cingulum is usually present. The majority of Old World monkeys display an edge-to-edge bite of the incisors, a notable exception being among the leaf-eating monkeys who occasionally display an underbite.

The maxillary canines are large, well developed teeth in most Old World monkeys. In both male and female they project beyond the occlusal plane, but the male canine is usually much larger than the female, particularly in baboons. A groove incises the mesial surface of the canine from the root to near the crown tip; it is deeper and more pronounced in males. The mesial surface is much wider than the blade-like distal surface which hones against P_3. Recently Zingeser (1969) and Every (1970) discussed the functional relationship of the canine/P_3 complex as an adaptation for honing the upper canine. The lower canines are smaller than the upper ones, and again more so in the female. A talonid is usually present along the distal aspect of the tooth. The crown curves labially and slightly distally from robust bases.

The upper premolars are quite variable among cercopithecids. P^3 may possess a single cusp (paracone) as in the majority of *Colobus polykomos* or it may display two cusps (paracone and protocone) as it does in most other cercopithecids. P^4, on the other hand, is generally more conservative, having two cusps as a general rule or occasionally three, particularly in baboons and mangabeys. The occlusal surfaces of P^{3-4} present a mesial fossa separated from a distal fossa (trigon basin) by an intervening preprotocrista. Concerning the smaller mesial fossa Delson (1973) felt that it was "perhaps slightly deeper in cercopithecines". The paracone is larger than the protocone. The lower premolars are heteromorphic, i.e. P_3 is unicuspid and sectorial (compressed buccolingually to form an oblique cutting edge which shears against the upper canine), while P_4 has two or more cusps. A sectorial P_3 is found in all extant cercopithecid monkeys. P_4 is more variable; it possesses anything from two to four cusps and indeed can be quite molariform in *Papio* (Hornbeck and Swindler, 1967). In general, the metaconid is always larger than the protoconid in the cercopithecines while it is reduced (smaller) than the protoconid in the colobines.

The molars are bilophodont, i.e. they possess four cusps, two mesial and two distal connected by transverse enamel ridges (lophs) resulting in a constriction or waisting of the crown between the mesial and distal pairs of cusps. On the maxillary molars the paracone is the highest cusp, followed by the metacone, while the protocone and hypocone are of nearly equal height. Three fossae are present on the occlusal surface. The central fossa is the trigon basin and the distal one is the talon basin. The mesial fossa is a cercopithecid novelty (Delson, 1973) and is present in all taxa. It is not reduced as in the lower molars of colobines, but instead is frequently quite large (Delson, 1973). The occlusal surface of each molar presents two developmental grooves, one passing mesiodistally through the center of the tooth and another perpendicular to the former, running buccolingually. The latter groove extends as a rather deep cleft onto both the buccal and lingual surfaces of the tooth. The lingual portion is usually wider and deeper than the buccal extension and in this manner each molar is divided into mesial and distal moieties. The upper molars flare both buccally and lingually.

On the mandibular molars the metaconid is the highest cusp, then the entoconid followed by the subequal protoconid and hypoconid. Three fossae occupy the occlusal surface of each lower molar: a mesial (trigonid basin), a central (talonid basin), which is the largest and deepest of the three, and a distal (post-talonid basin) which is probably homologous with the fovea posterior of hominids. Developmental grooves criss-cross the occlusal surface of each molar. The buccolingual groove is the deeper of the two and extends as a cleft onto both the buccal and lingual surfaces of the molars. The buccal surface of these molars flares outward giving the impression of semirounded columns when viewed from above, while the lingual side is almost vertical. The degree of flare is greater in *Cercocebus* and *Papio* and least in colobines according to Delson (1973), who indicated another structural difference useful for distinguishing between the lower molars of the two subfamilies, namely that the trigonid basin in colobines is generally nothing more than a narrow groove. The bilophodont molar is a hallmark of the dentition of Old World monkeys and the student should see Remane (1960), Kälin (1962) and Voruz (1970) regarding the development of bilophodonty in these monkeys. Voruz (1970) believes that the bilophodont pattern is not completed until the Pliocene. Hershkovitz (1971) using the varied molar patterns present in cebids, clearly demonstrated, in an analogous sense, the underlying mechanics of crista capture and conversion necessary during the formation of the bilophodont molar. It should be noted that the bilophodont molar is not peculiar to Old World monkeys, but is paralleled in other leaf-eating mammals, e.g. tapirs and kangaroos (Gregory, 1922).

The hypoconulid is absent on M_{1-2} and on M_3 it is present in all genera except *Cercopithecus* and *Erythrocebus.* The hypoconulid as well as many other dental features are variably expressed on cercopithecid molars and these will be considered at the appropriate time.

The dynamics of mastication in Old World monkeys is similar to that observed in many other primate taxa. A three-stroke masticatory cycle is present consisting of (a) a preparatory stroke which buccally aligns the teeth on the active side, (b) a power stroke which triturates the food and (c) a recovery stroke which returns the mandible to the beginning position. During this activity cycle the pabulum undergoes tooth–food–tooth contact (puncture-crushing) as well as tooth–tooth interdigitation (chewing).

Mills (1955) reported that Old World monkeys masticate differently than other primates in two ways: (a) during the buccal phase of occlusion there is an increased importance on the lingual cusps which are high relative to the buccal ones, affording continual contact throughout the buccal phase, and (b) during the lingual phase of occlusion the mandibular buccal cusps pass between the upper lingual ones resulting in two facets on each cusp. The obvious efficiency of this scissor-like mechanism is thought to be an adaptation to leaf and grass eating. As we mentioned earlier, the bilophodont molar is not indigenous to the primates, but among primates it is more or less unique in Old World monkeys.

Genus *Cercopithecus*

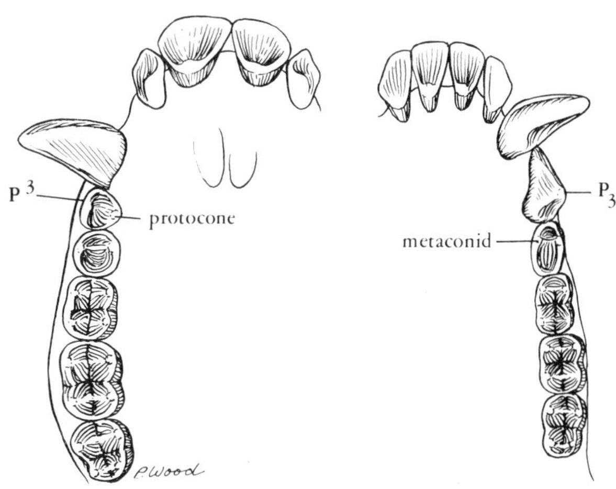

Plate 34. *Cercopithecus nictitans* male, occlusal view ×1·5.

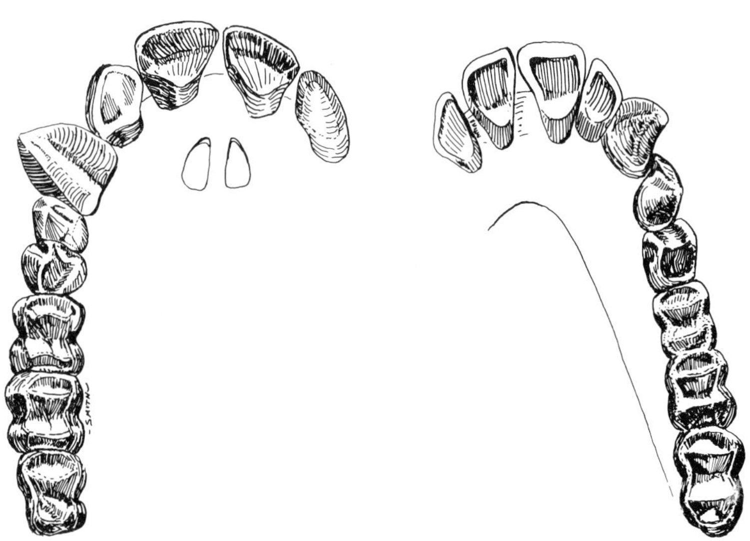

Plate 35. *Cercopithecus* (*Miopithecus*) *talapoin* female, occlusal view ×2.

Morphological Observations

	Sample	
	Male	Female
Cercopithecus aethiops (grivets)	35	20
Cercopithecus cephus (moustached monkey)	9	12
Cercopithecus nictitans (potnosed guenon)	11	7
Cercopithecus mona (mona monkey)	26	16
Cercopithecus mitis (blue monkey)	30	21
Cercopithecus l'hoesti (l'Hoests' monkey)	3	7
Cercopithecus neglectus (De Brazza's monkey)	13	7
Cercopithecus ascanius (redtail monkey)	30	18
Cercopithecus (Miopithecus) talapoin (talapoin)	2	3

The sequence of permanent tooth eruption in *Cercopithecus* is from Schultz (1935), who unfortunately does not mention the species examined. The only difference between this pattern and the one reported by Ockerse (1959) for *Cercopithecus aethiops* is the order of eruption of the premolars. In *Cercopithecus aethiops* P^4_4 appear before P^3_3, otherwise the pattern is identical.

Incisors

Upper. The upper incisors vary a great deal in appearance among the various species studied. In general I^1 is spatulate, having a rectangular labial surface and a rather triangular lingual surface. Mesial and distal marginal ridges are present as well as a lingual ridge. The latter structure varies among cercopithecids in general, and within *Cercopithecus* it may be absent, V-, U- or ⋃-shaped according to Sirianni (1974). For example, the lingual ridge is absent in over 25% of *Cercopithecus aethiops* and V-shaped in the remainder while in *Cercopithecus nictitans* it is ⋃-shaped. A median lingual sulcus is present, although it approaches more the configuration of a fossa in *Cercopithecus nictitans* and *Cercopithecus talapoin.* I^2 is much narrower mesiodistally than I^1; indeed, it is frequently truncated and rather pointed. Marginal ridges are present as well as a lingual ridge. The median lingual sulcus is often replaced by a stout median lingual ridge. A lingual tubercle is absent on I^{1-2} in the guenon species studied. I^{1-2} are heteromorphic as in the majority of Old World monkeys (Swindler, 1968; Swindler and Sirianni, 1969).

Lower. I_1 is larger than I_2 and the incisal border is horizontal, whereas this border tapers distally on I_2. Marginal ridges outline the lingual surfaces of both I_{1-2} although they are not as pronounced as on I^{1-2}. Both lower incisors have labial surfaces which are convex mesiodistally while the lingual surfaces tend to be triangular.

Canines

Upper. The canine is a large, prominent tooth projecting well above the occlusal plane. The base is strongly developed and the crown tapers off to end as a sharp point. The mesial surface is much wider than the blade-like distal surface which is maintained by occlusion with P_3. This honing mechanism is adequately described by Zingeser (1969). The most obvious feature on the mesial surface is the deep mesial groove extending vertically from near the crown tip onto the mesial surface of the root. In all species studied, the groove is much more pronounced in males than in females; in fact, in many females it is barely discernible.

Lower. The crown splays buccally as it passes above the occlusal plane. A narrow mesial longitudinal groove is present in most species, although it is absent in both sexes of *Cercopithecus ascanius.* A distostylid or heel is present in all species and in *Cercopithecus aethiops* it is extremely well developed, particularly in males.

Premolars

Upper. P^3 always has a large, well formed paracone; however, the number of lingual cusps varies from none to two. In *Cercopithecus neglectus* P^3 does not have a lingual cusp, whereas in *Cercopithecus aethiops* 65% of 40 animals have a protocone (Sirianni, 1974). In the majority of species, however, a protocone is present in less than 25% of the population. A narrow crista passes from the paracone to the protocone, or in its absence, the crista terminates on the primary cingulum. In any event, the crest separates the occlusal surface into mesial and distal portions. P^4 is bicuspid in the majority of animals in each species studied. As in P^3, a crista passes between the cusps demarcating mesial and distal occlusal areas.

Lower. P_3 is sectorial, i.e. it is compressed laterally to form an oblique cutting edge which shears against the upper canine. It varies in size among the species of *Cercopithecus* studied being quite small in talapoin and very elongated in the grivet monkey. It should also be noted that in body size, these two groups represent the smallest (talapoin) and among the largest (grivets) of extant guenons. P_4 always has two cusps and frequently there are one or two more adorning the talonid. A protocristid connects the protoconid with the metaconid and thereby forms the distal wall of a limited trigonid fossa. The trigonid is higher than the talonid.

Molars

The upper molars are bilophodont as in all Cercopithecidae. The buccal cusps are higher than the lingual ones, particularly as wear progresses. In all species of *Cercopithecus* studied, M^{1-2} possess four cusps, whereas M^3 may show less cusps. The missing cusp is almost always the hypocone; rarely, both the hypocone and metacone are absent. Indeed, among *Cercopithecus* Sirianni (1974) found the hypocone wanting in 8·3% of *Cercopithecus nictitans* samples and 46·2% of *Cercopithecus ascanius.* Remnants of a lingual cingulum (protostyle) are occasionally observed on the upper molars. They appear more frequently on M^2 (80% in *Cercopithecus neglectus*), although they may occur on all three molars (Sirianni, 1974). Of all species of *Cercopithecus* studied, the protostyle is least frequent in *Cercopithecus cephus* (2% on M^2).

Lower. The lower molars are bilophodont. M_{1-2} possess four cusps while the hypoconulid is lacking on M_3. The talapoin monkey has been reported to have three cusps on M_3 (Forbes, 1894; reported in James, 1960); however, Pocock (1925b) found four cusps on M_3, as we did in the five specimens presented here. The hypoconulid is absent on M_3 in all species studied. The lingual cusps are higher than the buccal ones. Finally, in general, there are very few extra cusplets or stylids adorning the lower molars. A protostylid is observable in over 85% of *Cercopithecus aethiops* lower molars and in all of the five talapoins studied; however, it is absent in the other species. The tuberculum intermedium and sextum are absent in the *Cercopithecus* material examined.

Odontometry (Tables 81–108, Appendix)

Sexual dimorphism in tooth size varied in degree among the different genera. In all species studied the upper canine and P_3 were significantly different in their mesiodistal dimensions and males were always larger than females. As we shall see later, this pattern obtains in all Old World monkeys studied. Regardless of the obvious sexual dimorphism in this trait, it seems that "within Old World monkeys the trend toward the development of the enlarged maxillary canine and sectorial third premolar occurred phylogenetically early and was expressed in both sexes" (Orlosky, Swindler and McCoy-Beck, 1974).

In general, sexual dimorphism in molar size is less in cercopithecids when the molars are small (Swindler, Gaven and Turner, 1963). Also, molar size relations express a great deal of sequential variability with no general pattern emerging.

Genus *Erythrocebus*

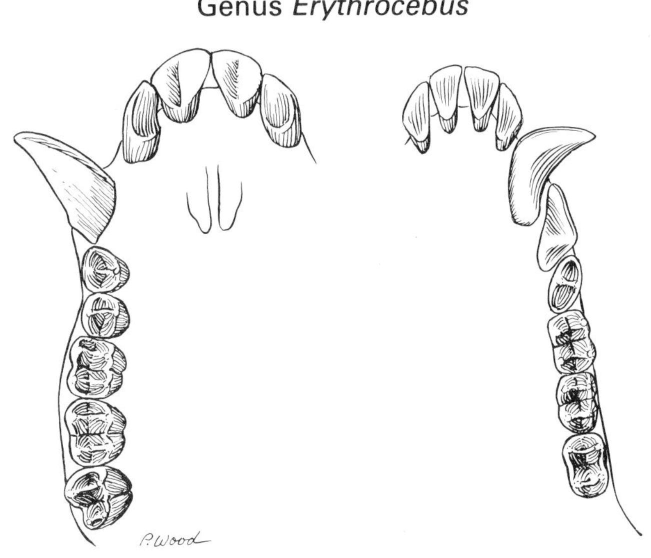

Plate 36. *Erythrocebus patas* male, occlusal view ×1·25.

Morphological Observations

	Sample	
	Male	Female
Erythrocebus patas (patas monkey)	5	2

Incisors

Upper. I^1 is a broad, quadrangular tooth. Thin mesial and distal marginal ridges are present, whereas the lingual cingulum is thick. A median lingual sulcus is present. I^2 is rounded mesiodistally and its incisal border is inclined distally. A slight median sulcus is present. I^{1-2} are heteromorphic.

Lower. I_1 presents a horizontal incisal border which in I_2 slopes distally. The lingual surface of both incisors is concave and in one specimen well defined tubercles are present at the cervical junction of the mesial and distal marginal ridges.

Canines

Upper. The upper canine is a large, long, pointed tooth. The distal border is narrow and sharp while the mesial border is broad and thick. A conspicuous deep mesial groove resides on the mesial surface. The groove is much deeper in males.

Lower. The lower canine is large and splays buccally as it passes well beyond the occlusal plane. A large, well formed heel dominates the distal portion of the tooth.

Premolars

Upper. P^3 is usually bicuspid, although in two of the seven specimens examined the protocone is extremely small. In all cases a primary cingulum swings around the lingual border of P^3. A narrow crista passes from the paracone to the protocone. P^4 is bicuspid: indeed, the protocone is nearly as large as the paracone. As in P^3, a crista connects the two cusps separating the occlusal surface into two uneven moieties. The mesial portion is more restricted than the expanded distal section.

Lower. P_3 is sectorial and the elongated crown lies buccal to the lower canine. P_4 is bicuspid with an expanded talonid. An incipient hypoconid and entoconid are present on P_4 in two of the seven specimens studied. P_{3-4} are heteromorphic.

Molars

Upper. M^{1-3} have four cusps and are bilophodont. The following variation is present in our seven animals: a protostyle occurs in one animal in M^2 and a distoconulus is present on M^3 in one specimen.

Lower. M_{1-3} are bilophodont. M_3 lacks a hypoconulid as did all species of *Cercopithecus*. No extra cusplets or stylids are present on the lower molars in our sample.

Odontometry

Because of the small number of females ($n = 2$) t-tests were not performed. The mesiodistal size gradient for the molars is $M^1_1 < M^2_2 > M^3_3$.

Genus *Cercocebus*

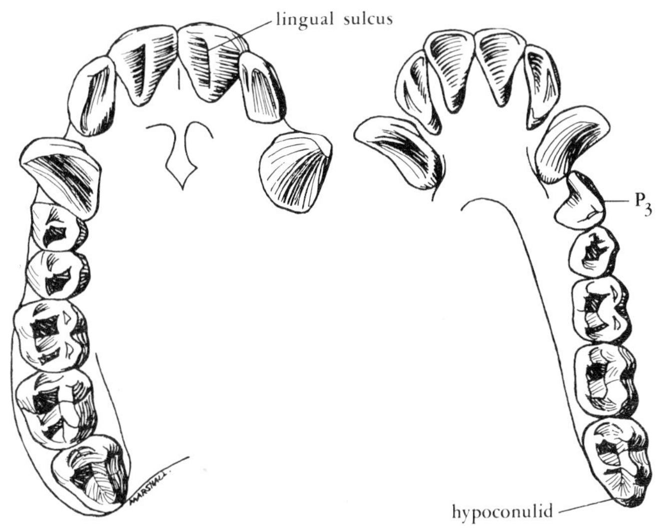

Plate 37. *Cercocebus albigena* female, occlusal view ×1·25.

Morphological Observations

	Sample	
	Male	Female
Cercocebus albigena (grey-cheeked mangabey)	30	29
Cercocebus torquatus (white-collared mangabey)	14	10
Cercocebus galeritus (agile mangabey)	10	10

Incisors

Upper. I^1 is a broad, spatulate tooth much larger than I^2. Mesial and distal marginal ridges outline the lingual surface of I^{1-2} although they are weakly developed on I^2. The surface within the marginal ridges is occupied by two enamel elevations which course upward toward the cervical portion of the tooth. A median lingual sulcus is formed between the two enamel pillars, a condition also observed in *Papio* and *Macaca*. I^2 shows little definition on its lingual surface. The median lingual sulcus is wanting on I^2, so that I^{1-2} are heteromorphic for this trait. They are also unequal in size.

Lower. The incisal border is horizontal on I_1 and slants distally on I_2. Narrow marginal ridges delineate smooth lingual surfaces on both I_{1-2}.

Canines

Upper. The upper canine is large and projects well above the occlusal plane. It is narrow distally and broad mesially. A mesial groove is present and as in most Old World monkeys, it is deeper and wider in males. A diastema separates the canine from I^2.

Lower. The lower canine is robust and inclines buccally. It has a well developed basal tubercle or heel.

Premolars

Upper. P^{3-4} are bicuspid; indeed, P^4 may present two lingual cusps (15% of the present sample). In both teeth a transverse crista connects the paracone and protocone thereby dividing the occlusal surface into two functional parts. The distal portion is more expanded than the mesial part, particularly on P^4.

Lower. P_3 is sectorial while P_4 is bicuspid, and the hypoconid and entoconid are frequently observable on the talonid. The talonid is always lower than the trigonid portion of P_4.

Molars

Upper. M^{1-3} possess four cusps and are bilophodont. A distoconulus (Plate 38) is present on M^3 in 10% of the present sample with no sexual difference observable. No other variation is present.

Lower. M_{1-3} are bilophodont. M_{1-2} have four cusps while M_3 displays a hypoconulid. A tuberculum intermedium is present on M_1 (7%), M_2 (9%) and M_3 (21%) of the time. M_3 in addition to possessing a hypoconulid may occasionally (20%) emblazon a tuberculum sextum between the hypoconulid and entoconid. The hypoconulid is absent 13% of the time.

Odontometry (Tables 109–120, Appendix)

The mangabeys display considerable sexual dimorphism in tooth size. Of the three species examined odontometrically, *Cercocebus albigena* had a significant difference for almost all teeth while *Cercocebus torquatus* had the least amount of sexual disparity. In all taxa, the mesiodistal diameters of the upper canine and P_3 are significant ($P < 0{\cdot}01$). The molar length relations are $M^1 < M^2 > M^3$ and $M_1 < M_2 < M_3$.

Genus *Macaca*

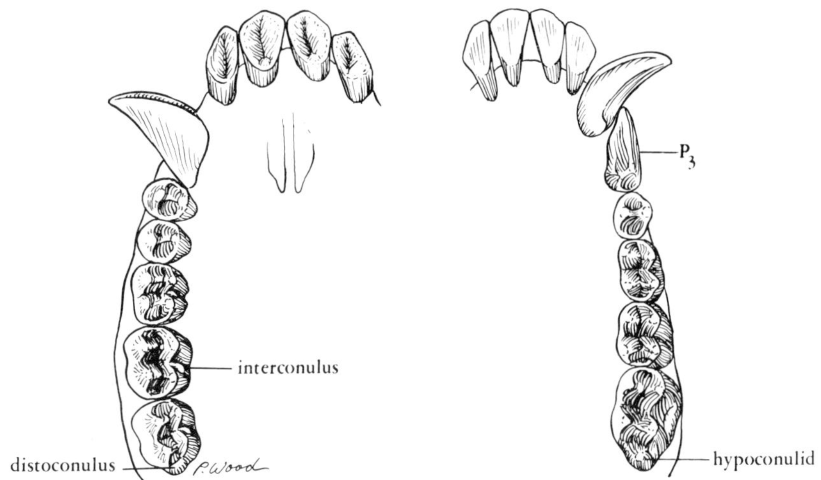

Plate 38. *Macaca nemestrina* male, occlusal view ×1.

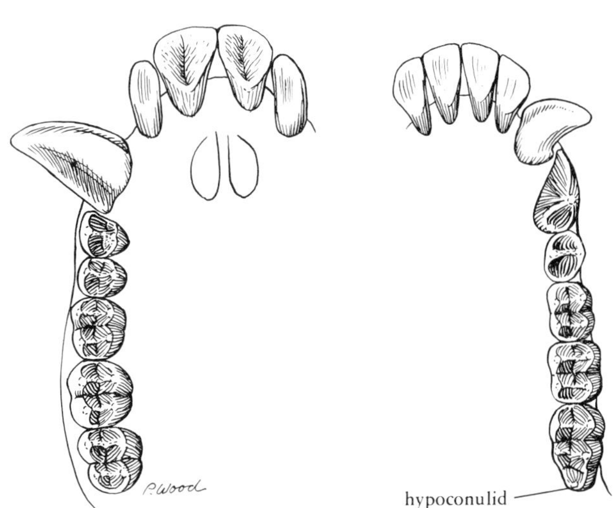

Plate 39. *Macaca niger* male, occlusal view ×1·25.

Morphological Observations

	Sample	
	Male	Female
Macaca mulatta (rhesus monkey)	90	83
Macaca fascicularis (crab-eating monkey)	60	55
Macaca nemestrina (pig-tailed macaque)	85	50
Macaca niger (Celebes black ape)	10	7

There is a sex difference in the eruption of the canine according to Hurme and Van Wagenen (1961). In the female the canine erupts before the premolars in the mandible and between the appearance of these two teeth in the maxillae. Also the odontiasis of the first permanent molar is more variable in males than females.

Incisors

Upper. I^1 is a broad, spatulate tooth. The labial surface is quadrilateral while the lingual surface is triangular. Mesial and distal marginal ridges are present with the latter better developed than the former. Within the marginal ridges there are two well formed enamel elevations with inclined surfaces near the mid-line of the tooth. The median lingual sulcus created by the enamel elevations expands toward the incisal border. The enamel elevations converge toward the cervical end of the lingual surface to form a "V". A lingual tubercle may be present at the base of the "V". In the present study a tubercle occurred in 43% of the animals examined. I^2 has a more pointed incisal border, a rounded lingual surface and the lingual tubercle is absent. If a median lingual sulcus is present, it is very shallow.

Lower. I_{1-2} have smooth, more or less triangular lingual surfaces. The incisal border of I_1 is horizontal while tapering distally on I_2. The labial surfaces of both I_{1-2} are smooth and convex mesiodistally.

Canines

Upper. The upper canines are large, robust teeth projecting well above the occlusal line. They display marked sexual dimorphism; indeed, in males when the teeth are occluded the crown tips of the upper canines surpass the necks of the lower canines. The blade-like distal surface shears against P_3 when the teeth are in occlusion. A deep mesial groove is present in both sexes but as in other cercopithecids, it is better developed in males.

Lower. The lower canines are much larger in males; in fact, in many females they barely surpass the occlusal plane. The crown curves distally and presents a well formed heel near its base.

Premolars

Upper. P^3 is bicuspid. The paracone is large and connects with the much smaller protocone by a narrow preprotocrista. The latter structure separates the occlusal surface into mesial and distal portions. P^4 is similar to P^3 except that it is larger and has an appreciably larger protocone.

Lower. P_3 is sectorial while P_4 is at least bicuspid and frequently small entoconids and hypoconids adorn the talonid rim. P_3 has a small indentation in its distal surface which is probably homologous with the talonid basin.

Molars

Upper. M^{1-3} are quadricuspid teeth and all are bilophodont. The lingual developmental groove is wider and deeper than the buccal extension and may present an interconulus, or what Saheki (1966) prefers to call a groove cusp (Plate 38). He found this cingular remnant on 20% M^1 ($n = 83$), 52% M^2 ($n = 76$) and 67% M^3 ($n = 36$) in *Macaca fuscata*. In our sample the highest incidence (38%) of the groove cusp appears in *Macaca nemestrina* and the lowest (4%) in *Macaca mulatta*. Another variable feature is the distoconulus on M^3 (Plate 38). Saheki (1966) found it in 38% of 49 *Macaca fuscata* while we identified it in 10% of *Macaca nemestrina* and 4% of *Macaca fascicularis*. It is absent in our series of *Macaca mulatta*.

Lower. The occlusal surfaces of M_{1-3} are rectangular and each has four cusps except M_3, which has a hypoconulid. All molars are bilophodont. The tuberculum sextum appears on M_3 in 61% of *Macaca nemestrina*, 68% of *Macaca fascicularis* and 26% of *Macaca mulatta*. Saheki (1966) reports a 39% incidence in *Macaca fuscata* ($n = 42$). *Macaca fuscata* and *Macaca mulatta* are significantly different ($P < 0{\cdot}01$) from *Macaca nemestrina* and *Macaca fascicularis*. The tuberculum intermedium is more frequent on M_2 in *Macaca fuscata* (38%, $n = 62$) than in *Macaca nemestrina* (11%), *Macaca fascicularis* (7%) or *Macaca mulatta* (3%). *Macaca fuscata* is quite different ($P < 0{\cdot}01$) from the other three macaque species.

Odontometry (Tables 121–140, Appendix)

The four macaque species studied present a great deal of sexual dimorphism in tooth size. In all species, the males are larger than the females and as in the majority of cercopithecids, the greatest sexual differences center around the upper canine/P_3 complex.

Genus *Papio*

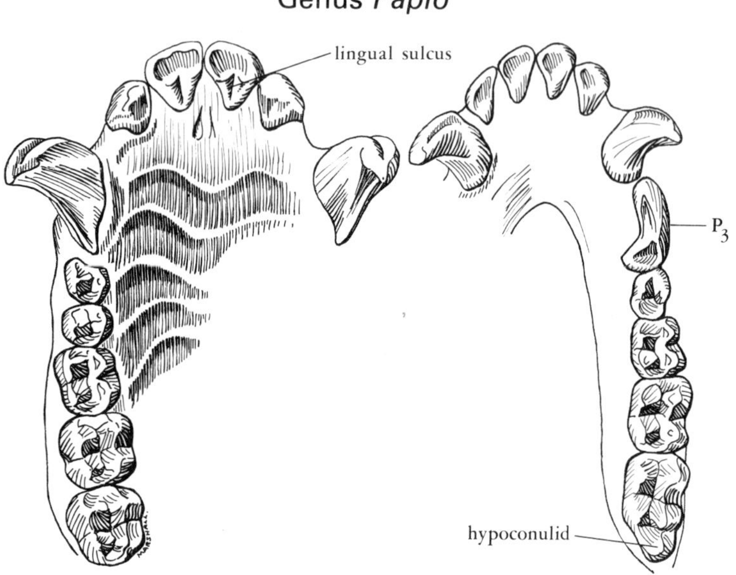

Plate 40. *Papio cynocephalus* male, occlusal view ×0·8.

Morphological Observations

	Sample	
	Male	Female
Papio cynocephalus (yellow baboon)	50	55

Incisors

Upper. The lingual surface of I^1 is essentially triangular with the apex toward the root. The labial surface is quadrilateral. Mesial and distal marginal ridges bound the lingual surface of I^1. The ridges merge cervically and have a V-shaped appearance which thickens to form a lingual cingulum. A pit is frequently present at the junction of the ridges. Between the marginal ridges are two rounded mesial and distal elevations of enamel running from inclined surfaces near the middle of the tooth, resulting in a deep median lingual sulcus. I^2 displays a rounded lingual surface upon which a slight median sulcus is present, always much more shallow than the median sulcus of I^1. The lingual surfaces of both I^{1-2} are better described as constituting a lingual sulcus rather than a lingual fossa which is occasionally present in hominoids (Swindler, McCoy and Hornbeck, 1967).

Lower. I_{1-2} have mesiodistally convex labial surfaces. Labial grooves are occasionally present passing from the incisal edge to the root. The lingual surfaces are flat or slightly convex mesiodistally. A median lingual tubercle may be present.

Canines

Upper. The upper canine is prominent, projecting well above the occlusal plane. In males it is exceedingly well developed and often passes for more than half its length below the occlusal line of the lower teeth. In cross-section, it is approximately triangular at the base from which it tapers off to end as a sharp point. The tooth curves mesially and this surface is much wider than the blade-like distal surface. The sharpness of the distal surface is maintained by its shearing (honing) contact with P_3. For an excellent description of the cercopithecid upper canine/P_3 complex the student should consult Zingeser (1969). A deep mesial groove occupies the mesial surface of the canine and it is always better developed in males than females.

Lower. The lower canines are much smaller than the uppers and their crowns twist or curve labially and distally from robust bases. A mesial groove is present but always shallow. A slight lingual cingulum and a well formed heel are discernible.

Premolars

Upper. P^3 is bicuspid, presenting both a paracone and a protocone. A preprotocrista forms a V-shaped nexus between the two cusps thereby separating the occlusal surface into a narrow mesial portion and a larger distal part. The paracone is always much larger than the protocone. P^4 is morphologically similar to P^3 except that it is larger and has a relatively more expanded distal portion.

Lower. P_3 is sectorial and its elongated crown shears against the upper canine. There is a fossa where the horizontal enamel ledge meets the inclined distal surface of the crown. Delson (1973) believes the fossa is homoplastic but probably not homologous to the talonid basin of the lower molars. P_4 is molariform in *Papio.* Of the two main cusps the metaconid is wider and higher than the protoconid in 100% of 21 animals (Hornbeck and Swindler, 1967). Because of eburnation the sample is reduced and in the study, sexes were combined. Also a variable number of cusplets adorn the talonid, thus three cusplets on 33%, four cusplets on 29%, while 38% had five cusplets (Hornbeck and Swindler, 1967). The occlusal surface is complex and greatly enlarged resulting in a highly molarized tooth.

Molars

Upper. The upper molars are approximately quadrilateral and of course they are high crowned and bilophodont. As is typical of *Papio,* the molars flare markedly between the cusp apexes and the cervical region. As mentioned earlier, molar flare is more pronounced among cercopithecids in *Papio* and *Cercocebus.*

Lower. The lower molars are subrectangular and bilophodont. M_3 presents a well formed hypoconulid. Molar flare is much more pronounced on the buccal aspect than on the lingual; indeed, the latter side of the tooth is nearly vertical. A mesiobuccal cusplet (protostylid) is present on the protoconid of all lower molars as well as P_4, thus demonstrating a morphogenetic field from P_4–M_3. In *Papio cynocephalus* a tuberculum intermedium is present on M_1 in 56·8% of 44 males and 71·7% of 46 females, on M_2 in 75% of 40 males and 75% of 32 females, and on M_3 in 70·8% of 24 males, 77·1% of 22 females; the tuberculum sextum appears on M_3 in 56·7% of 30 males and 25·9% of 7 females (Swindler, McCoy and Hornbeck, 1967). The hypoconulid was in a buccal position 82% of the time in both sexes, and central the remainder of the time.

Odontometry (Tables 141–144, Appendix)

Both I_1^1 are larger than I_2^2. In the premolars the size relations are reverse; $P_3 > P_4$, while $P^3 < P^4$. The molars increase in length in both jaws, $M_1^1 < M_2^2 < M_3^3$. Sequential relations are variable among cercopithecids and few if any clear patterns are demonstrable (Swindler, Gaven and Turner, 1963). Sexual dimorphism is apparent in every tooth, particularly in the upper canine/P_3 complex. It has been suggested that a canine "field" of sexual dimorphism in tooth size appears to characterize the permanent teeth of many primates, and that adjacent teeth are more influenced by the morphogenetic field than are the dental elements twice removed (Garn, Kerewsky and Swindler, 1966). This contention is substantiated by the present data on *Papio cynocephalus.*

Genus *Mandrillus*

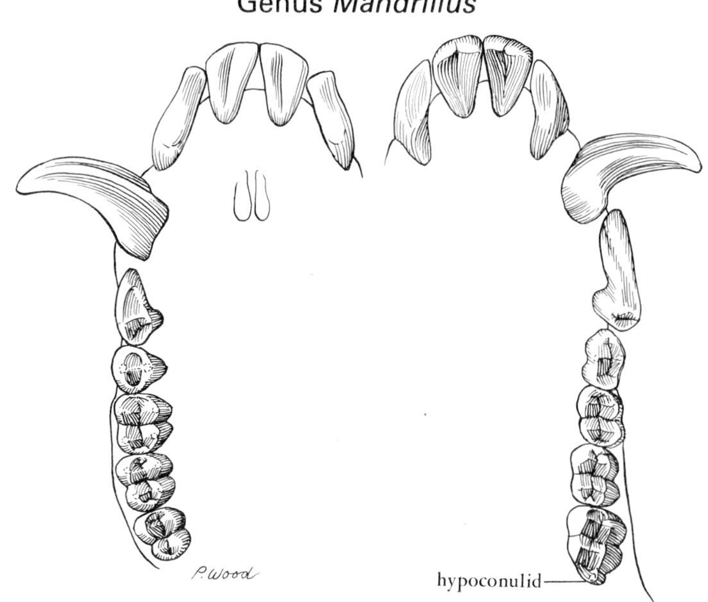

Plate 41. *Mandrillus sphinx* male, occlusal view ×0·65.

Morphological Observations

	Sample	
	Male	Female
Mandrillus sphinx (mandrill)	1	1

Incisors

Upper. I^{1-2} are heteromorphic and both incline mesiolingually. I^1 has a well formed median lingual sulcus while it is lacking on I^2. I^2 is separated by a diastema from the canine.

Lower. I_1 presents a convex buccal surface and a flatter, more triangular lingual face. A slight median lingual sulcus is present. I_2 has a mesially sloping crown which tilts toward I_1. The median lingual sulcus is absent.

Canines

Upper. The upper canine is a powerful tooth tapering from a triangular base to a sharp point. The mesial surface is broad and presents a deep groove while the distal surface is trenchant. An obvious broad groove occupies the lingual surface. A small heel is present distally.

Lower. The lower canine is large and projects well above the occlusal plane. The crown curves buccally and slightly distally to terminate as a rather sharp point. Two grooves are present, one on the mesial surface and another curving up along the lingual surface. A prominent heel is present distally at the base of the tooth.

Premolars

Upper. P^3 is bicuspid. The paracone is large and high compared with the small, low protocone. The two cusps are connected by a preprotocrista. The trigon basin is large and expanded in *Mandrillus* and perhaps the most characteristic feature of P^3 in the mandrill is the greater extension of the crown onto the mesiobuccal root. The trait appears in most cercopithecids and Delson (1973) notes its reliability in identifying isolated P^3's from P^4's. The morphology of P^3 therefore is much like that of P_3. P^4 is bicuspid and the paracone is still the largest cusp, albeit the protocone is almost equal in size to the paracone. The crown lacks the mesiodistal elongation of P^3.

Lower. P_3 is sectorial; indeed, it is the most specialized sectorial P_3 found among extant cercopithecids. The mesiobuccal flange is extremely elongated, passing nearly to the neck of the canine. The protoconid is distal on the buccal surface; from it passes a distomarginal crest which defines the buccal side of the deep talonid basin. Another crest courses lingually to complete the formation of this basin. P_4 is molariform as in *Papio.* The metaconid is wider and higher than the protoconid and the two are connected by a protocristid. The talonid basin is both elongated and expanded as in *Papio.*

Molars

Upper. M^{1-3} are bilophodont. They conform to the molar morphotype of cercopithecids described earlier. According to Jolly (1970a) the molars of *Mandrillus* have low, rounded cusps when compared with the large, high crowned molars of *Theropithecus.* At the same time, the cheek teeth of *Papio* are intermediate, tending to be more high crowned and angular than those of *Mandrillus*, but less than those of *Theropithecus.*

Lower. M_{1-3} are bilophodont. M_3 possesses a hypconulid and sometimes a tuberculum sextum (Hill, 1970). Our sample did not have a tuberculum sextum, but did display a tuberculum intermedium on M_{1-3}.

Odontometry

No measurements were taken of the two specimens. Mesiodistal size relations are $M^1 < M^2 > M^3$ while $M_1 < M_2 < M_3$. Jolly (1970a) found that the incisal breadth of the maxillary incisors (taken at the alveolar level) was greater than molar row length (M_{1-3}) in *Mandrillus*, while in *Theropithecus* the ratio is just the opposite, i.e. the incisal breadth is narrow and much less than molar row length. *Papio* was more or less intermediate. The implication is that incisal function is much more important in mandrillus and drills, less in *Papio* and least in *Theropithecus* when compared with molar function. The dietary differences obtaining between these three groups would seem to substantiate the functional hypothesis of Jolly (1970b) regarding small-object feeding in *Theropithecus*.

Genus *Theropithecus*

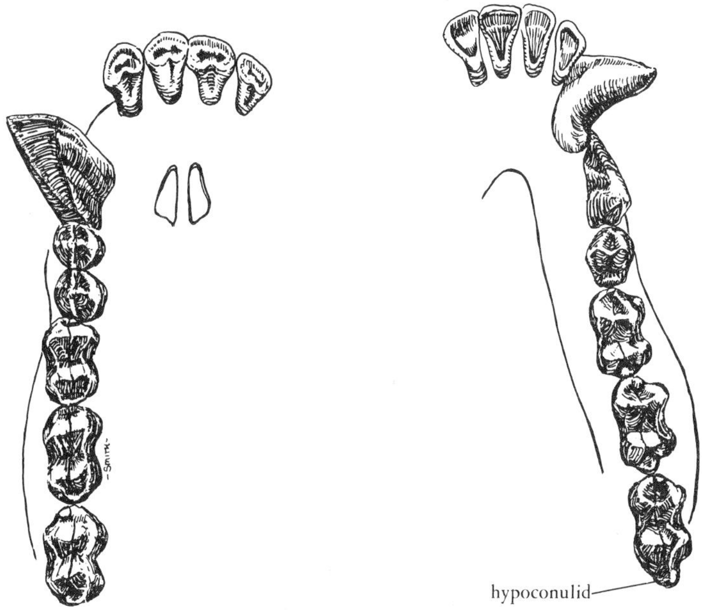

hypoconulid

Plate 42. *Theropithecus gelada* male, occlusal view ×1.

Morphological Observations

	Sample	
	Male	Female
Theropithecus gelada (gelada)	3	3

Incisors

Upper. I^{1-2} have convex labial surfaces. The lingual surface of I^1 is essentially triangular with the apex toward the root. Mesial and distal marginal ridges are present as well as thickened vertical ridges which delimit a median lingual sulcus. I^2 is more cylindrical with a convex lingual surface. A very shallow median lingual sulcus is present. Both I^{1-2} are similar to other Papionini except smaller.

Lower. I_1 has a broad, flat incisal border while in I_2 the border tapers distally. The labial face of I_{1-2} is convex and the lingual face is triangular. Delson (1973) says that the lingual surfaces of I_{1-2} lack enamel in all Papionini. During the course of this study we examined many lower incisors of *Papio*, *Cercocebus*, *Macaca* and *Theropithecus*. In our opinion, their lingual surfaces possess a thin layer of enamel, and therefore we cannot agree with Delson's observations.

Canines

Upper. The upper canine is robust in both sexes but as usual in cercopithecids the canine is appreciably smaller in females. The crown is convex mesiolabially and is much wider mesially than distally. A shallow median labial groove and a much deeper and wider mesial groove are present. Both pass onto the root. There is no indication of any cingular development along the lingual surface.

Lower. In the lower canines the crown passes labiolingually as it projects above the occlusal plane. The mesial surface is broad and convex mesiodistally. A slight groove passes vertically on the mesiolingual surface of the tooth, and a basilar heel projects distally.

Premolars

Upper. P^{3-4} are bicuspid. The paracone is larger than the protocone, particularly on P^4. The crown anatomy conforms to the morphotype described earlier. A mesiobuccal enamel extension or flange is obvious on P^3 but not to the degree that it is present in the mandrills (compare Plates 41 and 42). According to Delson (1973), P^{3-4} of *Theropithecus* display a mesiolingual cleft which is not usually seen in other taxa.

Lower. P_{3-4} are heteromorphic. P_3 is sectorial and P_4 is at least bicuspid since small cusps may also adorn the expanded talonid (Hill, 1970). The mesiobuccal extension of P_3 is more vertical and slightly less pronounced than in *Mandrillus* and *Papio*. As is characteristic of Cercopithecines, the metaconid of P_4 approximates the size of the protoconid.

Molars

Upper. M^{1-3} are bilophodont and high crowned, i.e. the cusps are high between the deep and widely separated fossae. The middle fossa is particularly wide and deep and a developmental groove passes transversely through it to cleft both the buccal and lingual surfaces, especially the latter. The mesial and distal fossae are expanded, particularly the mesial which appears more as an enamel ledge than a fossa. The molars flare as in other Papionini but the flare is lingual rather than buccal as discussed by Delson (1973) and thus differs from the other groups.

Lower. M_{1-3} are bilophodont and M_3 possesses a well formed hypoconulid while M_2 presents a distal cuspule instead of a hypoconulid. The deep, wide talonid basin opens as a deep groove or cleft onto the buccal surface of the tooth. The cleft is particularly well developed on M_3. A small mesiobuccal cuspule increases in size from M_{1-3}. The three molars flare buccally while the lingual face is nearly vertical. A tuberculum sextum may be present on M_3.

Odontometry (Tables 145–148, Appendix)

The data presented here were generously provided by Mr Gerald Eck. Mr Eck is preparing a detailed and systematic description of the teeth of *Theropithecus gelada* which should be ready in the near future from the University of California, Berkeley.

Genus *Colobus*

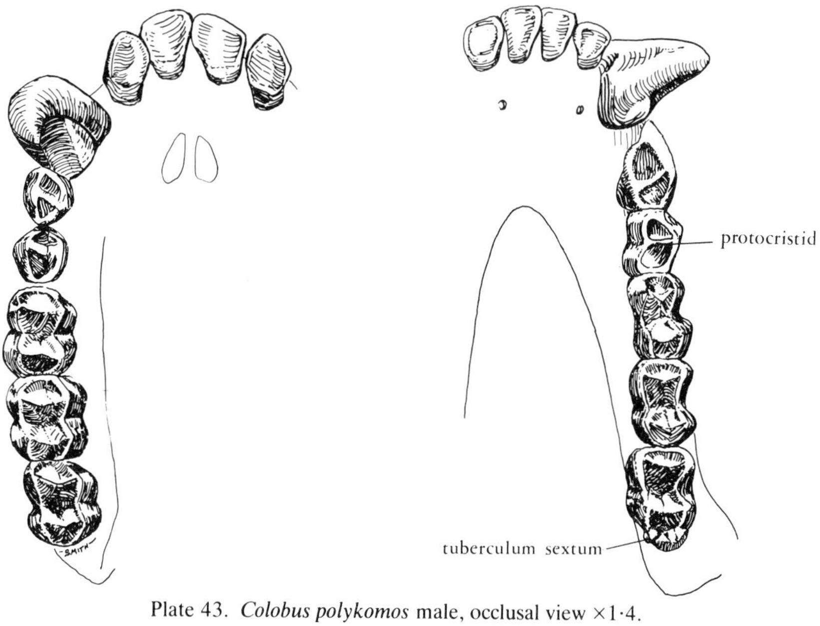

protocristid

tuberculum sextum

Plate 43. *Colobus polykomos* male, occlusal view ×1·4.

Morphological Observations

	Sample	
	Male	Female
Colobus polykomos (black and white colobus)	48	30
Colobus (Piliocolobus) badius (red colobus)	26	26

There is apparently a tendency among the colobines to erupt M_2^2 early, otherwise, the pattern is not different from that of other Old World monkeys.

Incisors

Upper. I^1 has a broad, flat incisal border. The buccal surface is quadrilateral and convex mesiodistally. Thin mesial and distal marginal ridges outline the lingual surface and pass onto the cervical portion as a lingual cingulum. The cingulum forms a broad "U" as it swings across this portion of I^1. A shallow lingual sulcus is observable just distal to the mesial marginal ridge. Of 52 *Colobus polykomos* satisfactory for study, 5·8% have a small lingual tubercle. I^2 is a narrow triangularly-shaped tooth with pointed incisal borders. Faint mesial and distal marginal ridges bound a lingual face which presents a vertical median ridge. The lingual tubercle is present in 3·8% of the 52 specimens. I^{1-2} are heteromorphic.

Lower. I_1 has a horizontal incisal border, while on I_2 it slopes distally. Narrow mesial and distal marginal ridges are present on both incisors. The lingual surfaces are slightly concave on I_{1-2}, more so on I_1. In the present material, underbite is more frequent in *Colobus polykomos* (78%) than in *Colobus badius* (57%). Schultz (1958) also found more underbite in *Colobus polykomos* (60%) than in *Colobus badius* (34%). The remaining incisor relations in our material are edge-to-edge while Schultz (1958) found a 5% incidence of this bite in *Colobus polykomos.*

Canines

Upper. The upper canine is robust, projecting well below the occlusal plane. It is broad mesially and narrow distally. A lingual cingulum skirts around the base of the tooth. The characteristic cercopithecid mesial groove is present.

Lower. The lower canine is large, rising well above the occlusal plane. From a subtriangular base the crown tapers to a rather sharp apex. An enamel heel is present distally.

Premolars

Upper. P^{3-4} possess the typical cercopithecid crown morphology outlined earlier. A notable difference between *Colobus* and the other leaf-eaters investigated concerns the development of the protocone on P^3. In *Colobus* there is a general reduction in the protocone, while in *Colobus polykomos* the protocone is absent in 93% of 60 males and 76% of 37 females (Swindler and Orlosky, 1972). This is also a significant sexual difference. Delson (1974) has also noticed the reduction of the protocone in several species of *Colobus.* P^4 is bicuspid although the paracone is the predominant cusp.

Lower. P_3 is a large sectorial tooth, particularly in males. P_4 is structurally more variable than P_3. The metaconid is variable: in *Colobus* it ranges from absent, to small, to an approximation of the size of the protoconid; in no case is the metaconid greater in height than the protoconid. Swindler and Orlosky (1972) found a small metaconid in 33% of 60 male *Colobus polykomos* but a large one in 62%. In the remainder (5%) the metaconid was absent. There was no significant sexual difference in the incidence of the trait. In an earlier study Hornbeck and Swindler (1967), stated that *Colobus badius* lacked a metaconid. That sample consisted of only seven specimens. Since then we have examined additional specimens and found a metaconid present. Delson (1973) also noted the presence of metaconids in a sample of 19 *Colobus badius.* In our opinion, such discrepancies are due primarily to two factors: small sample size and accurate definition of quantitative variables.

Molars

Upper. M^{1-3} are four-cusped bilophodont molars. As in colobines in general, these teeth tend to be high crowned and display little buccal flaring. The high cusps are in turn connected by tall, thin protocristids and entocristids. A distoconulus is present in 22% of 58 male *Colobus polykomos* (Swindler and Orlosky, 1972). No sexual dimorphism is present although females have a slightly higher frequency. The distoconulus does not articulate with the hypoconulid of M_3 and thus apparently has little, if any, role in mastication.

Lower. M_{1-2} have four cusps, M_3 has five, and all three are bilophodont. The position of the hypoconulid varies with respect to the long axis of M_3; it can be positioned centrally, buccally or lingually to the axis. In *Colobus polykomos* it is distributed about equally between the buccal (55%) and lingual (43%) positions (Swindler and Orlosky, 1972). The hypoconulid is absent in only 2% of the above animals. The variable tuberculum sextum is represented on M_3 in 63% of 57 male and 61% of 36 female *Colobus polykomos*.

Odontometry (Tables 149–156, Appendix)

Sexual dimorphism is greater in *Colobus polykomos* than in *Colobus badius*. However in both species the upper canine/P_3 complex is significantly dimorphic. Leutenegger (1971) found a highly significant sexual difference in the mesiodistal diameter of P_3 in both of these species. He also found more marked sexual differences in *Colobus polykomos* than in *Colobus badius* or *Colobus verus*. Mesiodistal molar size relations are $M^1 < M^2 \geq M^3$ and $M_1 < M_2 < M_3$.

Genus *Nasalis*

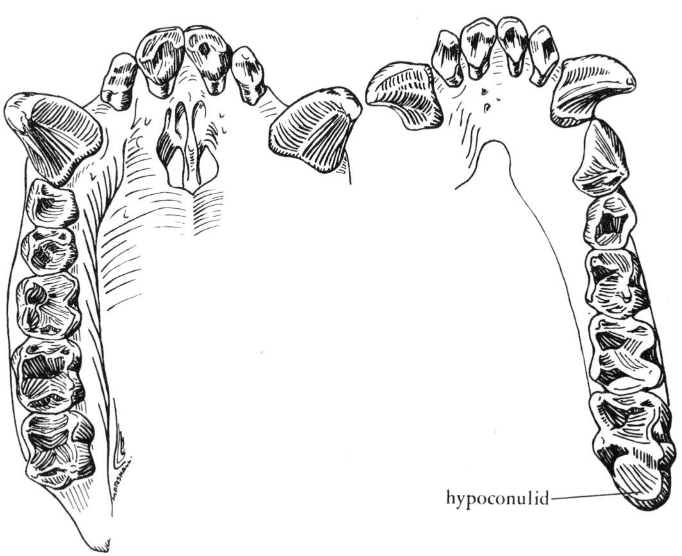

Plate 44. *Nasalis larvatus* male, occlusal view ×1·25.

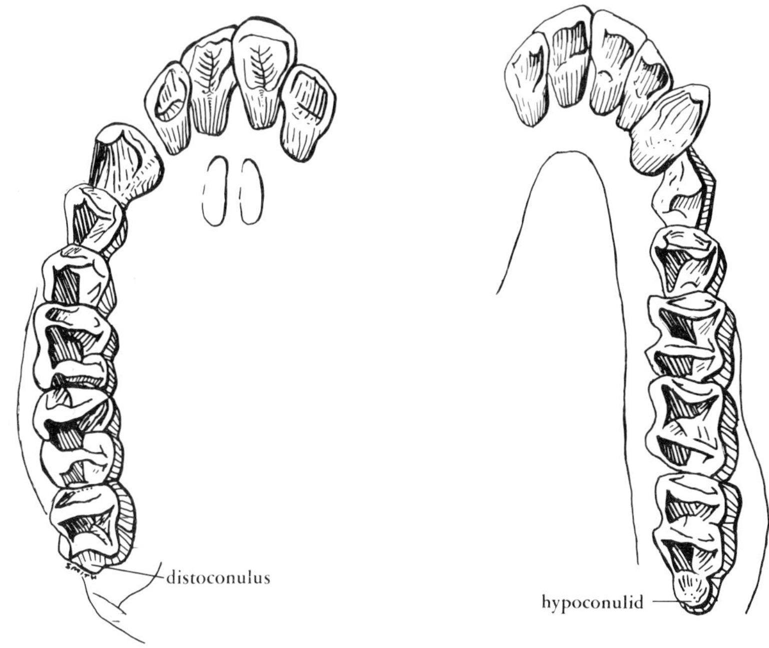

Plate 45. *Nasalis concolor* female, occlusal view ×2.

Morphological Observations

	Sample	
	Male	Female
Nasalis larvatus (proboscis monkey)	26	15
Nasalis concolor (Pagai Island langurs)	1	6

Nasalis differs from other leaf-eating monkeys in the eruption of M^2_2 after I^2_2 (Schultz, 1935). In this they agree with the cercopithecine pattern.

Incisors

Upper. I^{1-2} are heteromorphic: I^1 is broad with flat incisal borders while I^2 has sharp, pointed incisal tips. The lingual surface of I^1 exhibits narrow mesial and distal marginal ridges. Two enamel ridges with inclined surfaces are visible on the lingual surface of I^1. Between the two ridges is a median lingual sulcus, which is generally deeper and wider than in *Colobus.* A lingual tubercle, present in 87% of the sample near the middle of the lingual cingulum, is frequently well developed.

Lower. I_1 has a flat, horizontal border while on I_2 the incisal border tapers distally. Narrow distal and mesial marginal ridges are present and I_1 has a shallow median lingual sulcus. In *Nasalis larvatus* underbite appears in 65% of males and 76% of females. Schultz (1944) reports 77% (sexes pooled) underbite for the proboscis monkey.

Canines

Upper. The upper canines are large, projecting well below the occlusal plane. The mesial border is convex and presents the mesial groove. The distal border is narrow and sharp. A thin, narrow cingulum skirts the lingual surface near the cervix.

Lower. The lower canine rises above the other teeth and has a strong base from which the crown twists buccally and then distally. A lingual cingulum is present as well as a distal heel.

Premolars

Upper. P^3 is usually bicuspid although in *Nasalis larvatus* 11·5% of the males have only the paracone while all females possess two cusps. A secondary lingual cingulum is always present and in some cases is quite wide. P^4 is always bicuspid in the present sample. In *Nasalis concolor* P^3 has two cusps.

Lower. P_3 is sectorial while P_4 is bicuspid or more frequently tricuspid. Indeed, over 90% of *Nasalis larvatus* possess three cusps on P_4. The metaconid is usually larger in over 80% of *Nasalis larvatus*, while the small variety is present in less than 3% of the animals. In the remainder, the metaconid is larger than the protoconid. In *Nasalis concolor*, the metaconid is always as large as the protoconid and 80% of the animals possess three cusps on P_4. In both species, the metaconid lies opposite the protoconid in 93% of the samples and distal in the others.

Molars

Upper. M^{1-3} have four cusps and are bilophodont. There is usually some reduction in the size of M^3 and on occasion, the hypocone is small. As is typical in colobines, the molars display less flaring than the majority of cercopithecines. A distoconulus is present in about 33% of the *Nasalis larvatus* sample while absent in the small sample of *Nasalis concolor*.

Lower. M_{1-2} have four cusps and M_3 has five. A hypoconulid is present and is often quite large. In our sample its position varied from central (36%) to buccal (64%) in *Nasalis larvatus.* One male *Nasalid larvatus* lacked any development of the hypoconulid. A tuberculum sextum was present in 36% of the *Nasalis larvatus* sample but absent in the small *Nasalis concolor* sample. Groves (1970) reports only a 7–10% frequency for the genus. The molars of course are bilophodont.

Odontometry (Tables 157–162, Appendix)

The greatest degree of sexual dimorphism is present in the upper canine/P_3 complex. Although significant differences obtain in the premolars and molars, they are less intense. Regarding the magnitude of sexual dimorphism in tooth size correlated with body size, Swindler and Orlosky (1972) found that sexual dimorphism in tooth size was relatively less in *Nasalis larvatus* than in *Colobus polykomos* notwithstanding the much greater sexual dimorphism in body size in the former species. This finding suggests that there is little relationship between these two parameters among these two groups of leaf-eating Old World monkeys.

The mesiodistal molar relations are $M^1 < M^2 > M^3$ and $M_1 < M_2 < M_3$.

Genus *Pygathrix (Rhinopithecus)*

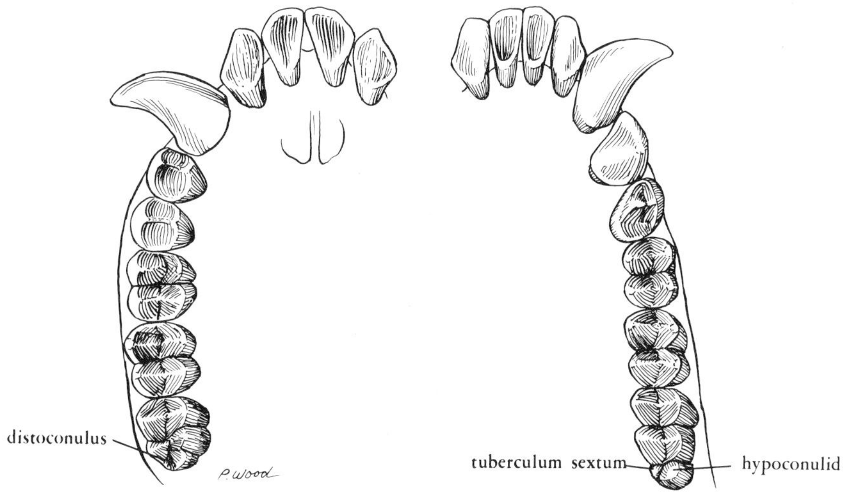

Plate 46. *Pygathrix namaeus* male, occlusal view ×1·3.

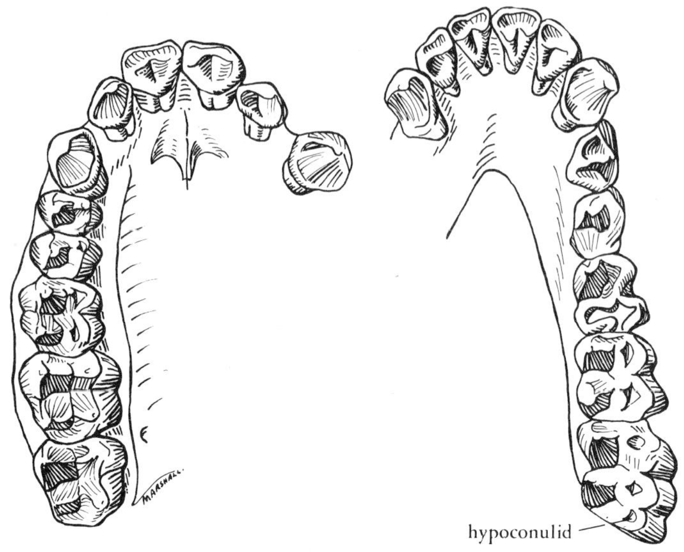

Plate 47. *Pygathrix (Rhinopithecus) roxellanae* female, occlusal view ×1·35.

Morphological Observations

	Sample	
	Male	Female
Pygathrix namaeus (douc langur)	7	4
Pygathrix (Rhinopithecus) roxellanae (golden snub-nosed langur)	6	11

According to Schultz (1935) *Pygathrix* is the only Old World monkey which erupts M^2_2 before the incisors and the other teeth have appeared.

Incisors

Upper. I^{1-2} are heteromorphic: I^1 is broad, while I^2 is more pointed. Mesial and distal marginal ridges delineate the lingual surface. A lingual cingulum is present on both I^{1-2}. A median lingual sulcus is present on I^1, whereas on I^2 a median ridge courses from the lingual cingulum to the incisal tip. Lingual tubercles are not present in our sample.

Lower. I_1 is slightly wider than I_2 and possesses a medial lingual sulcus while I_2 may display a slight sulcus or an elevated ridge dominating its lingual surface. Marginal ridges are better developed on I_1 than on I_2.

Canines

Upper. The upper canine is long and trenchant. The base is robust and presents a narrow lingual cingulum which is more pronounced in *Pygathrix (Rhinopithecus) roxellanae.* A deep groove incises the mesial border of the canine.

Lower. The lower canine twists buccally and slightly distally from its stout base to extend well above the occlusal plane. A lingual cingulum terminates as a distal heel.

Premolars

Upper. P^{3-4} are bicuspid. A single-cusped P^3 is not present in this sample. P^3 has two cusps in 74% of *Pygathrix namaeus*, three cusps in 13%, and four cusps in another 13%. On the other hand, it has only two cusps (33%) or three (67%) in *Pygathrix (Rhinopithecus) roxellanae.* P^4 is always bicuspid.

Lower. P_3 is sectorial. The buccal flange is well developed and lies to the buccal side of the canine. The tip of the compressed protoconid faces distalward. P_4 has a well formed protoconid and metaconid in both taxa, although it is larger relative to the protoconid in *Pygathrix (Rhinopithecus) roxellanae.* Thus, in this species the metaconid is the same size as the protoconid in 63% of our sample and larger in 37%, while in *Pygathrix namaeus* the metaconid is always smaller than the protoconid. At the same time, P_4 displays three cusps in *Pygathrix namaeus* and either three (13%) or four (87%) in *Pygathrix (Rhinopithecus) roxellanae.*

Molars

Upper. M^{1-3} have four well developed cusps and are bilophodont. As in other colobines, the molars have somewhat less flare than cercopithecines. The distoconulus enjoys a high frequency in both taxa; e.g. it is present on M^3 in 62% of *Pygathrix (Rhinopithecus) roxellanae* and 100% in *Pygathrix namaeus.*

Lower. The lower molars are bilophodont and M_3 has a hypoconulid. The fifth cusp is buccal in both species. The trigonid basin is narrow and shallow as in the majority of colobines. A tuberculum sextum is present on M_3 (89%) in *Pygathrix namaeus* while the tuberculum intermedium resides on M_1 (9%) and on M_3 (11%) in this species. In *Pygathrix (Rhinopithecus) roxellanae* the tuberculum sextum is present on M_3 (69%) while the tuberculum intermedium is variably expressed on M_1 (7%), M_2 (62%) and M_3 (46%).

Odontometry (Tables 163–168, Appendix)

The mesiodistal size gradient of the molars is $M^1_1 < M^2_2 \leq M^3_3$.

Genus *Presbytis*

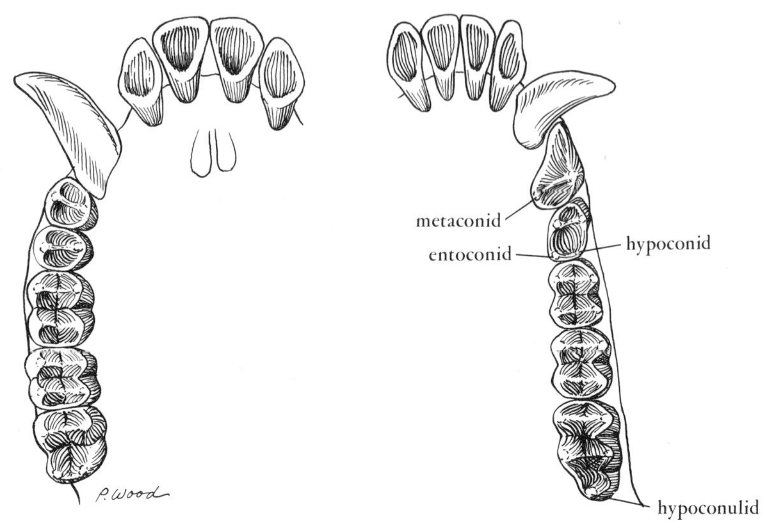

Plate 48. *Presbytis cristatus* male, occlusal view ×2.

Morphological Observations

	Sample	
	Male	Female
Presbytis pileatus (capped langur)	18	8
Presbytis cristatus (silvered leaf-monkey)	24	34
Presbytis aygula (Sunda Island leaf-monkey)	15	18
Presbytis phayrei (Phayre's leaf-monkey)	4	5
Presbytis johni (John's langur)	3	2

Incisors

Upper. I^{1-2} are heteromorphic. The labial surface of both incisors is convex mesiodistally. The lingual surface of I^1 is circumscribed by mesial and distal marginal ridges which converge to form a V-shape cervically. A shallow median lingual sulcus is present. I^2 has a triangular crown with a pointed incisal border. Both I^{1-2} have a lingual cingulum but not a lingual tubercle.

Lower. The lower incisors are similar to those in other colobines, i.e. narrow marginal ridges outline slightly concave lingual surfaces with I_2 more pointed than I_1. In the present sample, underbite is always more prevalent in females as well as displaying interesting differences between species. This is shown below.

Species	Sex	*n*	Underbite%
Presbytis cristatus	Male	21	33
	Female	32	50
Presbytis pileatus	Male	11	36
	Female	6	83
Presbytis aygula	Male	15	86
	Female	17	94

Colyer (1936) also found a high incidence of underbite in *Presbytis* (77%).

The explanation for these sexual and species differences is not known and the present data offer little clarification. What is required is extensive field studies correlated with detailed investigations of the developmental morphology of the animals.

Canines

Upper. The upper canine is a long, sabre-like tooth projecting well beyond the occlusal plane. It is broad mesially and narrow distally. It has a shallow groove along the distolingual face of the tooth as well as the characteristic cercopithecid mesial groove.

Lower. The lower canine is long and the crown extends beyond the occlusal plane from a rather stout base. A narrow lingual cingulum may be present and a distal heel projects from the base of the tooth.

Premolars

Upper. P^{3-4} are bicuspid and conform morphologically to the premolars described for other colobines. The number of cusps on P^3 ranges from one to three, the most common number being two as shown below (sexes pooled in brackets).

Species	1	2	3
Presbytis pileatus		100(21)	
Presbytis cristatus		91(45)	9(45)
Presbytis aygula		68(28)	32(28)
Presbytis phayrei	25(4)	86(7)	14(7)
Presbytis johni		50(4)	25(4)

Lower. P_3 is an elongated, sectorial tooth presenting a protoconid pointing distally and a shallow talonid fossa. A protocristid is normally present passing distolingually from the protoconid. P_4 is variable regarding the number of cusps adorning its crown as well as the size of the metaconid relative to the protoconid. The following outline below presents our findings (sexes pooled).

P_4 variability, metaconid to protoconid

Species	$M = P$	$M > P$	$M < P$	Number of cusps		
				2	3	4
Presbytis pileatus	74(19)	21(19)	5(19)	11(19)	42(19)	47(19)
Presbytis cristatus	86(49)	14(49)			22(51)	78(51)
Presbytis aygula	67(27)	26(27)	7(27)		48(27)	52(57)
Presbytis phayrei	57(7)	29(7)	14(7)		43(7)	57(7)
Presbytis johni	75(4)	25(4)			50(4)	50(4)

M = metaconid; P = protoconid

Molars

Upper. M^{1-3} have four cusps and are bilophodont. There is little if any reduction in M^3. The molar flare is normal for colobines. The distoconulus is variable among these species of langurs, as can be seen below (sexes pooled).

Species	M^3 distoconulus Present %
Presbytis pileatus	20(20)
Presbytis cristatus	36(44)
Presbytis aygula	24(25)
Presbytis phayrei	50(6)
Presbytis johni	50(4)

Lower. M_{1-3} have four cusps plus a well formed hypoconulid on M_3. All molars are bilophodont with some degree of buccal flaring. The trigonid basin is narrow as in colobines. The following summarizes our findings of lower molar plasticity in these species (sexes pooled).

Species	Tuberculum sextum M_3	Position of hypoconulid	
		Central	Buccal
Presbytis pileatus	17(18)	16(19)	84(19)
Presbytis cristatus	46(46)	30(46)	70(46)
Presbytis aygula	3(32)	96(24)	4(24)
Presbytis phayrei	20(5)	33(6)	67(6)
Presbytis johni	50(4)	25(4)	75(4)

A tuberculum intermedium is not present in these species of *Presbytis*. It is obvious from these data that a great deal of interspecific dental variability exists in *Presbytis*; indeed, earlier studies have consistently demonstrated the high degree of plasticity in the dentition of Old World monkeys (Freedman, 1957; Biggerstaff, 1966; Saheki, 1966; Swindler and Sirianni, 1969; Swindler and Orlosky, 1972; Delson, 1973; Sirianni, 1974).

Odontometry (Tables 169–184, Appendix)

In general, there are few sexual differences in the dentition of *Presbytis*. The obvious exception is the upper canine/P_3 complex which is always significantly greater in the mesiodistal dimension of the males. Among colobines it appears that the intensity of sexual dimorphism vacillates between genera; e.g. *Colobus polykomos* amd *Nasalis larvatus* display more differences than any of the *Presbytis* species studied. Differences also exist among species of the same genera: *Colobus polykomos* possesses more sexual dimorphism than *Colobus badius* or *Colobus verus*.

The mesiodistal molar relations are $M^1 < M^2 > M^3$ and $M_1 < M_2 < M_3$.

11

Family Hylobatidae

Present Distribution and Habitat

The family Hylobatidae is divided into two genera, common gibbons (*Hylobates*) and siamang gibbons (*Symphalangus*). The common gibbons are generally limited to the primary forests of South-East Asia, Sumatra, Borneo, Java and several of the smaller islands. Siamangs, on the other hand, are confined to the Malay peninsula and Sumatra. Both groups live in rain and montane forests, though the siamangs are found at high elevations (1,850 m) (Napier and Napier, 1967). They are primarily arboreal brachiating animals, preferring the middle to closed canopy forests. In the trees they are the most acrobatic of all primates, with the spider monkey of the New World a close second. On the ground they walk bipedally with arms raised above the heads for balance, although they rarely come to the ground naturally.

Dietary Habits

The gibbon diet is essentially frugivorous, being made up of as much as 80% fruit (Carpenter, 1940). The remainder includes leaves, insects, flowers and buds, birds' eggs and young birds. They are generally observed feeding at the ends of branches and it is interesting to note that Ellefson (1968) considers brachiation an adaptation for this mode of feeding.

General Dental Information

Permanent dentition: $I^2_2\ C^1_1\ P^2_2\ M^3_3$

Deciduous dentition: $i^2_2\ c^1_1\ m^2_2$

Sequence of eruption of permanent teeth (Schultz, 1935):

Hylobates

$M^1 \quad I^1 \quad I^2 \quad M^2 \quad P^3 \quad P^4 \quad C \quad M^3$

$M_1 \quad I_1 \quad I_2 \quad M_2 \quad P_3 \quad P_4 \quad C \quad M_3$

The eruption sequence for *Symphalangus* is not known

Both dental formulae are the same as those for the Old World monkeys. The upper incisors of gibbons are more heterodont than those of other hominoids; thus, the incisal border is broad and horizontal in I^1, more pointed in I^2. In both I^{1-2} the labial surfaces are convex mesiodistally. The lingual surface of I^1 is concave and presents a thickened lingual cingulum. A median lingual ridge divides I^2 into two nearly equal parts. The lower incisors are approximately equal in size. The incisal border of I_1 is more horizontal than that of I_2. The lingual surface of I_1 is usually indented by a shallow concavity which may take on the appearance of a lingual sulcus in *Symphalangus*. Gregory (1922, p. 315) believed that the incisors of extant gibbons were rather primitive chisel-like teeth "well adapted for holding and cutting fruits".

The upper canines are long, narrow, sabre-like teeth in both sexes. They curve buccally and then recurve lingually to terminate as sharp points. A vertical distolingual groove is more pronounced in females than males. The mesial groove is obvious in both sexes. The lower canines are large, projecting well above the occlusal plane. The bases are subtriangular in cross-section and present a well developed heel.

The upper premolars are bicuspid although the protocone is always smaller than the paracone, particularly on P^3. In *Symphalangus*, however, the protocone is relatively larger on both P^{3-4}. A preprotocrista passes down the lingual slope of the paracone and is interrupted by a mesiodistally-running developmental groove as it meets a crista from the paracone. The notch formed by the groove is more pronounced on P^4. P_{3-4} are heteromorphic: P_3 is sectorial while P_4 is bicuspid or multicuspid. Frisch (1963) notes the morphologic plasticity of P_3 in gibbons and defines two varieties: (a) the oblong type which possesses the common or elongated crown; and (b) the triangular type which displays a more developed distolingual portion resulting in a triangular outline. He found the triangular type more frequent in *Hylobates lar* from the Chiengmai district of Thailand. A cuspule is also occasionally present on the protocristid. P_4 possesses a well developed protoconid and metaconid; indeed, the latter cusp is frequently as large as the protoconid. The two cusps are connected by a V-shaped protocristid. The talonid is broad and slopes distally to lie at a lower level than the trigonid.

The upper molars have four cusps; the paracone and metacone are the highest, followed by the protocone and hypocone. The apex of each buccal cusp is mesial to those of the lingual cusps. A postprotocrista connects the protocone and metacone as in hominoids. Some authorities (e.g. Gregory, 1922) believe the upper molars are more primitive than those in other anthropoids since the trigonal pattern is only little modified by development of the hypocone and the lingual side of the molars is still narrow. The latter trait is variable in our sample, where the lingual side of the crown is frequently as wide as the buccal. A lingual cingulum is variably expressed in extant gibbons (Frisch, 1965; Kitahara-Frisch, 1973). Frisch (1965) found that the mode of reduction of the lingual cingulum differed greatly in the various species of Hylobates. Thus, there is almost no cingular reduction in *Hylobates concolor*, moderate reduction in *Hylobates lar* and *Hylobates moloch* and extreme reduction in *Hylobates agilis* and *Hylobates hoolock*. Remnants of a buccal cingulum are rare, appearing mainly in *Hylobates concolor* in the buccal groove between the paracone and metacone (Frisch, 1965). The siamang shows traces of a lingual cingulum but rarely does it exhibit a well developed one, e.g. only one of 36 specimens (Frisch, 1965).

The lower molars possess five cusps: the hypoconulid is added to each molar. The protoconid and metaconid are connected by a protocristid on M_{1-3}. The protocristid is better developed on M_1; indeed it is only slightly grooved at its center on M_1 but becomes progressively deeper grooved on M_{2-3}. The trigonid basin is extremely narrow and is little more than a narrow transverse slit along the mesial border of each molar. The central or talonid basin is greatly expanded and in occlusion it receives the protocone of the upper molars. The diverse anthropoid occlusal patterns resulting from these five cusps and the sulci separating them have been widely studied since Gregory (1916) pointed out the similarities between the patterns in fossil and extant hominoids. One of these, the fundamental hominoid arrangement, is called the *Dryopithecus* or Y-5 pattern and is defined as follows: the metaconid and hypoconid are in contact and the buccal groove is mesial to the lingual groove so that when viewed from the lingual side a Y-pattern is formed (Fig. VIII). During the course of evolution the Y-5 pattern has altered and gradually shifted to a plus pattern in which the protoconid touches the entoconid (Fig. VIII). As we shall see, this basic homonoid pattern obtains in all extant hominoid taxa although the frequency differs among and within the groups. Among extant hylobatids the most conservative group is *Symphalangus*. All of our specimens display a Y-5 pattern on M_{1-3}, as did those studied by Frisch (1965). In all other hylobatids, the patterns are more variable as discussed later.

In addition to these basic occlusal patterns the lower molars of hylobatids occasionally display extra cusplets.

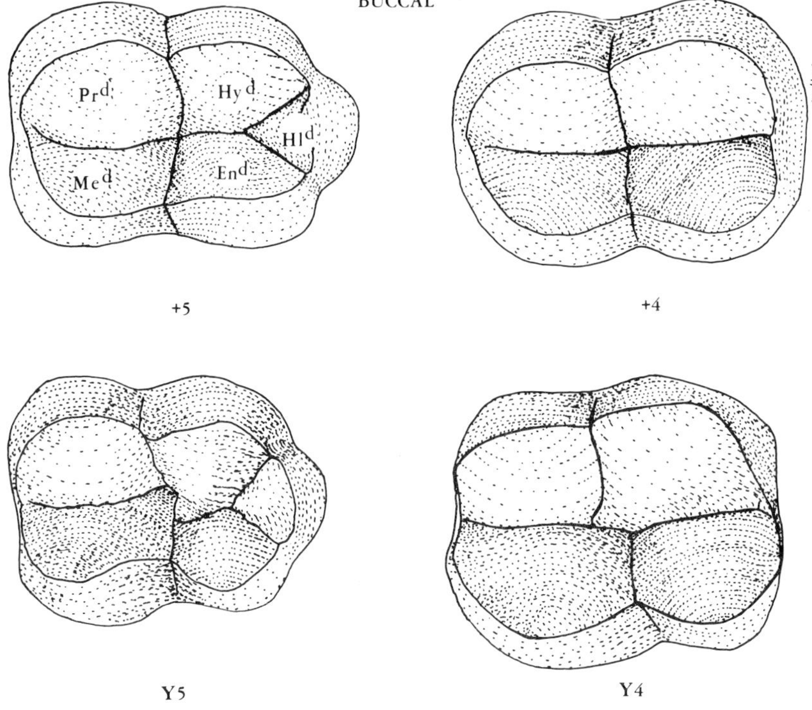

Fig. VIII. Lower molar patterns in Hominoidea.

Genus *Hylobates*

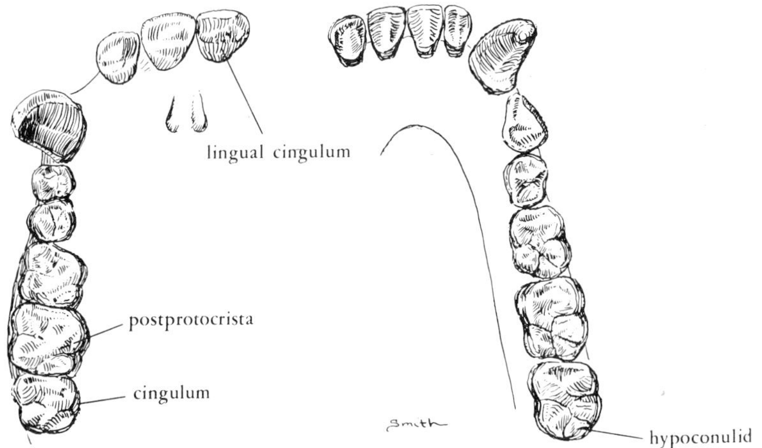

Plate 49. *Hylobates klossi* male, occlusal view ×1·65.

Morphological Observations

	Sample	
	Male	Female
Hylobates klossi (Kloss's gibbon)	18	15
Hylobates moloch (silvery gibbon)	5	4
Hylobates lar (whitehanded gibbon)	5	2

Incisors

Upper. I^{1-2} are heteromorphic. I^1 is broad and its lingual surface is excavated and presents a well developed lingual cingulum, which is thick and forms a ledge along the lingual surface. I^2 is narrow and pointed and has a lingual cingulum as well as a median lingual crest.

Lower. I_{1-2} are implanted vertically in the lower jaw. The incisal border is horizontal on I_1 and slopes distally on I_2. The lingual surfaces of both teeth are slightly concave, and mesial and distal marginal ridges lightly outline these surfaces. A lingual cingulum transverses the cervical region of each tooth.

Canines

Upper. The upper canines are long and trenchant in both sexes, and have both distolingual and mesial grooves; the former is better developed in females. These teeth curve buccally and then lingually to end as sharp points. These long sabre-like canines are probably a specialization for a frugivorous diet (Gregory, 1922).

Lower. The lower canines have broad bases from which the crown curves buccally and slightly distally. A cingulum curves around the lingual face of the tooth to terminate as a heel.

Premolars

Upper. P^{3-4} are bicuspid. The protocone on P^3 is frequently no more than a slight elevation, although it is always present in our sample. A small lingual cingulum is present on P^4 in only one of 33 specimens (*Hylobates klossi*). Frisch (1965) found a low frequency of lingual cingula on P^{3-4} in *Hylobates concolor* and on P^4 in *Hylobates moloch.* Buccal cingula are represented in the present sample on P^3 in two specimens while Frisch (1965) found a low incidence of the trait in *Hylobates concolor.*

Lower. P_3 is sectorial while P_4 is bicuspid. Buccal cingula are absent in the specimens examined and Frisch (1965) found only traces on the distal aspect of the protoconid of two P_4's in *Hylobates concolor*. As noted earlier, Frisch (1963) defined two types of P_3 in extant gibbons, oblong and triangular. In the present sample the oblong variety prevails since it is present in all but two of the specimens studied.

Molars

Upper. In general, the molars correspond to the gibbon morphotype described earlier. A lingual cingulum is present on M^1 in 30% of our sample, on M^2 in 24% and on M^3 in 9%. Nothing amounting to a Carabelli cusp exists in the present sample although it is described in *Hylobates lar* by Adloff (1908), Remane (1960) and Frisch (1965). No traces of a buccal cingulum are present in our material, although Frisch (1965) reports it in *Hylobates concolor*. M^3 displays a considerable degree of reduction in gibbons and the present sample is no exception. Variability ranges from an M^3 with a fully developed crown to one with both distal cusps absent. For a complete discussion of M^3 reduction in extant gibbons see Frisch (1965).

Lower. M_{1-3} possess five cusps. The Y-5 pattern is present on M_1 in 100%, on M_2 in 97% and on M_3 in 70% of our sample. On both M_2 and M_3 the other arrangement is the plus pattern. Frisch (1965) also found variability in the expression of the Y-5 pattern; as in our study, M_3 was the most labile and M_1 the most conservative regarding the development of the Y-5 pattern. The position of the hypoconulid relative to the long axis of the tooth can be lingual, central or buccal. According to Gregory and Hellman (1926), the primitive position is central in hominoids. The trend has been for the hypoconulid to shift buccally. In recent hylobatids the position of the hypoconulid is variable, being mainly distributed between the central and buccal positions, although Frisch (1965) did find a few in a lingual position in *Hylobates lar*. In the present sample they are predominantly buccal (*c*. 80%) on all three molars. The hypoconulid may be absent on all three molars but is more often missing on M_3 (10% of the present sample). The position of the metaconid relative to the protoconid is of phylogenetic interest. Originally (tribosphenic molars) the metaconid was distal to the protoconid, but Frisch (1965) has described a trend to shift mesially so as to face the protoconid. In hylobatids the metaconid on M_1 is more frequently distal to the protoconid although in *Hylobates lar* (Chiengmai) it is almost always opposite the protoconid (Frisch, 1965). The relationship tends to reverse itself distally since the two cusps face each other more often on M_{2-3}, particularly on M_3. On M_3 the metaconid may even come to lie mesial to the protoconid. A similar trend is present in our sample. A tuberculum intermedium is observable on M_1 in 0·07%, on M_2 in 18% and on M_3 in 37% of our sample.

Odontometry (Tables 185–198, Appendix)

Extant gibbons display few sexual differences in tooth size, even for the upper canine/P_3 complex. This is also true for the molars of several species of gibbons (Kitahara-Frisch, 1973). The most common mesiodistal size relations are $M^1 < M^2 > M^3$ and $M_1 < M_2 = M_3$; however, Kitahara-Frisch (1973, p. 131) is quick to point out that they are far from being universal in gibbons and that variability in molar length (particularly in lower molars) suggests that this region of the dentition "has probably undergone metrical changes in the not-too-distant past".

Genus *Symphalangus*

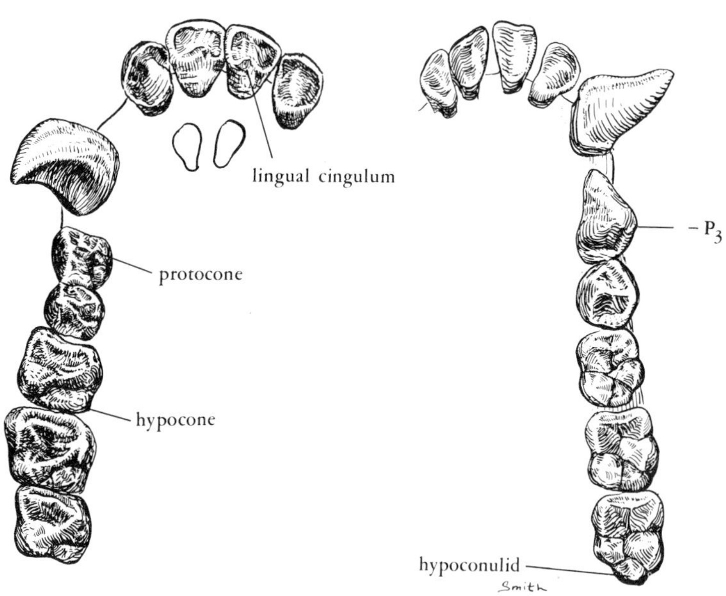

Plate 50. *Symphalangus syndactylus* male, occlusal view ×1·5.

Morphological Observations

	Sample Male	Sample Female
Symphalangus syndactylus (siamang)	5	

Incisors

Upper. I^{1-2} are heteromorphic: I^1 is broad with a concave lingual surface while I^2 is pointed with a median vertical convex ridge occupying the lingual surface. A well developed lingual cingulum is present in both incisors; however, it constitutes a veritable ledge on I^1.

Lower. I_1 is smaller than I_2 and both present rather horizontal incisal borders. A narrow lingual sulcus is present on I_{1-2}.

Canine

Upper. The upper canines are long, blade-like teeth projecting well beyond the occlusal plane. The mesial surface is broad and contains the mesial groove. The distal surface is sharp and narrow. In general, the canines have less lingual curve than observed in *Hylobates.* In other words, they pass nearly vertically from the upper jaw.

Lower. The lower canines are robust teeth very similar to those in *Hylobates.* They curve buccally and have a well formed distal heel.

Premolars

Upper. P^{3-4} are bicuspid. The protocone of P^3 is well developed and when compared with *Hylobates,* the relative sizes of the protocone and paracone are immediately obvious. The mesial face of P^3 is practically vertical and lacks a mesial fossa. The protocone is large on P^4 although never attaining the dimensions of the paracone. A V-shaped preprotocrista is present on P^{3-4}. Neither buccal nor lingual cingula are present in our sample, nor are they reported by Frisch (1965).

Lower. P_{3-4} are heteromorphic: P_3 is sectorial and presents a distal depression or fossa, while P_4 is bicuspid. The protoconid and metaconid are close together and connected by a V-shaped protocristid. The metaconid approximates the protoconid in size.

Molars

Upper. The upper molars possess four cusps although the size of the hypocone is variable, particularly on M^3. As in *Hylobates,* there is evidence of M^3 reduction. A postprotocrista is present on all three molars in the present sample. Lingual and buccal cingula are wanting, and Frisch (1965) mentions the general lack of developed lingual cingula in *Symphalangus.* Indeed, only one of his 37 specimens displayed lingual cingula on all three molars.

Lower. The lower molars have five cusps and each presents a Y-5 occlusal pattern. Frisch (1965) notes a Y-5 pattern on all lower molars of *Symphalangus* and attributes it to the small size and distal position of the entoconid. A hypoconulid is present on M_{1-3} and its position is about equally distributed between central and buccal. The metaconid varies in its position relative to the protoconid as observed in *Hylobates*. In general, it is distal on M_1, opposite or slightly distal on M_2 and opposite or even slightly mesial on M_3. In the present sample, a tuberculum intermedium resides on M_{1-3} in one specimen.

Odontometry

No measurements were taken of these five males.

12

Family Pongidae

Present Distribution and Habitat

The family Pongidae embraces the three great apes: gorillas, chimpanzees and orang-utans. The gorilla lives in two major areas of Equatorial Africa: the western gorilla inhabits the western parts of the Congo basin, and the eastern gorilla ranges from the eastern lowlands of the Upper Congo to the mountains east of Lake Kivu. There is an eastern and western lowland gorilla as well as an eastern highland gorilla. The highland gorilla has been found in bamboo forests up to 3,000 m (10,000 ft) (Napier and Napier, 1967). The gorilla prefers rain forests, either lowland or montane.

The chimpanzee is usually separated into two species, *Pan troglodytes* and the celebrated pygmy chimpanzee, *Pan paniscus. Pan troglodytes* is widely distributed from West Africa through parts of Central Africa to as far east as Lake Victoria and Lake Tanganyika. The pygmy chimpanzee is restricted to an area bordered by the Congo and Lualaba Rivers. Chimpanzees live in rain forests, savannas and montane forests as high as 3,000 m.

The orang-utan is limited to Sumatra and Borneo where it occupies tropical rain forests, particularly the peatswamp forests.

Of the three great apes, the orang-utan is predominantly arboreal and rarely comes to the ground, while the gorilla is more often on the ground—particularly the large males. Chimpanzees are more or less intermediate since they may spend as much as one-third of the daylight hours on the ground. The arboreal locomotion of all three great apes is brachiation or what Napier has called modified brachiation (Napier and Napier, 1967). It should be pointed out, however, that on the ground the African species display a specialized quadrupedal locomotion known as knuckle walking (Tuttle, 1967). The great apes are diurnal.

Dietary Habits

The gorillas are vegetarian. Lowland gorillas eat some fruit while highland gorillas evidently eat little fruit, preferring instead bamboo shoots and roots, stalks, bark and vines. The chimpanzee is vegetarian, i.e. eats fruits, leaves and seeds, although it supplements this diet with meat in the form of termites, ants and an occasional baboon (Goodall, 1963). The orang-utan prefers fruit; however, as with many primates, it changes somewhat with the seasons, eating bark, birds, birds' eggs and leaves at certain times of the year.

General Dental Information

Permanent dentition: $I^2_2\ C^1_1\ P^2_2\ M^3_3$

Deciduous dentition: $i^2_2\ c^1_1\ m^2_2$

These dental formulae are the same as in the Old World monkeys, gibbons and man.
Sequence of eruption of permanent teeth (Schultz, 1935):

Pongo

$$\frac{M^1 \quad I^1 \quad I^2 \quad M^2 \quad P^3 \quad P^4 \quad C \quad M^3}{M_1 \quad I_1 \quad I_2 \quad M_2 \quad P_4 \quad P_3 \quad C \quad M_3}$$

Gorilla

$$\frac{M^1 \quad I^1 \quad I^2 \quad M^2 \quad P^3 \quad P^4 \quad C \quad M^3}{M_1 \quad I_1 \quad I_2 \quad M_2 \quad P_4 \quad P_3 \quad C \quad M_3}$$

Pan

$$\frac{M^1 \quad I^1 \quad I^2 \quad M^2 \quad P^3 \quad P^4 \quad C \quad M^3}{M_1 \quad I_1 \quad I_2 \quad M_2 \quad P_3 \quad P_4 \quad C \quad M_3}$$

The upper central incisors are broad, spatulate teeth with straight incisal borders, while the lateral incisors are not as wide and their incisal borders tend to be more oblique, particularly in the orang-utan. With the exception of the orang-utan, I^{1-2} tend to be more homomorphic than in the cercopithecids. The central incisors are especially hypertrophied in gorillas and orang-utans. Well developed mesial and distal marginal ridges are present on both I^{1-2} and the lingual surface is frequently concave. An obvious lingual cingulum is present on I^{1-2} in all great apes. The lingual cingulum frequently has a central enlargement and one to several vertical enamel ridges may arise from it to pass to the incisal border. This is the tuberculum dentale.

Another common condition, particularly in *Pan*, is a median lingual ridge. When present, it separates the lingual fossa into approximately equal portions and when well developed it overshadows the fossa as well as the marginal ridges.

The lower incisors are also large in the great apes and are generally similar in shape and size. Their lingual surfaces are subrectangular and present mesial and distal marginal ridges. A lingual cingulum is present and is sometimes quite pronounced as it traverses the cervical portion of both I_{1-2}. A true lingual tubercle is absent although a swelling of the basal part of the tooth is not uncommon (see also Weidenreich, 1937). A median lingual ridge may be present on both I_{1-2}; it seems to be more frequent in *Pongo* and *Pan* and when present it tends to obliterate the shallow fossa.

The upper canines are large, conical teeth displaying a great amount of sexual dimorphism. The mesial border is convex while the distal border is either straight or slightly concave from the crown tip to the base. The lingual surface differs somewhat among the three genera. In the orang-utan a mesiolingual groove follows the curvature of the mesial border from the base to near the tip. The chimpanzee and gorilla possess a lingual groove but it is median in position. It should be noted, however, that in addition to the mesiolingual groove the orang-utan may display a slight median lingual groove. A lingual cingulum is present and is particularly well developed in the orang-utan. In some orang-utan specimens it is serrated.

The lower canines are also large and more or less conical, although the shape varies somewhat in the different taxa. In all three pongids they are more or less triangular when viewed occlusally; however, in the chimpanzee the borders are rounder and less obtuse than in the other two. In all three species the canines splay buccally, more so in the gorilla. A lingual cingulum is present as well as a basal heel.

The upper premolars are bicuspid. The paracone is larger and higher than the protocone although in the orang-utan P^4 the protocone nearly approximates the paracone in size. In all species, a well formed mesiodistal developmental groove separates the two cusps. It is particularly wide and deep in the gorilla. Mesial and distal marginal ridges define these respective borders of each premolar. Although protocristae are absent, the occlusal surface of each premolar presents mesial and distal faces inclining cervically, with the former face more steeply inclined than the latter. There are occasional traces of both lingual and buccal cingula on the upper premolars; they are more frequent in the gorilla than in the other forms.

The lower premolars are heteromorphic. P_3 is sectorial, having but a single compressed protoconid although metaconids are not uncommon. If present, the two cusps are connected by a strong protocristid, or if the metaconid is absent, the protocristid courses distolingually to terminate at the lingual cingulum. The protocristid effectively divides the lingual occlusal surface into mesial and distal parts. These two portions are usually somewhat depressed and are probably homoplastic if not homologous with the trigonid and talonid basins of the more distal teeth. P_4 has two cusps, a protoconid and a metaconid connected by a protocristid which is cut into by the mesiodistal sulcus. In unworn specimens, the metaconid may be more prominent than the protoconid and the former cusp lies either opposite or distal to the latter one. The trigonid basin varies in shape and depth in the different anthropoid apes, as does the wider, more excavated talonid basin. A hypoconid or entoconid may adorn the talonid. Buccal cingula are occasionally seen on the lower premolars.

The upper molars possess four cusps which vary in size in recent pongids. In general, the paracone and protocone are larger than the metacone and hypocone. This is always true regarding the hypocone, but the protocone and metacone may be subequal. There is a progressive diminution in the size of the hypocone from M^{1-3} and in the orang-utan this is accompanied by reduction of the metacone (Hooijer, 1948). There is less hypocone reduction in gorilla. Thus, a general trend of M^3 reduction in extant pongids runs from *Gorilla* to *Pongo* to *Pan* (Frisch, 1965). A cruciform design of developmental grooves separates the four cusps occlusally. The buccolingual groove overflows onto the buccal and lingual surfaces of the molars. In all forms, the buccal sulcus is more pronounced than the lingual one; it is deepest in gorilla and shallowest in orang-utan. A postprotocrista connects the protocone with the metacone. The occlusal surface is usually divided into three fossae: a mesial fossa (fovea anterior) is delimited by a preprotocrista distally and a mesial marginal ridge mesially, a central fossa or trigon basin lies between the preprotocrista mesially and the postprotocrista distally, and the distal fossa (fovea posterior) lies between the postprotocrista and the distal marginal ridge. Variability of the three fossae is generally limited to the formation of the mesial and distal fossae.

The lower molars have five cusps on M_{1-3}. The most prevelant occlusal configuration on M_1 is the Y-5 pattern which may be modified on M_{2-3}. The lingual cusps are higher and more pointed than the buccal ones, particularly in the gorilla. The metaconid on M_1 is either opposite the protoconid or distal, whereas on the majority of M_{2-3}, the metaconid is opposite or mesial to the protoconid. The hypoconulid of M_3 is variably expressed in all pongids, particularly in the chimpanzee where it may be absent in 21% of M_3 (Schuman and Brace, 1954). The protoconid and metaconid are connected by a protocristid which forms the distal boundary of a narrow transverse valley, the trigonid basin. Immediately distal to the protocristid the central part of the crown widens to form the basin-shaped talonid fossa. A narrow postentocristid connects the entoconid with the hypoconulid, thereby forming a narrow, oblique transverse depression, the post-talonid basin. A buccal cingulum is often expressed on the lower molars, most frequently in gorilla, least in orang-utan.

The masticatory habits of the Pongidae have been investigated by many students through the years and little can be added at this time. From the work of Mills (1955, 1973) it is seen that two phases of occlusion are represented in pongids: (a) the buccal phase, resulting in mandibular rotation of the ipsilateral condyle and (b) the lingual phase, corresponding to rotation of the contralateral condyle. Both of these motions occur simultaneously on opposite sides of the jaw resulting in a balanced occlusion. This is a common pattern of chewing throughout the Primates, even in *Homo.*

Genus *Pongo*

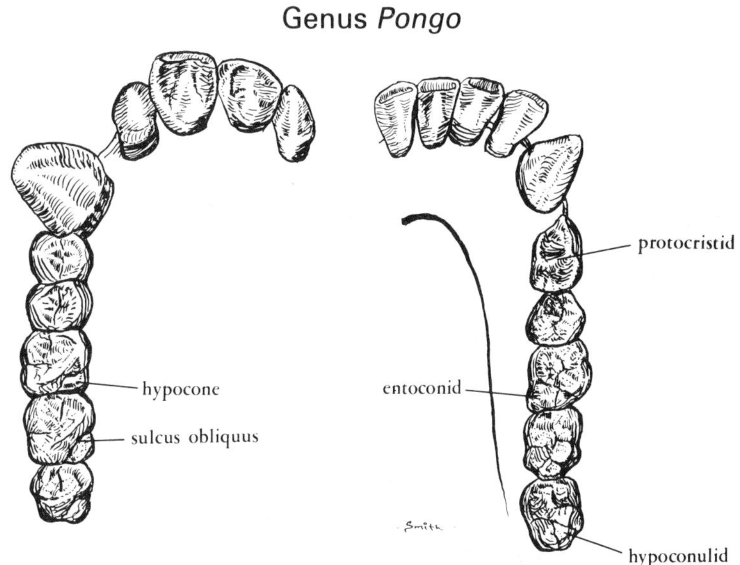

Plate 51. *Pongo pygmaeus* female, occlusal view ×0·8.

Morphological Observations

	Sample	
	Male	Female
Pongo pygmaeus (orang-utan)	8	11

Incisors

Upper. I^{1-2} are more heteromorphic in *Pongo* than in the other pongids; I^1 is very large with a broad, flat incisal border, while I^2 is much smaller and has a pointed incisal border. The labial surface of I^1 forms a gentle curve mesiodistally while the convexity of I^2 is much more apparent. The lingual surfaces of both I^{1-2} present a complex series of thin crenulations running vertically across the face of the tooth. There are mesial and distal marginal ridges on I^{1-2}. The marginal ridges become confluent with the lingual cingulum which on I^1 is quite thick centrally. Emanating from the median lingual thickening—particularly on I^1—are one, or more commonly two, median lingual ridges that stand out in more relief than the crenulations. These ridges ascend toward the incisal border. The lingual surface of I^1 is concave while that of I^2 is slightly convex mesiodistally.

Lower. The labial surface is flat on I_1, convex mesiodistally on I_2. The incisal border is horizontal on I_1 and tapers distally on I_2. Mesial and distal marginal ridges as well as a lingual cingulum outline the lingual face of each tooth. In addition, I_{1-2} present a median lingual ridge which can be quite well developed. A median labial ridge is infrequently present on I_{1-2} in *Pongo.*

Canines

Upper. The upper canine is a large, robust tooth with a mesiodistally elongated base. It is much larger in all dimensions in the male. The mesial border is convex and rounded while the distal surface is either straight or slightly concave from tip to base. The lingual surface usually presents two vertical grooves, one mesially and the other distally; between them a vertical enamel pillar is observable. The pillar and groove morphology is always better developed in males than in females. A cingulum skirts around the lingual base of the canine; it is much more pronounced in the female, a character also mentioned by Hooijer (1948). Indeed, Hooijer even found an occasional central tubercle developed on the cingular ridge in females, a condition observed only once in our sample of 11 females.

Lower. The lower canine is much larger in the male than in the female and as in the upper canine, the major nonmetric sexual dimorphism is the better developed lingual cingulum in the female. The crown rises from a stout base to curve distally and slightly lingually, particularly near the tip. A shallow groove is present mesiolingually on the lingual surface of the tooth and, incidentally, is better developed in males. A distal prominence is present at the base of the tooth; like the cingulum, it is more pronounced in females.

Premolars

Upper. P^{3-4} are bicuspid. The paracone is always the largest cusp and on P^3 the protocone is somewhat blunter than it is on P^4. Well developed mesial and distal marginal ridges are present on P^{3-4} and two other cristae are usually visibly associated with the protocone and metacone. On P^{3-4} a mesial crista passes from the paracone or just mesial to it to end at the marginal ridge just mesial to the protocone; and a distal crista runs from the protocone to the distal aspect of the paracone. Several small crenulations are also obvious on the occlusal surface. The buccal and lingual cusps are separated by a deep mesiodistal developmental groove. The mesiobuccal surface on P^3 is more prominent and projects further mesially than on P^4, so that isolated P^{3-4}'s are readily distinguished from one another (Hooijer, 1948). Buccal cingula are rare on P^3 but do appear on the mesiobuccal surface of the paracone on P^3 in one of our eight males. None is found on P^4 in our collection. Hooijer (1948) reports one specimen with a distinct hypocone on P^4 and another one with a small metacone. The latter cusp is also noted by Remane (1921). Neither of these cusps can be identified in our sample ($n = 18$). Finally Hooijer (1948) mentions a minute protoconule present in five specimens. This trait is absent in our collection.

Lower. P_3 is sectorial, possessing a large, elongated protoconid projecting slightly buccal to the canine. Commensurate with this compression the enamel extends mesiobuccally for some distance toward the root. A small metaconid is present lying just distolingual to the apex of the protoconid. The two cusps are connected by a high transverse crest, the protocristid. Buccally two crests arise from the protoconid; one passes mesially, the other distally, while a well formed primary cingulum outlines the lingual surface of P_3. In this manner, two fossae are created, a mesial trigonid basin and a distal talonid basin. Hooijer (1948) describes two different morphologies for P_3. In one the cingulum and protocristid are virtually absent and the trigonid area is nothing more than a flat triangular region; in the other the protocristid and cingulum are pronounced, resulting in a markedly depressed trigonid basin. P_3 is variable in our sample but never to these two extremes. Hooijer (1948) found both an entoconid and a hypoconid in one specimen of orang-utan. In our study an entoconid is present in one animal, but the hypoconid is absent. Crenulations are obvious in the fossae. P_4 is essentially bicuspid, having a large protoconid and frequently an even larger metaconid. In addition, a hypoconid and entoconid are reported by Hooijer (1948) with the entoconid less frequent that the hypoconid. This is reversed in our study, with entoconids occurring in 85% and hypoconids in 18% of the sample. Hooijer (1948) even records a P_4 with a hypoconulid, but we failed to find a hypoconulid in our sample. A V-shaped protocristid forms the distal boundary of the trigonid basin. The trigonid basin varies mainly in shape from a wide, deep fossa to a narrow, shallow depression. Distally, P_4 widens to form the talonid basin which is circumscribed by a well developed marginal ridge. Enamel crenulations radiate from the center of both fossae. Buccal cingula are absent.

Molars

Upper. M^{1-3} possess four cusps and are more oval in occlusal outline than are the upper molars of the other pongids. In this *Pongo* agrees more with *Homo* than do the other extant anthropoid apes (Korenhof, 1960). Of the four principal cusps, the metacone is usually the highest on M^1 while the paracone may assume this rank on M^{2-3}. The hypocone is always the lowest cusp, which correlates with its late phylogenetic and ontogenetic development. (The student should see the excellent work of Butler, 1956, regarding cusp formation in mammals.) M^3 is the smallest upper molar and in some cases both hypocone and metacone are lacking, or at best vestigial. The most pronounced crest on the occlusal surface is the postprotocrista, which is always present in our sample. As noted by Korenhof (1960), it is frequently (66%) interrupted by the mesiodistal developmental groove. Additional crests separate the occlusal surface into narrow fovea anterior and posterior, and a broad central or talon basin. Both the mesial and distal fossae may be absent, particularly the mesial one. In orang-utans the occlusal surface is always covered by a series of enamel wrinkles radiating from the center of all fossae obscuring the general picture to some degree. A sulcus obliquus (distolingual groove) is present, isolating the hypocone from the other cusps. A metaconule is absent in our collection and has not been reported by other authors. A protoconule, however, is frequently present (80%) on the three molars of our sample. A high incidence of the protoconule is reported by Hooijer (1948) and Korenhof (1960). Remnants of a small lingual cingulum are present on our sample on M^1 (50%), M^2 (51%) and M^3 (20%), and Korenhof (1960) found traces of a cingulum in a series of 41 specimens. A narrow ledge-like buccal cingulum is rarely (<35%) observed in the present sample. According to Frisch (1965), the cingulum is more reduced in *Pongo* than in the other pongids, a conclusion supported by the present data.

Lower. As in all pongids, the lower molars of *Pongo* possess five cusps and the most frequent occlusal pattern is the *Dryopithecus* (Y-5) pattern. Thus M_1 (94%), M_2 (77%) and M_3 (64%) display this pattern in our sample. The other pattern represented is the "plus" arrangement. The lingual cusps are higher and more pointed than the buccal ones. On M_1 the metaconid is either slightly distal to or opposite the protoconid, whereas on M_{2-3} the metaconid more frequently faces the protoconid. In the present sample the hypoconulid is invariably buccal in position on all three molars, a situation also noted for *Pongo* (Frisch, 1965). The fossae pattern is similar to the pongid morphotype described earlier. The trigonid basin varies in shape from a narrow transverse slit to one of slightly wider dimensions. The trigonid basin is usually larger on M_1 than on M_2 or particularly on M_3 where it is frequently only a slit, or even wanting. In the latter condition, the protocristid is absent. Similar variability is manifested by the post-talonid fossa. Buccal cingula are rare in extant orang-utans (Frisch, 1965) and the present sample is no exception. A small ridge-like structure is found on the buccal surface of the protoconid or in the groove between the protoconid and hypoconid on M_1 (22%), M_2 (22%) and M_3 (27%). A tuberculum intermedium appears in our collection as follows: M_1 (47%), M_2 (63%) and M_3 (50%).

Odontometry (Tables 199–202, Appendix)

There are 22 significant sexual differences in the tooth dimensions listed in the Appendix. The upper canine is much larger in males in both dimensions, M–D ($P<0{\cdot}02$) and B–L ($P<0{\cdot}03$). The lower canine does not show significant sexual dimorphism in the M–D diameter, but does in the B–L dimension ($P<0{\cdot}005$). In contrast to Old World monkeys, *Pongo* does not display significant sexual dimorphism in P_3. Molar length relations are $M^1 \leq M^2 > M^3$ and $M_1 < M_2 > M_3$. M_2 is usually longer than M_3 as reported here, although the relationship is obviously quite variable in extant orangs (see Gregory and Hellman, 1926; Hooijer, 1948; Ashton and Zuckerman, 1950).

Genus *Gorilla*

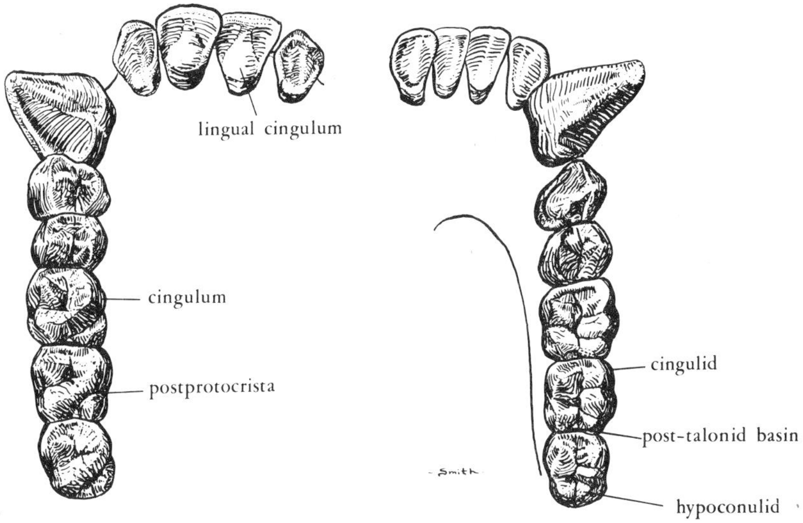

Plate 52. *Gorilla gorilla* male, occlusal view ×0·6.

Morphological Observations

	Sample Male	Sample Female
Gorilla gorilla (western lowland gorilla)	6	9

The eruption of P_4 before P_3 is similar to that in the orang and differs from that in *Pan*; otherwise the sequence of eruption is the same in extant pongids (Schultz, 1935).

Incisors

Upper. I^1 is large compared with I^2. The incisal border of I^1 is horizontal and chisel-like while in I^2 it slopes off distally. The labial surface of both teeth is convex and on I^1 a median vertical groove is frequently observed. The lingual surface of I^1 is circumscribed on three sides by marginal ridges while a cingulum defines the fourth or cervical border of the tooth. The cervical region constitutes the thickest portion of the tooth from which a tuberculum dentale usually extends finger-like processes toward the incisal border. The large tuberculum and the ridge or ridges arising from it tend to mask the depth of the lingual fossa. I^2 is similar to I^1 but smaller and the tuberculum usually lacks any extensions.

Lower. I_1 is smaller than I_2. The incisal border of I_1 is flat while that of I_2 slopes off distally. Both teeth possess well formed mesial, distal and incisal marginal ridges as well as a transverse lingual cingulum. The lingual surface of both teeth may be slightly concave, particularly the distal portion of I_2.

Canines

Upper. The upper canines are large, robust teeth projecting well beyond the occlusal plane, particularly in the male. The base is elongated mesiodistally from which the crown curves distally. The mesial border is broad and thick while the distal border is narrow. A thin groove incises the mesial border from the base to near the tip. A median lingual groove is also present as well as a lingual cingulum. The latter structure is not sexually dimorphic as in the orang-utan.

Lower. Like the upper canines, the lower ones display a high degree of sexual dimorphism. Indeed, in females, the canines are not much higher than the incisors. The canine projects buccally and presents three surfaces separated by three ridges. A narrow groove runs up the mesiolingual surface of the tooth. A lingual cingulum is present as well as a basal heel.

Premolars

Upper. P^{3-4} are bicuspid. The paracone is always much larger and more pointed than the protocone. A cleft-like developmental groove separates the cusps longitudinally. The occlusal surface presents two major cristae: one passes from the mesiolingual side of the paracone transversely to end just mesial to the protocone, while the other courses from the distolingual side of the paracone to terminate at the distal surface of the protocone. In this manner, these diverging cristae separate the occlusal surface of each premolar into three parts. Each premolar presents well defined mesial and distal ridges. As in the orang-utan, the enamel on the mesiobuccal surface of P^3 extends toward the root much more than on P^4. A small hypocone is present on P^4 in one specimen. A lingual cingulum is present on P^3 in 85% of our sample and on P^4 in over 90%. It usually manifests as a narrow band along the distolingual surface of the protocone, although on occasion it is limited to the mesiolingual aspect of this cusp. In one male, a buccal cingulum is present on P^{3-4} and in addition, a distostyle arises from the cingulum.

Lower. P_3 is sectorial although in the present collection a metaconid is always present as a light elevation on the distolingual aspect of the protocristid. The latter structure divides the lingual surface of P_3 into mesial and distal parts. A fossa or depression is observable on both parts; however, the distal fossa is broader and more excavated than the mesial one. It should be noted that this spacious distal fossa receives the protocone of P^3 when the teeth are occluded. A well developed lingual cingulum defines the entire lingual surface of P_3. P_4 is bicuspid with an expanded talonid which frequently displays a hypoconid and entoconid. Remane (1921) identifies a hypoconulid on P_4 in his gorilla sample, but we are unable to detect it in our specimens. The protoconid and metaconid are connected by a V-shaped protocristid which forms the distal boundary of the trigonid basin. The configuration of this basin varies in *Gorilla* as it did in *Pongo*. A narrow, band-like buccal cingulum is present on P_4 in over 90% of the present animals and in most cases it is complete.

Molars

Upper. M^{1-3} possess four cusps. The highest cusp is usually the metacone although the paracone, which is a close second, is almost always the highest cusp on M^3. The largest cusp is usually the protocone. The smallest is the hypocone, which, however, is larger relative to the other cusps in *Gorilla* than in the other pongids (Korenhof, 1960). The occlusal surface conforms to the pongid morphotype outlined earlier, i.e. three fossae defined by cristae. In *Gorilla*, the mesial fossa is usually quite narrow while the distal fossa is rather wide. As in *Pongo*, a sulcus obliquus surrounds the hypocone and a postprotocrista passes between the protocone and metacone. A protoconule, or a remnant thereof, is usually described on gorilla molars (see Korenhof, 1960 for an excellent discussion of this feature in the gorilla). Slight wear quickly obliterates the protoconule and in the present sample we were able to identify it in only two animals. Lingual cingula are common in gorillas (Korenhof, 1960) and the present sample attests to this: they are present on M^1 and M^2 in all of our specimens, and on M^3 in 71%. Buccal cingula however are rare, being present on M^1 in 13% and on M^2 in 8% of our sample, and completely absent on M^3. In all cases, the buccal cingula are limited to the mesiobuccal surface of the paracone or to the groove between the paracone and metacone.

Lower. M_{1-3} have five cusps and the most common groove and cusp arrangement in the present collection is the Y-5 pattern, thus: M_1 (91%), M_2 (67%) and M_3 (71%). The other pattern represented is the cruciform configuration. The metaconid and entoconid are the highest cusps, especially after attrition commences. The groove and fossa systems are as discussed earlier for the pongids. The buccal groove is deep while its lingual counterpart is hardly discernible, particularly on M_{2-3}. However, the occlusal separation between the metaconid and entoconid is very wide and deep on all three molars, a feature unique in gorillas among all pongids. Another peculiarity at this location on gorilla molars is that the lingual marginal ridge passes up along the lingual side of the metaconid, creating a narrow ledge separated from the metaconid by a slit-like fissure. On some molars, the ascending ledge nearly achieves contact with the lingual part of the mesial marginal ridge. Even in extreme wear the groove and ledge are perceptible. The result of the wide cleft is a commodious talonid basin which probably developed *pari passu* with the voluminous protocone of the upper molars, since the latter cusp is accommodated by the talonid basin when the teeth are in occlusion. The hypoconulid is usually buccal in position (90% of the present sample). In the other cases it is central. A tuberculum intermedium appears on M_1 (43%), M_2 (100%) and M_3 (100%). A buccal cingulum appears on M_1 (80%), M_2 (50%) and M_3 (15%). It is variable in position, being complete or limited to either the protoconid or hypoconid. The buccal cingulum is much more common in *Gorilla* than in the other pongids and is also better developed; e.g. in the present study only *Gorilla* possesses a complete buccal cingulum. For these and other reasons, many students believe the dentition of *Gorilla* represents the most primitive among extant Pongidae (Remane, 1921, 1960; Korenhof, 1960; Frisch, 1965).

Odontometry (Tables 203–206, Appendix)

The dentition of the lowland gorilla emblazons a high degree of sexual dimorphism, particularly in the upper canine where both dimensions are significantly different ($P<0{\cdot}01$). A similar magnitude of canine sexual dimorphism is reported for the lowland gorilla (Greene, 1973). Like Greene, we found much greater canine sexual dimorphism in the gorilla than in the chimpanzee. Notwithstanding the hypertrophied upper central incisors of both sexes, our findings indicate a significant sexual difference in both dimensions (M–D: $P<0{\cdot}002$ and B–L: $P<0{\cdot}02$).

The molar length relations are $M^1<M^2>M^3$ and $M_1<M_2\geq M_3$.

Genus *Pan*

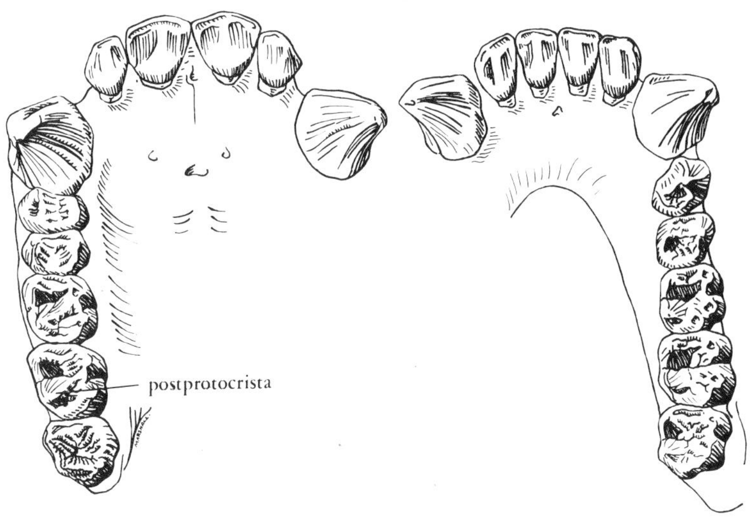

Plate 53. *Pan troglodytes* male, occlusal view ×0·8.

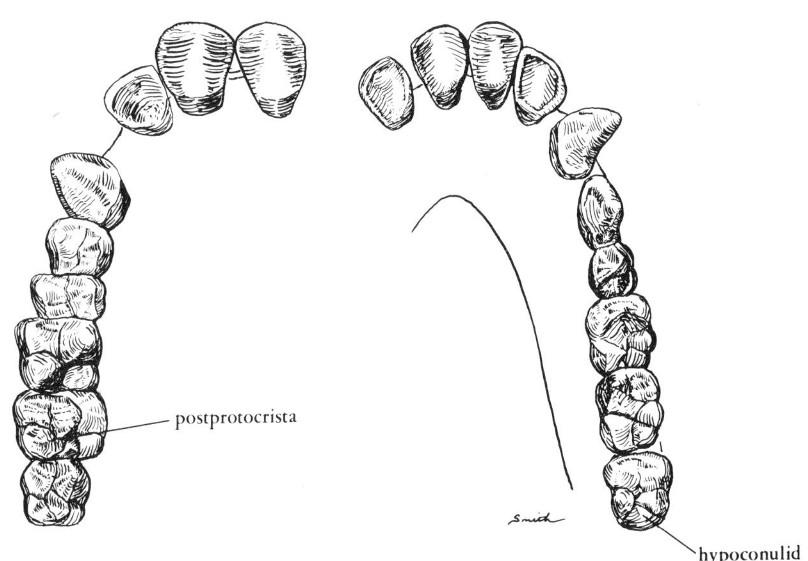

Plate 54. *Pan paniscus* male, occlusal view ×1·15.

Morphological Observations

	Sample	
	Male	Female
Pan troglodytes (chimpanzee)	32	61
Pan paniscus (pygmy chimpanzee)	2	2

Incisors

Upper. I^1 is broad compared with I^2. The labial surfaces are convex mesiodistally and a median vertical ridge is occasionally observed on I^1. The lingual surfaces of both I^{1-2} present well defined mesial and distal marginal ridges. A transverse lingual cingulum resides on both I^{1-2} from which a median tuberculum dentale arises to pass to the incisal border. In *Pan* it most often possesses a single enamel extension although multiple projections are observed. The median tubercle tends to mask the concave nature of the lingual surface in both incisors.

Lower. I_{1-2} are similar morphologically but differ slightly in size ($I_1 < I_2$). Both possess marginal ridges as well as lingual cingula. The median part of the cingulum is frequently somewhat swollen, suggesting a lingual tubercle. A median enamel ridge occupies the lingual surface of both incisors, passing from the cingulum to the incisal border.

Canines

Upper. The upper canine is larger in males than in females of both species. The base is stout and elongated mesiodistally. The mesial border is more convex in *Pan troglodytes* and possesses a mesial groove which is absent in our sample ($n = 4$) of *Pan paniscus.* The lingual surface presents a median longitudinal groove in *Pan troglodytes* which is absent in the pygmy chimpanzee. A lingual cingulum is present in both species.

Lower. The lower canine is large, particularly in males. The base is stout and presents a lingual cingulum. A true basilar heel is lacking. The crown tapers to a point; mesiolingual and distolingual grooves are observable along its sides. Neither is well developed and occasionally the mesiolingual groove is lacking.

Premolars

Upper. P^{3-4} are bicuspid. The protocone is always smaller than the paracone and is somewhat more pointed on P^4. The enamel extends onto the mesiobuccal surface of the root as in other pongids. Mesial and distal marginal ridges are present as well as several small cristae on the occlusal surface. Of the latter, two are more prominent than the others; both emanate from the side of the paracone and pass transversely lying mesially and distally to the protocone. They are usually more distinguishable on P^4 than on P^3. There are no cingular remnants on either premolar in the present sample.

Lower. P_3 is sectorial. A metaconid is frequently observable as a small cusp upon the distolingual surface of the protocristid. The protocristid divides the lingual surface of P_3 into mesial and distal parts. In addition, well marked ridges pass down the mesial and distal slopes of the protoconid to become confluent with the lingual cingulum. Thereby the mesial part of P_3 is a shallow triangular design while the distal portion is a rather deep triangular fossa (talonid basin). Remane (1960) discusses the talonid basin in P_3 which he calls the fovea posterior, noting its large size and depth in the chimpanzee and orang-utan. P_4 is bicuspid and an entoconid (93%) and a hypoconid (51%) are frequently present in our sample. Remane (1921) describes a hypoconulid as a rare feature on P_4 in chimpanzees. The hypoconulid is absent in our collection. The metaconid approximates the size of the protoconid and frequently surpasses it. The position of the metaconid to the protoconid is quite variable in our sample, ranging from distal (21%) to opposite (69%) to mesial (10%). A V-shaped protocristid connects the protoconid and metaconid and at the same time forms the distal boundary of the moderately deep trigonid fossa. The mesial border is constituted by the mesial marginal ridge. The talonid basin is broad and flat and completely circumscribed by the distal marginal ridge upon which the hypoconid and entoconid develop. The talonid is on a lower level than the trigonid. The size and configuration of both fossae are variable: i.e. they may be broad and expanded in the same tooth or one or the other may be narrow and constricted. The various combinations vary a great deal in the present sample. Cingular remnants are absent in this sample of chimpanzees.

Molars

Upper. Of all the teeth in the extant Ponginae, the molars of *Pan* most resemble those of *Homo* both morphologically and metrically. The four principal cusps vary in height: para- and metacone are higher than the proto- and hypocone; the hypocone is smallest. Indeed, Schuman and Brace (1955, p. 83) consider the progressive reduction (M^{1-3}) of the hypocone "to be another hallmark of the later phases of dentition evolution". In their study of the Liberian chimpanzee, they found the hypocone largest on M^1 (61%), medium on M^2 (55%) and smallest on M^3 (52%). Also, they never identified a three-cusped molar. The present data agree with the morphogenetic reduction gradient recorded by Schuman and Brace, but there are three-cusped M^3's in our sample. The crista pattern is similar to that discussed for the gorilla. The preprotocrista (crista transversa anterior) is less often interrupted by the mesiodistal developmental groove than in *Gorilla.* Korenhof (1960) found it uninterrupted on M^1 (87%), M^2 (78%) and M^3 (66%) and took this to indicate its stronger development in *Pan.* The postprotocrista is quite apparent on these molars and as with the preprotocrista, it is generally uninterrupted on M^{1-2}, while on M^3 it may be interrupted as much as 50% of the time (Korenhof, 1960). An oblique sulcus is present although it is not usually as distinct as in *Gorilla* and *Pongo.* In addition to these crests the occlusal surface of the molars in *Pan* have secondary wrinkles or crenulations, but in our experience never to the extent found in *Pongo.* A metaconule is not reported in *Pan,* though a vestigial protoconule is recorded as a variety of the lingual end of the preprotocrista (Korenhof, 1960). Korenhof reports the protoconule is variably expressed on M^{1-3} in seven of 58 skulls. We have examined our material for the protoconule and confess that we must remain dubious about its presence in *Pan.* An occasional wrinkle is found at this location but to homologize this as a remnant of the protoconule is in our opinion rather unlikely, though it remains a possibility.

A lingual cingulum is a rather common occurrence in *Pan*; it usually appears on the mesiolingual aspect of the protocone in M^1, being more extensive on M^{2-3} so as to include the protocone. On M^3 the cingulum may even pass distally around the hypocone to become confluent with the distal marginal ridge. These findings are not unlike those of Schuman and Brace (1955). The incidence of the lingual cingulum in the present sample is M^1 (80%, $n=92$), M^2 (80%, $n=85$) and M^3 (63%, $n=56$). Since there is no sexual dimorphism, the sexes are combined. Unfortunately neither Schuman and Brace (1955) nor Korenhof (1960) present frequencies for the trait. It is interesting to note that the lingual cingulum occurs less frequently on M^3, but when it does it is generally larger and more extensive than it is on M^1 or M^2. A buccal cingulum is less frequent than the lingual one, M^1 (33%, $n=91$), M^2 (13%, $n=80$) and M^3 (33%, $n=43$). In most cases it is limited to the paracone or bridges the buccal groove between the two buccal cusps.

A final comment is necessary regarding the lingual cingulum. A controversy appears in the literature regarding this trait in primates, i.e. does the tumescence appearing on the lingual side of the protocone represent the same developmental phenomenon in all primates, or should it be considered one thing in nonhuman primates and another (unrelated) condition in hominids? The divisionists believe the term "Carabelli's cusp" should be limited to hominids and not used for the cingulum in nonhuman primates. In this study we have seen a protoconal cingulum in prosimians, New and Old World monkeys, pongids and *Homo.* It is also found in a wide variety of fossil forms. In our opinion the protoconal complex is an old (primitive), well entrenched feature in the primates and it matters little what appellation is attached to it. There are differences in its expression in the primates but this is true of any quantitative dental trait that we are familiar with. In summary, we believe that simply because a quantitative feature does not extend as far distally or protrude as high on the crown in one taxon as it does in another does not mean that the structures are not necessarily homologous.

Lower. The lower molars generally possess five cusps, of which the metaconid and entoconid are the highest. The expression of the *Dryopithecus* pattern is variable in *Pan*. In the present sample it is present on M_1 (95%, $n=74$), M_2 (62%, $n=66$) and M_3 (47%, $n=43$). In the other cases it is cruciform with either four, five or six cusps represented. Schuman and Brace (1955) found many fewer Y-5 patterns on the lower molars: M_1 (54%, $n=96$), M_2 (16%, $n=99$) and M_3 (1%, $n=114$). It should be mentioned that their *n*'s refer to number of teeth while our samples are number of animals studied. Also, they separated the Y-5 and Y-6 patterns which we did not do; thus when combined on M_1 the frequency is 79% Y-configurations, a figure somewhat more compatible with our findings. The almost complete lack of Y-patterns on M_3 is difficult to explain, unless it is another manifestation of the large interpopulational variability of the dentition in extant chimpanzees. The metaconid on M_1 is most frequently distal to the protoconid; when it is not, it is opposite the protoconid. The latter condition is more prevalent on $M_{2\text{-}3}$. A protocristid is well developed between the protoconid and metaconid. The size and shape of the trigonid basin are variable as in other extant pongids. The talonid basin is spacious but the separation between the metaconid and entoconid is narrow compared with the wide gap between these cusps in *Gorilla*. There is more flare to the buccal side of the molars than exemplified by the lingual face and also, the buccal groove is more incisive than the lingual one. The hypoconulid is most often in a buccal position, M_1 (100%), M_2 (94%) and M_3 (76%). A buccal cingulum is a rare trait in extant *Pan*, M_1 (10%, $n=89$), M_2 (5%, $n=85$) and M_3 (10%, $n=49$) in our sample while a tuberculum intermedium appears on M_1 (12%, $n=83$), M_2 (29%, $n=89$) and M_2 (37%, $n=60$).

Odontometry (Tables 207–214, Appendix)

As is true in all anthropoid apes, the upper canines are markedly sexually dimorphic ($P<0{\cdot}01$). This finding contradicts that of Schuman and Brace (1955) for the Liberian chimpanzee but corroborates the data of Ashton and Zuckerman (1950), and Pilbeam (1969) in supporting male superiority in canine tooth size. Our data substantiate the observation by Korenhof (1960) that there is little if any metric increase in M^2 over M^1 in *Pan* and when these figures are compared with *Gorilla* the reverse ratio exists. Additionally, M^3 is frequently smaller than M^{1-2}, giving molar length relations of $M^1 \geqslant M^2 > M^3$ and $M_1 < M_2 > M_3$, although variations of these formulae are to be anticipated in *Pan*.

13

Family Hominidae

Present Distribution and Habitat

Modern *Homo sapiens* lives in all of the habitable parts of the earth and in many regions that are not so hospitable. The only major land mass not conquered by *Homo* until quite recently was Antarctica. Many people live below sea level and there are a few groups found living at more than 4,500 m (15,000 ft) in the Andes of Peru. For the most part, mankind is an erect bipedal creature although other modes of locomotion may be resorted to when the occasion demands. The activity rhythm of *Homo* is diurnal.

Dietary Habits

Omnivorous.

General Dental Information

Permanent dentition: $I^2_2\ C^1_1\ P^2_2\ M^3_3$

Deciduous dentition: $i^2_2\ c^1_1\ m^2_2$

Sequence of eruption of permanent teeth:

Homo sapiens

	M^1		I^1		I^2	P^3		C	P^4		M^2	M^3
M_1		I_1		I_2		C	P_3		P_4	M_2		M_3

Genus *Homo*

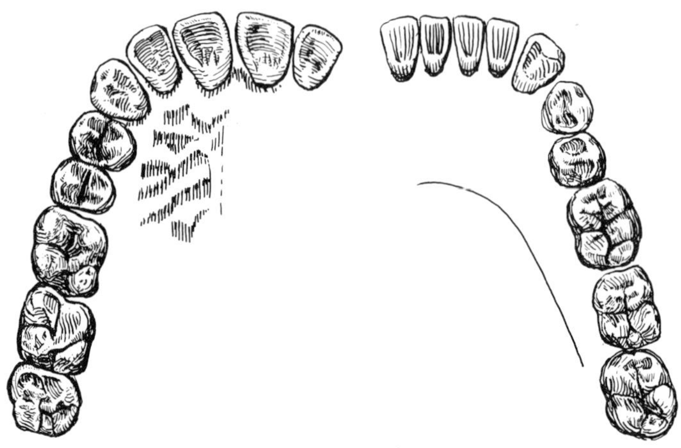

Plate 55. *Homo sapiens* male, occlusal view, ×1.

Morphological Observations

	Sample	
	Male	Female
Homo sapiens (man)	40	20

The central incisor is larger than the lateral one. Both have rather broad, flat incisal borders and I^{1-2} are usually homomorphic. The labial surfaces of both I^{1-2} are convex, I^2 more so than I^1. A mesial and distal marginal ridge and a transverse cingulum outline the lingual face of each tooth. The lingual topography of I^{1-2} gives them a scoop-like appearance which when well developed is termed "shovel shaped" (Hrdlička, 1920). Since then numerous students have studied the distribution of shovel-shaped incisors in *Homo* and found them to be more common in populations of the Mongoloid stock (Nelson, 1938; Dahlberg, 1949; Pedersen, 1949; Moorrees, 1957). The trait is more frequent in the maxillary incisors than in the mandibular incisors. A tuberculum dentale, or what some students call a lingual tubercle, probably represents the hypertrophied lingual cingulum. The lingual tubercle on occasion sends enamel extensions into the fossa. The tubercle may be present on either I^1 or I^2 and it varies in frequency among the different populations. Robinson (1956) distinguishes a lingual cingulum from a lingual tubercle on the basis of the latter feature being marked off from the marginal ridges rather than having the ridges pass across the gingival eminence as they do with the cingulum. He believes that the cingulum is a primitive and widely spread trait among nonhuman primates and that the tubercle is phylogenetically later, occurring "only in the more recent hominids" i.e. Neanderthal and modern man. In our opinion, a lingual cingulum is present in extant pongids, however a lingual tubercle is frequently present in Old World monkeys, e.g. *Macaca*, *Colobus* and *Nasalis*, as well as in recent hominids. The lateral incisor is more variable than the central incisor, ranging from normal size to absent. It may be barrel-shaped, peg-shaped or caniniform.

The lower central incisor is somewhat smaller than the lateral incisor. The labial face of both I_{1-2} are ordinarily very smooth while the lingual surface displays a slight concavity. The marginal ridges are usually slight, although they can form into prominent rolls of enamel near the incisal edges. The cingular area of I_{1-2} is generally not well marked, but lingual tubercles do occur.

The maxillary canine is very much reduced in all dimensions in extant *Homo* when compared with the canines of recent pongids. However, whether hominids evolved from a large or small canine Miocene ancestor continues to be debatable. Recently Kinzey (1971) presented some very convincing evidence that the orthodox view deriving living man from an antecedent possessing large canines is no longer tenable. It is quite conceivable that man's immediate ancestor had small canines. The labial surface is convex while the lingual face presents thin marginal ridges and a lingual cingulum. A median lingual ridge is frequently present dividing the shallow lingual fossa into two parts. Sometimes, however, the lingual surface is so smooth the fossae and ridges are practically imperceptible.

The lower canines are usually not much larger than the adjacent teeth. Surface markings are slight: even the marginal ridges and cingular regions are difficult to identify.

The maxillary premolars are usually bicuspid although a small third cusp occurs. In both P^{3-4} the paracone is larger than the protocone. In all modern populations P^3 is consistently larger than P^4. The principal two cusps are connected by a V-shaped preprotocrista which is divided by a central developmental groove. In modern dentistry, this crest is separated into two sections and called the lingual and buccal triangular ridges. Other topographic features include marginal ridges enclosing the entire occlusal surface as well as mesial and distal fossae. P^{3-4} are very similar to each other morphologically.

The lower premolars are usually described as possessing a large protoconid and a small metaconid united by a transverse ridge. Kraus and Furr (1953) and Ludwig (1957) demonstrated that this description does not apply to human lower premolars. Indeed, these investigators pointed out the great variability in these teeth in modern populations. For example, Kraus and Furr (1953) identified 16 variable traits on P_3 and Ludwig (1957) outlined seven variable features on P_4. These characters include numerical differences in cusp number, ridge and crista variability, position of the metaconid relative to the protoconid, the independence of the metaconid, etc. There is apparently also a certain degree of sexual dimorphism regarding the reduction of cusp numbers. According to Moorrees (1957), females express a greater frequency of two-cusp premolars than males. Such detailed, systematic studies alert us to the magnitude of premolar variability in extant *Homo*, a fact known for many years by students of human dentition.

The maxillary molars are characterized as possessing four cusps with a mesiodistal reduction gradient in the hypocone as well as in the absolute size of the molars. M^3 is particularly variable: it may have a normal morphology or a peg-shape, or it may be completely absent. The metacone is also frequently reduced, resulting in a change in occlusal outline from a parallelogram to a more triangular form. The parallelogram configuration is much more common in pongids. A more or less cruciform groove pattern separates the four main cusps and the hypocone is separated by an oblique groove from the trigon. A mesial and distal fossae (fovea anterior and posterior) are present although the former depression may be poorly developed or absent. Crenulations may appear on the occlusal surface but these are rarely pronounced. The Carabelli complex is commonly present on M^1 in *Homo*. Excellent studies are available on the Carabelli complex in *Homo* (Dietz, 1944; Pedersen, 1949; Kraus, 1951; Moorrees, 1957). It varies in its incidence on M^{1-3} as well as between human populations. The trait is more common in white ethnic groups (70%), least prevalent in Mongoloid populations (10%).

The lower molars have five cusps and in general decrease in size from M_1 to M_3. The cusps are more bunodont than in the pongids and the metaconid is usually the largest cusp although the protoconid may be larger. The hypoconulid is normally the smallest cusp. The typical *Dryopithecus* cusp and groove pattern expresses a differential gradient from M_{1-3}, i.e. the Y-5 configuration is more common on M_1, less on M_2 and least on M_3 in all ethnic groups. The tendency is to transform the Y-5 to a +4-type and this development also shows different frequencies among human populations (Hellman, 1928; Nelson, 1938; Goldstein, 1948; Moorrees, 1957). A mesial and distal fossa are present and both display developmental plasticity, i.e. they range in size from small slits to rather well formed depressions. In modern dentistry, they are referred to as the mesial and distal triangular fossae. A true protocristid seems to be lacking in modern forms. The central fossa (talonid basin) is roughly circular and not as commodious as in the pongids. The tuberculum sextum and intermedium are present in modern hominids and the former trait increases in frequency from M_{1-3}. Finally, Dahlberg (1950) mentions the presence of protostylids on the mandibular molars of modern man; indeed, in a population of 80 Pima Indians he found a 31% incidence of the trait.

14

Summary

Incisors

In primates the incisor teeth are more variable (less conservative) in both size and morphology than are the posterior teeth. This is particularly so for the prosimians. The majority of living primates possess two upper and two lower incisors in each quadrant of the jaw. The tree shrews, on the other hand, have retained three incisors in the lower jaw. In *Daubentonia*, a single pair of incisors (canines, see Schwartz, 1974a) are present in both upper and lower jaws, and they grow continuously throughout the animal's life. The two maxillary incisors are generally absent in *Lepilemur* although small stubs are occasionally observable. The upper incisors are broad and more spatulate among the higher primates and in general they are more heteromorphic in the New and Old World monkeys than in the hominoids. A sulcus occupies the lingual surface of the upper incisors in the majority of Old World monkeys, while a fossa is more frequently present in hominoids. In *Homo* the fossa may be broad and deep, forming a shovel-shaped incisor.

The lower incisors are long and procumbent in extant prosimians (except for *Daubentonia* and *Tarsius*) and form, along with the procumbent canines, the characteristic dental comb. *Tarsius* possesses a single, erect incisor in each quadrant of the lower jaw.

Canines

The size, shape and height of the canine varies tremendously among primates and in many groups sexual dimorphism is common. The upper canines are more projecting than the lower ones and frequently display great amounts of distal curvature. The degree of sexual dimorphism is small among prosimians, particularly in *Tarsius*, *Propithecus* and *Lepilemur*, while among New World monkeys little sexual differences are noted for *Saguinus* and *Callicebus*. The most projecting, dagger-like canines, as well as the greatest degree of sexual dimorphism, are found among the cercopithecids, e.g. *Papio* and *Mandrillus.* The canines are thick and massive among the male pongids, particularly in *Gorilla*, whereas among the hylobatids there is virtually no sexual dimorphism in canine size. It has been suggested by several authorities that the explanation for this sexual dimorphism is not completely alimentary but rather is to be found in the animals' aggression patterns and social organization.

Premolars

The first premolar was lost early in primate history. Among extant primates, Indriidae, Old World monkeys and hominoids possess two premolars (P_3^3 and P_4^4); all other living primates display three premolars (P_2^2, P_3^3 and P_4^4) with the exception of *Daubentonia* which possesses a single upper premolar (P^4). Among prosimians P_2^2 is frequently caniniform and in *Phaner* it is particularly hypertrophied and caniniform; in *Hapalemur* and *Cheirogaleus* P^4 is molariform, as is P_4 in *Hapalemur.* In the Lorisidae P_4 is usually quite molarized, particularly in *Loris* and *Galago.* Among the New World monkeys P_2 is somewhat caniniform in the Callithricidae while the upper premolars are generally bicuspid in this family. In cebids the upper premolars are usually bicuspid, especially in *Cebus*, and the lower premolars tend to add cusps from P_{2-4}. The lower premolars are heteromorphic (P_3 sectorial, P_4 multicuspid) in Old World monkeys and Great Apes, while in *Homo* they are homomorphic (P_{3-4} possess two or more cusps). The upper premolars in these three groups display anywhere from one cusp to four, although two is the most common number.

Molars

Among the primates the trend is to develop a hypocone on the upper molars, making them quadritubercular. The three-cusped upper molar persists in the Cheirogalinae although an incipient hypocone is present in *Lemur.* A four-cusped molar is common in the Indriidae and Lorisidae. Among the Anthropoidea the four-cusped molar prevails even in *Callicebus*, where the molar outline is somewhat reminiscent of the tribosphenic design. In cercopithecids both upper and lower molars display the bilophodont pattern, while bilophodonty is absent in the hominoids. In hominoids as well as in some prosimians (*Propithecus*, *Tarsius* and the Lorisidae) and New World monkeys (*Ateles*, *Alouatta*, *Aotus*, etc.) the upper molars possess an oblique crest, the postprotocrista connecting the protocone and metacone.

The lower molars lack the paraconid (except in *Tarsius* and in other rather rare cases); the trigonid and talonid become equal in height and each possesses four cusps. The fifth cusp, hypoconulid, is variably present throughout the primates, e.g. it is present on M_3 in cercopithecids except for *Cercopithecus* and *Erythrocebus* while in hominoids it may be found on all three lower molars. The Y-5 and +4 cusp and groove patterns are also variably developed among the hominoids.

Odontometric Appendix

The measurements presented in the Appendix of original data were taken following the methods and definitions presented in Chapter 1. The odontometric data taken from the work of other students however, may vary somewhat from our work, but in general, we believe the measuring techniques to be more or less comparable.

T-tests of tooth size sexual dimorphism were performed on the majority of species and this information appears in the male tables.

Table 1
Tupaia javanica male

Maxillary teeth		n	X̄±s.e.	s.d.	Range	P
I^1	M–D					
	B–L					
I^2	M–D					
	B–L					
C	M–D					
	B–L					
P^2	M–D					
	B–L					
P^3	M–D	13	1·8±0·07	0·27	1·2–2·2	
	B–L	13	1·3±0·05	0·19	1·1–1·6	
P^4	M–D	19	2·1±0·04	0·17	1·9–2·5	*
	B–L	18	2·2±0·05	0·20	1·7–2·6	
M^1	M–D	19	2·8±0·03	0·13	2·5–3·0	
	Ant B	19	2·7±0·04	0·16	2·3–3·0	
	Post B	18	2·8±0·05	0·21	2·5–3·4	
M^2	M–D	17	2·5±0·03	0·14	2·3–2·9	
	Ant B	17	3·0±0·04	0·17	2·7–3·4	
	Post B	17	2·9±0·05	0·19	2·5–3·2	
M^3	M–D	15	2·0±0·05	0·19	1·7–2·5	
	Ant B	15	2·4±0·06	0·21	2·1–2·9	
	Post B					

* $P<0·05$

Table 2
Tupaia javanica male

Mandibular teeth		n	X̄±s.e.	s.d.	Range
I_1	M–D				
	B–L				
I_2	M–D				
	B–L				
C	M–D				
	B–L				
P_2	M–D				
	B–L				
P_3	M–D				
	B–L				
P_4	M–D	16	1·9±0·04	0·14	1·7–2·1
	B–L	16	1·4±0·05	0·22	1·0–2·0
M_1	M–D	17	2·6±0·03	0·13	2·3–2·9
	Tri B	17	1·8±0·06	0·25	1·5–2·6
	Tal B	17	1·9±0·06	0·23	1·6–2·7
M_2	M–D	18	2·6±0·02	0·10	2·4–2·7
	Tri B	18	1·8±0·05	0·22	1·6–2·6
	Tal B	18	1·8±0·05	0·22	1·6–2·6
M_3	M–D	17	2·2±0·03	0·14	2·0–2·6
	Tri B	17	1·4±0·02	0·09	1·3–1·6
	Tal B	17	1·0±0·03	0·11	0·9–1·3
	Hypo-conulid				

Table 3
Tupaia javanica female

Maxillary teeth		n	$\bar{X} \pm$ s.e.	s.d.	Range
I^1	M–D				
	B–L				
I^2	M–D				
	B–L				
C	M–D				
	B–L				
P^2	M–D				
	B–L				
P^3	M–D	11	1·9±0·05	0·18	1·5–2·2
	B–L	11	1·3±0·03	0·10	1·2–1·5
P^4	M–D	12	2·2±0·05	0·18	1·7–2·4
	B–L	12	2·2±0·05	0·18	1·8–2·5
M^1	M–D	13	2·7±0·04	0·16	2·5–3·0
	Ant B	13	2·8±0·04	0·15	2·5–3·0
	Post B	13	2·8±0·05	0·17	2·4–3·0
M^2	M–D	14	2·5±0·03	0·11	2·2–2·6
	Ant B	14	3·1±0·04	0·16	2·7–3·2
	Post B	14	2·8±0·04	0·14	2·5–3·1
M^3	M–D	13	2·1±0·04	0·15	1·8–2·3
	Ant B	13	2·5±0·03	0·12	2·4–2·7
	Post B				

Table 4
Tupaia javanica female

Mandibular teeth		n	$\bar{X} \pm$ s.e.	s.d.	Range
I_1	M–D				
	B–L				
I_2	M–D				
	B–L				
C	M–D				
	B–L				
P_2	M–D				
	B–L				
P_3	M–D				
	B–L				
P_4	M–D	13	1·8±0·04	0·16	1·6–2·1
	B–L	13	1·4±0·09	0·32	1·1–2·4
M_1	M–D	13	2·6±0·03	0·12	2·4–2·8
	Tri B	13	1·8±0·04	0·13	1·4–1·9
	Tal B	13	1·8±0·04	0·13	1·5-2·0
M_2	M–D	14	2·6±0·03	0·11	2·5–2·8
	Tri B	14	1·8±0·03	0·12	1·5–2·0
	Tal B	11	1·8±0·03	0·11	1·7–2·0
M_3	M–D	12	2·3±0·03	0·12	2·1–2·5
	Tri B	11	1·4±0·03	0·08	1·3–1·5
	Tal B	11	1·1±0·04	0·12	0·9–1·2
	Hypo-conulid				

Table 5
Tupaia minor male

Maxillary teeth		n	$\bar{X}\pm$s.e.	s.d.	Range
I^1	M–D				
	B–L				
I^2	M–D				
	B–L				
C	M–D				
	B–L				
P^2	M–D				
	B–L				
P^3	M–D				
	B–L				
P^4	M–D	7	2·0±0·05	0·13	1·8–2·2
	B–L	7	2·1±0·05	0·13	2·0–2·3
M^1	M–D	7	2·5±0·06	0·17	2·2–2·7
	Ant B	7	2·5±0·04	0·10	2·4–2·7
	Post B	7	2·5±0·04	0·10	2·3–2·6
M^2	M–D	7	2·4±0·07	0·20	2·2–2·7
	Ant B	7	2·7±0·06	0·15	2·5–2·9
	Post B	7	2·5±0·08	0·21	2·2–2·8
M^3	M–D	6	1·8±0·03	0·08	1·7–1·9
	Ant B	6	2·1±0·07	0·18	1·8–2·2
	Post B				

Table 6
Tupaia minor male

Mandibular teeth		n	$\bar{X}\pm$s.e.	s.d.	Range
I_1	M–D				
	B–L				
I_2	M–D				
	B–L				
C	M–D				
	B–L				
P_2	M–D				
	B–L				
P_3	M–D				
	B–L				
P_4	M–D	7	1·8±0·15	0·40	1·5–2·7
	B–L	7	1·3±0·16	0·41	1·0–2·2
M_1	M–D	7	2·4±0·04	0·11	2·3–2·6
	Tri B	7	1·5±0·03	0·07	1·4–1·6
	Tal B	7	1·6±0·04	0·11	1·4–1·7
M_2	M–D	7	2·4±0·03	0·08	2·4–2·6
	Tri B	7	1·4±0·03	0·07	1·3–1·5
	Tal B	7	1·5±0·05	0·13	1·3–1·7
M_3	M–D	6	2·0±0·05	0·11	1·9–2·2
	Tri B	6	1·2±0·03	0·06	1·1–1·3
	Tal B	6	0·9±0·03	0·08	0·8–1·0
	Hypoconulid				

Table 7
Tupaia minor female

Maxillary teeth		n	$\bar{X}\pm$s.e.	s.d.	Range
I^1	M–D				
	B–L				
I^2	M–D				
	B–L				
C	M–D				
	B–L				
P^2	M–D				
	B–L				
P^3	M–D				
	B–L				
P^4	M–D	3	1·8±0·15	0·25	1·5–2·0
	B–L	3	1·9±0·17	0·30	1·6–2·2
M^1	M–D	4	2·5±0·10	0·21	2·3–2·8
	Ant B	4	2·5±0·15	0·30	2·2–2·9
	Post B	4	2·5±0·18	0·36	2·2–3·0
M^2	M–D	4	2·4±0·10	0·19	2·1–2·5
	Ant B	4	2·8±0·15	0·29	2·2–3·0
	Post B	4	2·6±0·11	0·21	2·4–2·9
M^3	M–D	4	1·8±0·08	0·15	1·7–2·0
	Ant B	4	2·1±0·33	0·66	1·2–2·8
	Post B				

Table 8
Tupaia minor female

Mandibular teeth		n	$\bar{X}\pm$s.e.	s.d.	Range
I_1	M–D				
	B–L				
I_2	M–D				
	B–L				
C	M–D				
	B–L				
P_2	M–D				
	B–L				
P_3	M–D				
	B–L				
P_4	M–D	2	1·7±0·15	0·21	1·5–1·8
	B–L	2	1·2±0·15	0·21	1·0–1·3
M_1	M–D	2	2·5±0·15	0·21	2·4–2·7
	Tri B	2	1·6±0·20	0·28	1·4–1·8
	Tal B	2	1·8±0·15	0·21	1·6–1·9
M_2	M–D	2	2·6±0·20	0·28	2·4–2·8
	Tri B	2	1·6±0·20	0·28	1·4–1·8
	Tal B	2	1·8±0·15	0·21	1·6–1·9
M_3	M–D	2	2·1±0·05	0·07	2·0–2·1
	Tri B	2	1·4±0·15	0·21	1·2–1·5
	Tal B	2	1·0±0·05	0·07	0·9–1·0
	Hypoconulid				

Table 9
Tupaia glis male

Maxillary teeth		*n*	$\bar{X}\pm$s.e.	s.d.	Range
I^1	M–D				
	B–L				
I^2	M–D				
	B–L				
C	M–D				
	B–L				
P^2	M–D				
	B–L				
P^3	M–D	26	2·4±0·03	0·16	2·1–2·7
	B–L	27	1·9±0·03	0·18	1·7–2·5
P^4	M–D	25	2·7±0·04	0·21	2·4–3·1
	B–L	24	2·8±0·04	0·19	2·5–3·2
M^1	M–D	27	3·5±0·05	0·23	3·2–4·0
	Ant B	27	3·4±0·05	0·24	2·8–3·8
	Post B	27	3·4±0·04	0·19	3·0–3·7
M^2	M–D	23	3·1±0·04	0·17	2·8–3·3
	Ant B	26	3·6±0·05	0·26	3·1–4·0
	Post B	26	3·3±0·04	0·22	2·8–3·7
M^3	M–D	20	2·4±0·04	0·16	2·0–2·7
	Ant B	20	2·7±0·04	0·19	2·3–3·0
	Post B				

Table 10
Tupaia glis male

Mandibular teeth		*n*	$\bar{X}\pm$s.e.	s.d.	Range
I_1	M–D				
	B–L				
I_2	M–D				
	B–L				
C	M–D				
	B–L				
P_2	M–D				
	B–L				
P_3	M–D				
	B–L				
P_4	M–D	25	2·4±0·03	0·14	2·1–2·7
	B–L	25	1·7±0·04	0·21	1·5–2·5
M_1	M–D	27	3·4±0·04	0·19	3·0–3·7
	Tri B	27	2·2±0·03	0·17	1·8–2·5
	Tal B	27	2·3±0·03	0·17	2·0–2·6
M_2	M–D	25	3·3±0·03	0·16	3·0–3·6
	Tri B	25	2·1±0·03	0·15	1·8–2·4
	Tal B	25	2·1±0·03	0·16	1·9–2·4
M_3	M–D	25	2·7±0·02	0·12	2·5–2·9
	Tri B	26	1·7±0·04	0·21	1·5–2·5
	Tal B	25	1·3±0·06	0·30	0·9–2·5
	Hypoconulid				

Table 11
Tupaia glis female

Maxillary teeth		n	$\bar{X} \pm$ s.e.	s.d.	Range
I^1	M–D				
	B–L				
I^2	M–D				
	B–L				
C	M–D				
	B–L				
P^2	M–D				
	B–L				
P^3	M–D	20	2·5 ± 0·04	0·16	2·2–2·7
	B–L	19	2·0 ± 0·05	0·20	1·7–2·3
P^4	M–D	21	2·6 ± 0·04	0·17	2·4–3·0
	B–L	21	3·0 ± 0·24	1·09	2·6–7·7
M^1	M–D	21	3·5 ± 0·04	0·17	3·2–3·7
	Ant B	21	3·4 ± 0·05	0·21	3·2–3·8
	Post B	20	3·4 ± 0·05	0·20	3·1–3·8
M^2	M–D	19	3·0 ± 0·03	0·13	2·8–3·0
	Ant B	18	3·6 ± 0·06	0·25	3·2–4·2
	Post B	18	3·3 ± 0·05	0·21	2·9–3·6
M^3	M–D	18	2·4 ± 0·06	0·25	2·0–2·9
	Ant B	17	2·8 ± 0·06	0·25	2·3–3·4
	Post B	19	2·0 ± 0·06	0·24	1·6–2·4

Table 12
Tupaia glis female

Mandibular teeth		n	$\bar{X} \pm$ s.e.	s.d.	Range
I_1	M–D				
	B–L				
I_2	M–D				
	B–L				
C	M–D				
	B–L				
P_2	M–D				
	B–L				
P_3	M–D				
	B–L				
P_4	M–D	19	2·5 ± 0·04	0·19	1·9–2·7
	B–L	18	1·7 ± 0·03	0·12	1·5–2·0
M_1	M–D	19	3·4 ± 0·04	0·16	3·1–3·6
	Tri B	19	2·1 ± 0·04	0·19	1·9–2·5
	Tal B	20	2·3 ± 0·04	0·17	2·0–2·7
M_2	M–D	20	3·3 ± 0·03	0·15	2·9–3·5
	Tri B	20	2·1 ± 0·05	0·20	1·8–2·5
	Tal B	20	2·1 ± 0·04	0·19	1·8–2·5
M_3	M–D	19	2·7 ± 0·05	0·20	2·3–3·3
	Tri B	19	1·7 ± 0·03	0·13	1·5–2·0
	Tal B	19	1·3 ± 0·04	0·16	1·1–1·7
	Hypo-conulid				

Table 13
Tupaia tana male

Maxillary teeth		n	$\bar{X}\pm$s.e.	s.d.	Range	P
I^1	M–D					
	B–L					
I^2	M–D					
	B–L					
C	M–D					
	B–L					
P^2	M–D					
	B–L					
P^3	M–D	3	2·7±0·17	0·29	2·5–3·0	
	B–L	3	1·7±0·24	0·42	1·2–2·0	
P^4	M–D	3	2·7±0·10	0·17	2·5–2·8	
	B–L	3	2·8±0·21	0·36	2·5–3·2	
M^1	M–D	4	3·6±0·05	0·10	3·5–3·7	
	Ant B	4	3·3±0·11	0·22	3·0–3·5	
	Post B	4	3·3±0·11	0·22	3·0–3·5	
M^2	M–D	4	3·3±0·00	0·00	3·3–3·3	*
	Ant B	4	3·4±0·10	0·21	3·2–3·6	
	Post B	4	3·2±0·12	0·23	3·0–3·4	
M^3	M–D	4	2·6±0·10	0·19	2·3–2·7	
	Ant B	4	2·5±0·10	0·21	2·3–2·8	
	Post B					

* $P<0·05$

Table 14
Tupaia tana male

Mandibular teeth		n	$\bar{X}\pm$s.e.	s.d.	Range
I_1	M–D				
	B–L				
I_2	M–D				
	B–L				
C	M–D				
	B–L				
P_2	M–D				
	B–L				
P_3	M–D				
	B–L				
P_4	M–D	4	2·8±0·12	0·24	2·5–3·0
	B–L	4	1·6±0·13	0·26	1·3–1·9
M_1	M–D	4	3·6±0·11	0·22	3·4–3·9
	Tri B	4	2·0±0·16	0·13	1·8–2·1
	Tal B	4	2·2±0·06	0·13	2·0–2·3
M_2	M–D	4	3·6±0·06	0·13	3·4–3·7
	Tri B	4	2·1±0·09	0·17	1·9–2·3
	Tal B	4	2·2±0·10	0·21	1·9–2·4
M_3	M–D	4	3·0±0·03	0·06	2·9–3·0
	Tri B	4	1·7±0·08	0·15	1·5–1·8
	Tal B	3	1·5±0·03	0·06	1·4–1·5
	Hypo-conulid				

Table 15
Tupaia tana female

Maxillary teeth		n	$\bar{X}\pm$s.e.	s.d.	Range
I^1	M–D				
	B–L				
I^2	M–D				
	B–L				
C	M–D				
	B–L				
P^2	M–D				
	B–L				
P^3	M–D	7	2·6±0·11	0·29	2·0–2·9
	B–L	7	1·7±0·10	0·27	1·2–2·0
P^4	M–D	8	2·7±0·06	0·16	2·5–3·0
	B–L	8	2·6±0·08	0·23	2·2–2·9
M^1	M–D	9	3·7±0·05	0·16	3·5–4·0
	Ant B	9	3·4±0·06	0·16	3·2–3·7
	Post B	9	3·2±0·04	0·11	3·0–3·3
M^2	M–D	9	3·4±0·04	0·11	3·2–3·5
	Ant B	9	3·6±0·09	0·26	3·2–4·0
	Post B	8	3·3±0·07	0·20	3·1–3·7
M^3	M–D	5	2·6±0·07	0·15	2·4–2·8
	Ant B	6	2·9±0·12	0·30	2·6–3·4
	Post B				

Table 16
Tupaia tana female

Mandibular teeth		n	$\bar{X}\pm$s.e.	s.d.	Range
I_1	M–D				
	B–L				
I_2	M–D				
	B–L				
C	M–D				
	B–L				
P_2	M–D				
	B–L				
P_3	M–D	1	2·2±0·00	0·00	2·2–2·2
	B–L	1	1·4±0·00	0·00	1·4–1·4
P_4	M–D	9	2·6±0·07	0·20	2·4–2·9
	B–L	9	1·7±0·08	0·25	1·4–2·0
M_1	M–D	9	3·6±0·08	0·24	3·2–4·0
	Tri B	9	2·1±0·09	0·26	1·8–2·6
	Tal B	9	2·3±0·09	0·28	2·0–2·8
M_2	M–D	9	3·5±0·07	0·21	3·2–3·8
	Tri B	9	2·2±0·08	0·24	1·9–2·6
	Tal B	9	2·3±0·08	0·23	1·9–2·6
M_3	M–D	7	3·0±0·06	0·17	2·8–3·3
	Tri B	6	1·9±0·10	0·25	1·6–2·2
	Tal B	7	1·5±0·05	0·13	1·4–1·8
	Hypo-conulid				

Table 17
Lemur variegatus male

Maxillary teeth		n	$\bar{X}$ ± s.e.	s.d.	Range
I^1	M–D	3	2·1 ± 0·07	0·12	2·0–2·2
	B–L	2	1·5 ± 0·00	0·00	1·5–1·5
I^2	M–D	3	2·1 ± 0·17	0·29	1·8–2·3
	B–L	2	1·5 ± 0·05	0·07	1·4–1·5
C	M–D	5	6·3 ± 1·04	0·33	4·2–7·7
	B–L	5	3·0 ± 0·10	0·23	2·7–3·3
P^2	M–D	4	4·9 ± 0·17	0·34	4·6–5·3
	B–L	4	2·8 ± 0·19	0·38	2·2–3·0
P^3	M–D	5	6·7 ± 0·16	0·35	6·3–7·2
	B–L	4	4·6 ± 0·25	0·49	3·9–5·0
P^4	M–D	5	6·3 ± 0·12	0·27	6·0–6·6
	B–L	4	7·0 ± 0·17	0·33	6·5–7·3
M^1	M–D	6	7·3 ± 0·19	0·46	6·5–7·8
	Ant B	6	7·6 ± 0·62	1·53	4·5–8·4
	Post B	6	7·8 ± 0·56	1·38	5·0–8·5
M^2	M–D	5	7·1 ± 0·08	0·18	6·8–7·2
	Ant B	5	8·0 ± 0·09	0·20	7·7–8·2
	Post B	5	7·8 ± 0·16	0·36	7·4–8·3
M^3	M–D	6	5·3 ± 0·23	0·56	4·6–6·2
	Ant B	5	5·6 ± 0·10	0·21	5·3–5·8
	Post B	5	5·0 ± 0·34	0·76	3·7–5·6

Table 18
Lemur variegatus male

Mandibular teeth		n	$\bar{X}$ ± s.e.	s.d.	Range
I_1	M–D				
	B–L				
I_2	M–D				
	B–L				
C	M–D	3	2·1 ± 0·07	0·12	2·0–2·2
	B–L	3	2·8 ± 0·12	0·21	2·6–3·0
P_2	M–D	2	6·9 ± 0·80	1·13	6·1–7·7
	B–L	3	2·9 ± 0·17	0·29	2·7–3·2
P_3	M–D	4	5·4 ± 0·17	0·35	4·9–5·7
	B–L	5	3·1 ± 0·11	0·26	2·9–3·5
P_4	M–D	4	6·7 ± 0·13	0·25	6·4–7·0
	B–L	4	4·1 ± 0·12	0·25	3·9–4·4
M_1	M–D	5	7·8 ± 0·07	0·16	7·6–8·0
	Tri B	5	4·6 ± 0·14	0·32	4·2–5·0
	Tal B	4	5·0 ± 0·09	0·17	4·8–5·2
M_2	M–D	4	7·0 ± 0·07	0·14	6·8–7·1
	Tri B	4	4·7 ± 0·17	0·34	4·3–5·0
	Tal B	3	4·7 ± 0·03	0·06	4·6–4·7
M_3	M–D	4	5·5 ± 0·19	0·39	5·0–5·9
	Tri B	2	3·7 ± 0·15	0·21	3·5–3·8
	Tal B	2	3·6 ± 0·25	0·35	3·3–3·8
	Hypo-conulid				

Lemur variegatus female measurements not available

Table 19
Lepilemur mustelinus male

Maxillary teeth		n	$\bar{X}\pm$s.e.	s.d.	Range	P
I^1	M–D					
	B–L					
I^2	M–D					
	B–L					
C	M–D	6	3·13±0·13	0·32	2·6–3·5	
	B–L	6	1·62±0·11	0·28	1·3–2·0	
P^2	M–D	6	3·00±0·07	0·17	2·8–3·2	
	B–L	6	1·62±0·09	0·21	1·4–2·0	
P^3	M–D	6	2·85±0·10	0·25	2·6–3·2	
	B–L	6	2·17±0·08	0·19	2·0–2·4	
P^4	M–D	6	2·65±0·11	0·26	2·3–3·0	
	B–L	6	2·73±0·03	0·08	2·6–2·8	
M^1	M–D	6	3·50±0·11	0·27	3·2–4·0	
	Ant B	6	3·27±0·08	0·20	3·0–3·6	*
	Post B					
M^2	M–D	6	3·40±0·11	0·28	3·1–3·9	
	Ant B	6	3·42±0·06	0·15	3·2–3·6	*
	Post B					
M^3	M–D	6	2·67±0·06	0·14	2·5–2·9	
	Ant B	6	3·05±0·09	0·21	2·7–3·3	
	Post B					

* $P<0·05$

Table 20
Lepilemur mustelinus male

Mandibular teeth		n	$\bar{X}\pm$s.e.	s.d.	Range
I_1	M–D				
	B–L				
I_2	M–D				
	B–L				
C	M–D				
	B–L				
P_2	M–D	4	2·70±0·07	0·14	2·5–2·8
	B–L	4	1·75±0·07	0·13	1·6–1·9
P_3	M–D	6	2·93±0·09	0·23	2·7–3·2
	B–L	6	1·63±0·08	0·20	1·4–1·9
P_4	M–D	6	2·75±0·10	0·24	2·4–3·0
	B–L	6	1·78±0·15	0·38	1·1–2·2
M_1	M–D	6	3·50±0·08	0·19	3·3–3·7
	Tri B	6	2·25±0·07	0·16	2·0–2·5
	Tal B				
M_2	M–D	6	3·58±0·07	0·16	3·4–3·8
	Tri B	6	2·38±0·09	0·21	2·0–2·6
	Tal B				
M_3	M–D	6	3·97±0·06	0·14	3·8–4·2
	Tri B	6	2·12±0·07	0·16	1·9–2·3
	Tal B				
	Hypoconulid				

Table 21
Lepilemur mustelinus female

Maxillary teeth		n	$\bar{X} \pm$ s.e.	s.d.	Range
I^1	M–D				
	B–L				
I^2	M–D				
	B–L				
C	M–D	10	3·05 ± 0·13	0·40	2·3–3·7
	B–L	10	1·78 ± 0·10	0·32	1·4–2·3
P^2	M–D	11	3·13 ± 0·15	0·50	2·5–4·0
	B–L	11	1·67 ± 0·06	0·21	1·2–2·0
P^3	M–D	11	2·91 ± 0·03	0·11	2·7–3·0
	B–L	11	2·29 ± 0·04	0·13	2·0–2·5
P^4	M–D	11	2·65 ± 0·07	0·22	2·3–3·0
	B–L	11	2·78 ± 0·07	0·22	2·3–3·1
M^1	M–D	11	3·40 ± 0·06	0·18	3·2–3·7
	Ant B	11	3·46 ± 0·05	0·16	3·2–3·7
	Post B				
M^2	M–D	11	3·29 ± 0·06	0·18	3·0–3·5
	Ant B	11	3·63 ± 0·06	0·20	3·4–4·0
	Post B				
M^3	M–D	11	2·75 ± 0·06	0·20	2·4–3·0
	Ant B	11	3·11 ± 0·06	0·18	2·7–3·4
	Post B				

Table 22
Lepilemur mustelinus female

Mandibular teeth		n	$\bar{X} \pm$ s.e.	s.d.	Range
I_1	M–D				
	B–L				
I_2	M–D				
	B–L				
C	M–D				
	B–L				
P_2	M–D	9	2·80 ± 0·13	0·39	2·2–3·6
	B–L	9	1·87 ± 0·12	0·36	1·5–2·5
P_3	M–D	11	2·71 ± 0·09	0·30	2·3–3·2
	B–L	11	1·63 ± 0·05	0·17	1·3–1·9
P_4	M–D	11	2·58 ± 0·09	0·29	2·2–3·1
	B–L	11	2·00 ± 0·05	0·16	1·6–2·2
M_1	M–D	11	3·56 ± 0·06	0·21	3·2–3·8
	Tri B	11	2·36 ± 0·05	0·17	2·0–2·6
	Tal B				
M_2	M–D	11	3·47 ± 0·04	0·14	3·2–3·7
	Tri B	11	2·41 ± 0·05	0·17	2·2–2·7
	Tal B				
M_3	M–D	11	3·95 ± 0·04	0·14	3·6–4·1
	Tri B	11	2·13 ± 0·05	0·17	2·0–2·5
	Tal B				
	Hypoconulid				

Table 23
Propithecus verreauxi male

Maxillary teeth		n	$\bar{X} \pm$ s.e.	s.d.	Range	P
I^1	M–D	3	4·23 ± 0·09	0·15	4·0–4·4	
	B–L	3	2·87 ± 0·03	0·06	2·8–3·0	
I^2	M–D	3	3·00 ± 0·12	0·20	2·7–3·7	
	B–L	3	2·10 ± 0·21	0·36	1·6–2·5	
C	M–D	2	6·0 ± 0·00	0·00	6·0–6·0	
	B–L	2	3·80 ± 0·30	0·42	3·3–4·3	
P^3	M–D	3	6·30 ± 0·15	0·27	5·6–6·6	
	B–L	3	3·77 ± 0·09	0·15	3·6–4·0	
P^4	M–D	3	5·83 ± 0·07	0·12	5·0–6·0	*
	B–L	3	4·77 ± 0·03	0·06	4·6–4·8	*
M^1	M–D	3	7·13 ± 0·19	0·32	6·7–7·5	
	Ant B	3	6·67 ± 0·09	0·15	6·4–6·9	*
	Post B	3	6·60 ± 0·15	0·27	6·3–6·9	†
M^2	M–D	3	7·43 ± 0·12	0·21	6·7–7·7	†
	Ant B	3	7·43 ± 0·09	0·15	7·2–7·6	†
	Post B	2	7·00 ± 0·10	0·14	6·8–7·2	†
M^3	M–D	3	5·80 ± 0·15	0·27	5·0–6·1	†
	Ant B	3	6·10 ± 0·10	0·17	5·9–6·3	†
	Post B	2	4·40 ± 0·30	0·42	3.9–4·9	†

* $P<0·05$
† $P<0·01$

Table 24
Propithecus verreauxi male

Mandibular teeth		n	$\bar{X} \pm$ s.e.	s.d.	Range	P
I_1	M–D	3	1·47 ± 0·03	0·06	1·3–1·5	
	B–L	2	2·75 ± 0·05	0·07	2·7–2·8	
I_2	M–D	3	2·77 ± 0·15	0·25	2·5–3·6	
	B–L	3	3·53 ± 0·15	0·25	3·2–3·8	
C	M–D					
	B–L					
P_3	M–D	3	6·03 ± 0·41	0·71	5·2–6·8	
	B–L	3	3·13 ± 0·12	0·21	2·9–3·3	
P_4	M–D	3	5·40 ± 0·27	0·46	4·9–5·9	
	B–L	3	2·87 ± 0·12	0·21	2·5–3·2	
M_1	M–D	3	7·00 ± 0·06	0·10	6·7–7·2	
	Tri B	3	4·10 ± 0·15	0·27	3·8–4·4	
	Tal B	3	4·97 ± 0·03	0·06	4·9–5·1	*
M_2	M–D	3	7·20 ± 0·10	0·17	6·9–7·4	†
	Tri B	3	4·97 ± 0·03	0·06	4·9–5·0	†
	Tal B	3	5·43 ± 0·03	0·06	5·3–5·5	
M_3	M–D	2	7·30 ± 0·10	0·14	6·9–7·7	*
	Tri B	2	5·30 ± 0·10	0·14	5·1–5·5	*
	Tal B	3	4·67 ± 0·13	0·23	4·3–5·0	
	Hypo-conulid					

* $P<0·05$
† $P<0·01$

Table 25
Propithecus verreauxi female

Maxillary teeth		n	$\bar{X} \pm$ s.e.	s.d.	Range
I^1	M–D	5	3·66±0·19	0·43	3·0–4·1
	B–L	5	2·62±0·12	0·27	2·2–2·9
I^2	M–D	5	2·72±0·09	0·19	2·5–3·0
	B–L	5	1·86±0·16	0·35	1·5–2·5
C	M–D	7	5·77±0·19	0·51	5·2–6·5
	B–L	7	3·57±0·15	0·39	3·1–4·3
P^3	M–D	7	6·27±0·19	0·50	5·7–7·0
	B–L	7	3·56±0·10	0·27	3·2–4·0
P^4	M–D	7	5·29±0·13	0·35	4·8–5·9
	B–L	7	4·09±0·14	0·36	3·5–4·8
M^1	M–D	7	6·70±0·12	0·31	6·1–7·5
	Ant B	6	5·83±0·17	0·41	5·4–6·5
	Post B	7	5·87±0·11	0·30	5·5–6·9
M^2	M–D	7	6·43±0·14	0·36	5·9–7·5
	Ant B	7	6·33±0·18	0·48	5·6–7·6
	Post B	6	6·10±0·10	0·24	5·8–7·1
M^3	M–D	7	4·44±0·13	0·34	4·0–5·5
	Ant B	6	4·70±0·19	0·37	4·2–6·0
	Post B	5	3·14±0·15	0·34	2·8–4·9

Table 26
Propithecus verreauxi female

Mandibular teeth		n	$\bar{X} \pm$ s.e.	s.d.	Range
I_1	M–D	5	1·38±0·06	0·13	1·2–1·6
	B–L	7	2·43±0·10	0·26	2·1–2·8
I_2	M–D	6	2·62±0·21	0·52	2·1–3·2
	B–L	8	3·21±0·12	0·33	2·7–3·8
C	M–D				
	B–L				
P_3	M–D	8	6·50±0·16	0·44	5·0–7·0
	B–L	8	2·81±0·08	0·23	2·6–3·3
P_4	M–D	8	5·58±0·20	0·57	4·7–6·4
	B–L	8	2·80±0·09	0·25	2·3–3·1
M_1	M–D	8	6·69±0·13	0·35	6·1–7·3
	Tri B	8	3·83±0·10	0·29	3·4–4·3
	Tal B	7	4·66±0·07	0·18	4·3–5·0
M_2	M–D	8	6·44±0·13	0·38	6·0–7·3
	Tri B	8	4·39±0·11	0·30	4·0–4·7
	Tal B	8	4·69±0·07	0·19	4·5–5·4
M_3	M–D	7	6·47±0·14	0·37	5·9–7·4
	Tri B	7	4·50±0·15	0·39	4·1–5·2
	Tal B	7	4·16±0·16	0·41	3·7–4·8
	Hypo-conulid	2	1·90±0·10	0·14	1·8–2·0

Table 27
Nycticebus coucang male

Maxillary teeth		n	$\bar{X}\pm$s.e.	s.d.	Range	P
I^1	M–D	7	1·4±0·06	0·15	1·2–1·6	
	B–L	9	1·4+0·05	0·14	1·2–1·6	
I^2	M–D	3	1·2±0·17	0·29	0·9–1·4	
	B–L	2	0·9±0·10	0·14	0·8–1·0	
C	M–D	18	3·6±0·08	0·34	3·2–4·3	*
	B–L	18	2·7±0·06	0·26	2·4–3·2	
P^2	M–D	20	3·1±0·08	0·35	2·0–3·6	
	B–L	21	2·4±0·05	0·23	2·0–2·9	
P^3	M–D	20	2·3±0·07	0·31	1·6–2·9	
	B–L	18	2·6±0·09	0·38	1·6–3·0	
P^4	M–D	20	2·4±0·08	0·36	1·9–3·2	
	B–L	19	3·1±0·11	0·50	1·8–4·0	
M^1	M–D	21	3·5±0·09	0·40	2·9–4·5	
	Ant B	17	4·3±0·12	0·48	3·6–5·0	
	Post B	17	4·4±0·13	0·54	3·6–5·6	
M^2	M–D	21	3·2±0·07	0·32	2·6–3·8	
	Ant B	19	4·2±0·10	0·44	3·5–5·0	
	Post B	19	4·1±0·11	0·48	3·3–5·2	
M^3	M–D	18	2·5±0·08	0·33	1·8–2·9	
	Ant B	18	3·5±0·08	0·35	2·9–4·3	
	Post B	2	3·8±0·75	1·06	3·0–4·5	

* $P<0{\cdot}05$

Table 28
Nycticebus coucang male

Mandibular teeth		n	$\bar{X}\pm$s.e.	s.d.	Range
I_1	M–D				
	B–L	14	1·6±0·06	0·21	1·2–2·0
I_2	M–D				
	B–L	14	1·6±0·06	0·21	1·2–2·0
C	M–D	16	1·4±0·06	0·23	1·0–1·8
	B–L	14	2·3±0·08	0·29	1·9–2·9
P_2	M–D	20	3·5±0·11	0·50	1·8–4·2
	B–L	20	3·1±0·09	0·39	2·0–3·8
P_3	M–D	17	2·2±0·07	0·29	1·8–2·8
	B–L	17	2·3±0·08	0·31	2·0–3·1
P_4	M–D	18	2·4±0·09	0·38	1·7–3·0
	B–L	17	2·5±0·08	0·31	2·0–3·0
M_1	M–D	21	3·4±0·07	0·33	3·0–4·1
	Tri B	20	2·7±0·07	0·29	2·3–3·4
	Tal B	19	2·9±0·06	0·25	2·5–3·5
M_2	M–D	20	3·4±0·06	0·28	2·9–3·9
	Tri B	19	2·8±0·07	0·30	2·4–3·5
	Tal B	18	2·8±0·06	0·27	2·3–3·3
M_3	M–D	16	3·4±0·11	0·44	2·5–4·1
	Tri B	17	2·5±0·07	0·27	1·9–2·9
	Tal B	14	2·3±0·05	0·19	1·9–2·6
	Hypo-conulid	13	1·3±0·05	0·18	1·0–1·6

Table 29
Nycticebus coucang female

Maxillary teeth		n	$\bar{X}\pm$s.e.	s.d.	Range
I^1	M–D	6	1·5±0·08	0·21	1·2–1·8
	B–L	6	1·4±0·06	0·14	1·2–1·6
I^2	M–D	3	1·2±0·17	0·29	0·9–1·4
	B–L	2	0·9±0·10	0·14	0·8–1·0
C	M–D	7	3·3±0·05	0·14	3·2–3·5
	B–L	8	2·5±0·09	0·26	2·2–3·0
P^2	M–D	9	3·0±0·14	0·42	2·3–3·6
	B–L	9	2·3±0·11	0·32	1·6–2·8
P^3	M–D	9	2·2±0·14	0·42	1·7–3·0
	B–L	9	2·6±0·11	0·33	1·9–3·0
P^4	M–D	9	2·5±0·17	0·52	2·0–3·8
	B–L	9	3·1±0·14	0·42	2·4–3·6
M^1	M–D	9	3·5±0·08	0·25	3·2–4·0
	Ant B	8	4·2±0·11	0·32	3·7–4·6
	Post B	8	4·4±0·10	0·29	4·0–4·7
M^2	M–D	9	3·2±0·08	0·25	2·9–3·6
	Ant B	9	4·1±0·08	0·23	3·7–4·4
	Post B	9	4·2±0·09	0·25	3·8–4·6
M^3	M–D	9	2·4±0·06	0·18	2·3–2·8
	Ant B	9	3·5±0·05	0·15	3·3–3·7
	Post B	2	3·0±0·05	0·07	2·9–3·0

Table 30
Nycticebus coucang female

Mandibular teeth		n	$\bar{X}\pm$s.e.	s.d.	Range
I_1	M–D				
	B–L	8	1·6±0·05	0·15	1·4–1·9
I_2	M–D				
	B–L	8	1·7±0·07	0·18	1·5-2·0
C	M–D	8	1·3±0·04	0·11	1·2–1·5
	B–L	6	2·4±0·13	0·32	2·1–3·0
P_2	M–D	8	3·2±0·17	0·49	2·2–3·6
	B–L	8	2·9±0·12	0·34	2·3–3·3
P_3	M–D	8	2·0±0·12	0·35	1·3–2·5
	B–L	9	2·3±0·14	0·41	1·7–2·9
P_4	M–D	9	2·2±0·12	0·35	1·8–2·7
	B–L	9	2·4±0·08	0·25	1·9–2·8
M_1	M–D	9	3·4±0·09	0·28	2·8–3·8
	Tri B	8	2·7±0·09	0·25	2·3–2·9
	Tal B	8	2·9±0·08	0·23	2·5–3·2
M_2	M–D	8	3·3±0·09	0·25	2·9–3·6
	Tri B	6	2·8±0·07	0·17	2·6–3·0
	Tal B	7	2·8±0·07	0·17	2·6–3·0
M_3	M–D	8	3·5±0·09	0·26	3·1–3·9
	Tri B	7	2·3±0·09	0·23	2·0–2·6
	Tal B	7	2·2±0·09	0·23	2·0–2·6
	Hypoconulid	8	1·4±0·05	0·14	1·2–1·6

Table 31
Perodicticus potto male

Maxillary teeth		n	$\bar{X} \pm$ s.e.	s.d.	Range	P
I^1	M–D					
	B–L					
I^2	M–D					
	B–L					
C	M–D	8	4·3 ± 0·15	0·43	3·5–4·9	
	B–L	8	3·1 ± 0·11	0·31	2·5–3·5	*
P^2	M–D	11	2·8 ± 0·05	0·17	2·6–3·2	
	B–L	9	2·5 ± 0·03	0·09	2·3–2·6	
P^3	M–D	8	2·5 ± 0·11	0·30	2·2–3·0	
	B–L	6	2·3 ± 0·10	0·25	2·0–2·7	
P^4	M–D	8	2·3 ± 0·08	0·24	2·0–2·7	
	B–L	7	2·3 ± 0·10	0·26	3·0–3·8	
M^1	M–D	11	3·5 ± 0·08	0·27	2·8–3·8	
	Ant B	10	4·0 ± 0·11	0·34	3·5–4·7	
	Post B					
M^2	M–D	10	3·3 ± 0·05	0·15	3·1–3·5	
	Ant B	8	4·3 ± 0·10	0·28	3·7–4·6	
	Post B					
M^3	M–D	6	2·3 ± 0·13	0·31	1·7–2·5	
	Ant B	6	3·5 ± 0·17	0·42	2·9–4·1	
	Post B					

* $P < 0{\cdot}05$

Table 32
Perodicticus potto male

Mandibular teeth		n	$\bar{X} \pm$ s.e.	s.d.	Range
I_1	M–D				
	B–L				
I_2	M–D				
	B–L				
C	M–D				
	B–L				
P_2	M–D	10	2·9 ± 0·12	0·38	2·3–3·7
	B–L	10	2·7 ± 0·09	0·28	2·3–3·2
P_3	M–D	7	2·3 ± 0·11	0·29	1·7–2·6
	B–L	7	2·4 ± 0·04	0·10	2·3–2·5
P_4	M–D	9	2·3 ± 0·06	0·18	2·0–2·5
	B–L	8	2·5 ± 0·07	0·19	2·2–2·7
M_1	M–D	12	3·2 ± 0·08	0·29	2·8–3·6
	Tri B	10	2·7 ± 0·10	0·32	2·2–3·3
	Tal B	10	2·6 ± 0·08	0·26	2·4–3·2
M_2	M–D	12	3·3 ± 0·09	0·29	2·9–4·0
	Tri B	11	2·9 ± 0·09	0·31	2·4–3·7
	Tal B	11	2·8 ± 0·10	0·32	2·5–3·7
M_3	M–D	8	3·0 ± 0·09	0·26	2·6–3·4
	Tri B	8	2·5 ± 0·13	0·36	2·0–3·1
	Tal B	5	2·2 ± 0·08	0·18	2·0–2·5
	Hypoconulid				

Table 33
Perodicticus potto female

Maxillary teeth		n	$\bar{X} \pm$ s.e.	s.d.	Range
I^1	M–D				
	B–L				
I^2	M–D				
	B–L				
C	M–D	9	4·1 ± 0·14	0·42	3·5–4·8
	B–L	10	2·7 ± 0·09	0·28	2·2–3·2
P^2	M–D	9	2·9 ± 0·10	0·31	2·4–3·3
	B–L	9	2·5 ± 0·09	0·27	2·2–3·0
P^3	M–D	9	2·4 ± 0·07	0·21	2·0–2·7
	B–L	8	2·2 ± 0·07	0·20	1·9–2·4
P^4	M–D	8	2·4 ± 0·02	0·05	2·3–2·5
	B–L	9	3·3 ± 0·11	0·33	2·9–3·7
M^1	M–D	7	3·4 ± 0·07	0·19	3·2–3·7
	Ant B	8	3·8 ± 0·10	0·28	3·6–4·3
	Post B				
M^2	M–D	7	3·4 ± 0·06	0·17	3·1–3·6
	Ant B	8	4·4 ± 0·12	0·33	3·9–4·9
	Post B				
M^3	M–D	6	2·0 ± 0·12	0·30	1·7–2·4
	Ant B	8	3·4 ± 0·08	0·23	2·9–3·7
	Post B				

Table 34
Perodicticus potto female

Mandibular teeth		n	$\bar{X} \pm$ s.e.	s.d.	Range
I_1	M–D				
	B–L				
I_2	M–D				
	B–L				
C	M–D				
	B–L				
P_2	M–D	8	2·8 ± 0·10	0·28	2·4–3·2
	B–L	9	2·9 ± 0·11	0·32	2·4–3·5
P_3	M–D	9	2·2 ± 0·09	0·26	1·9–2·7
	B–L	10	2·3 ± 0·06	0·18	2·0–2·6
P_4	M–D	8	2·4 ± 0·06	0·17	2·0–2·5
	B–L	10	2·4 ± 0·06	0·20	2·1–2·7
M_1	M–D	7	3·1 ± 0·14	0·38	2·4–3·4
	Tri B	10	2·5 ± 0·08	0·26	2·0–2·8
	Tal B	10	2·4 ± 0·07	0·23	2·0–2·7
M_2	M–D	9	3·4 ± 0·05	0·16	3·2–3·7
	Tri B	10	2·8 ± 0·09	0·29	2·5–3·2
	Tal B	10	2·7 ± 0·08	0·26	2·2–3·0
M_3	M–D	4	3·3 ± 0·10	0·19	3·0–3·4
	Tri B	5	2·3 ± 0·10	0·23	2·0–2·6
	Tal B	5	2·1 ± 0·06	0·12	2·0–2·3
	Hypoconulid				

Table 35
Galago crassicaudatus male

Maxillary teeth		n	$\bar{X}\pm$s.e.	s.d.	Range	P
I^1	M–D					
	B–L					
I^2	M–D					
	B–L					
C	M–D	15	4·6±0·15	0·56	3·9–5·7	*
	B–L	13	3·0±0·09	0·32	2·5–3·8	*
P^2	M–D	18	3·9±0·13	0·55	3·2–5·7	
	B–L	19	2·4±0·09	0·37	1·9–3·5	
P^3	M–D	20	3·1±0·08	0·37	2·5–3·7	
	B–L	20	2·7±0·08	0·37	2·1–3·7	
P^4	M–D	19	3·6±0·06	0·25	3·2–4·0	
	B–L	19	4·1±0·10	0·45	3·0–5·0	
M^1	M–D	19	4·2±0·06	0·26	3·7–4·5	*
	Ant B	18	5·0±0·11	0·47	3·6–5·6	
	Post B	18	5·3±0·09	0·37	4·3–5·8	
M^2	M–D	18	3·9±0·04	0·17	3·6–4·3	
	Ant B	17	5·4±0·07	0·28	4·8–5·8	
	Post B	17	5·1±0·07	0·28	4·7–5·7	
M^3	M–D	15	3·0±0·07	0·28	2·3–3·3	
	Ant B	13	4·2±0·09	0·32	3·7–4·7	
	Post B	5	3·6±0·23	0·52	2·7–4·0	

* $P<0·05$

Table 36
Galago crassicaudatus male

Mandibular teeth		n	$\bar{X}\pm$s.e.	s.d.	Range
I_1	M–D				
	B–L	10	1·9±0·08	0·26	1·5–2·2
I_2	M–D				
	B–L	10	1·9±0·08	0·26	1·5–2·2
C	M–D	10	1·2±0·04	0·14	0·9–1·3
	B–L	10	2·4±0·08	0·26	1·9–2·6
P_2	M–D	17	3·4±0·11	0·46	2·2–4·0
	B–L	17	2·7±0·08	0·31	2·3–3·3
P_3	M–D	18	3·5±0·16	0·66	1·6–4·3
	B–L	18	2·4±0·07	0·28	1·7–2·9
P_4	M–D	19	3·4±0·10	0·45	1·8–3·8
	B–L	17	2·8±0·09	0·35	2·2–3·8
M_1	M–D	18	3·9±0·04	0·18	3·5–4·2
	Tri B	18	3·3±0·04	0·17	3·0–3·7
	Tal B	17	3·3±0·05	0·21	2·9–3·6
M_2	M–D	19	3·9±0·07	0·29	3·5–5·0
	Tri B	19	3·4±0·04	0·19	2·9–3·7
	Tal B	18	3·2±0·05	0·20	2·7–3·4
M_3	M–D	18	4·3±0·07	0·31	3·7–4·8
	Tri B	18	2·8±0·06	0·23	2·4–3·3
	Tal B	18	2·4±0·04	0·19	2·1–2·8
	Hypo-conulid	15	1·4±0·06	0·25	0·9–1·8

Table 37
Galago crassicaudatus female

Maxillary teeth		n	$\bar{X} \pm$ s.e.	s.d.	Range
I^1	M–D				
	B–L				
I^2	M–D				
	B–L				
C	M–D	7	4·1 ± 0·19	0·51	3·1–4·7
	B–L	6	2·9 ± 0·15	0·36	2·1–3·2
P^2	M–D	8	4·0 ± 0·08	0·23	3·6–4·3
	B–L	8	2·5 ± 0·10	0·27	1·9–2·8
P^3	M–D	8	3·1 ± 0·11	0·30	2·5–3·5
	B–L	8	2·6 ± 0·11	0·31	2·1–3·0
P^4	M–D	10	3·7 ± 0·07	0·23	3·3–4·2
	B–L	11	4·3 ± 0·06	0·19	3·9–4·6
M^1	M–D	11	4·4 ± 0·07	0·23	3·8–4·6
	Ant B	11	4·9 ± 0·08	0·25	4·6–5·5
	Post B	11	5·3 ± 0·06	0·20	5·0–5·6
M^2	M–D	11	4·1 ± 0·07	0·24	3·8–4·7
	Ant B	10	5·3 ± 0·07	0·23	4·9–5·5
	Post B	10	5·0 ± 0·07	0·22	4·7–5·3
M^3	M–D	10	3·2 ± 0·04	0·14	3·0–3·4
	Ant B	10	4·3 ± 0·03	0·10	4·2–4·5
	Post B	6	2·8 ± 0·15	0·37	2·1–3·2

Table 38
Galago crassicaudatus female

Mandibular teeth		n	$\bar{X} \pm$ s.e.	s.d.	Range
I_1	M–D				
	B–L	6	1·8 ± 0·04	0·10	1·8–2·0
I_2	M–D				
	B–L	6	1·9 ± 0·04	0·10	1·8–2·0
C	M–D	6	1·3 ± 0·06	0·15	1·0–1·4
	B–L	6	2·4 ± 0·09	0·22	2·0–2·6
P_2	M–D	8	3·3 ± 0·06	0·17	3·0–3·5
	B–L	8	2·7 ± 0·15	0·41	1·7–3·0
P_3	M–D	7	3·6 ± 0·08	0·20	3·3–3·9
	B–L	6	2·6 ± 0·06	0·15	2·4–2·8
P_4	M–D	8	3·6 ± 0·07	0·19	3·3–3·9
	B–L	8	2·7 ± 0·06	0·18	2·3–2·8
M_1	M–D	10	4·0 ± 0·07	0·21	3·7–4·3
	Tri B	10	3·2 ± 0·06	0·18	3·0–3·5
	Tal B	11	3·2 ± 0·05	0·18	2·9–3·5
M_2	M–D	11	3·9 ± 0·05	0·18	3·5–4·2
	Tri B	11	3·3 ± 0·05	0·16	3·0–3·6
	Tal B	10	3·2 ± 0·03	0·08	3·0–3·3
M_3	M–D	11	4·3 ± 0·05	0·15	4·2–4·7
	Tri B	10	2·9 ± 0·04	0·12	2·7–3·0
	Tal B	10	2·5 ± 0·02	0·07	2·4–2·6
	Hypo-conulid	10	1·4 ± 0·07	0·22	0·9–1·7

Table 39
Galago senegalensis male

Maxillary teeth		n	$\bar{X}\pm$s.e.	s.d.	Range	P
I^1	M–D					
	B–L					
I^2	M–D					
	B–L					
C	M–D	16	2·3±0·05	0·19	2·0–2·6	*
	B–L	17	1·6±0·08	0·32	1·4–2·6	*
P^2	M–D	20	2·2±0·04	0·16	1·9–2·5	*
	B–L	21	1·2±0·03	0·12	1·0–1·4	
P^3	M–D	20	1·8±0·04	0·17	1·5–2·1	
	B–L	21	1·7±0·07	0·31	1·2–2·6	
P^4	M–D	22	2·2±0·04	0·17	1·9–2·5	
	B–L	22	2·6±0·05	0·25	2·2–3·5	
M^1	M–D	24	2·5±0·03	0·14	2·1–2·8	
	Ant B	23	3·0±0·04	0·17	2·6–3·2	
	Post B	23	3·3±0·04	0·21	2·7–3·7	
M^2	M–D	23	2·4±0·03	0·12	2·2–2·6	
	Ant B	22	3·3±0·03	0·16	3·0–3·6	
	Post B	19	3·4±0·04	0·16	3·1–3·7	
M^3	M–D	19	1·9±0·05	0·23	1·6–2·7	
	Ant B	19	2·7±0·05	0·21	2·2–3·2	
	Post B					

* $P<0·05$

Table 40
Galago senegalensis male

Mandibular teeth		n	$\bar{X}\pm$s.e.	s.d.	Range	P
I_1	M–D					
	B–L					
I_2	M–D					
	B–L					
C	M–D					
	B–L					
P_2	M–D	10	1·8±0·08	0·24	1·4–2·2	*
	B–L	10	1·4±0·04	0·14	1·2–1·6	*
P_3	M–D	21	2·1±0·07	0·30	1·5–2·6	
	B–L	17	1·3±0·05	0·19	1·0–1·6	
P_4	M–D	21	2·1±0·04	0·16	1·8–2·4	
	B–L	21	1·6±0·04	0·20	1·4–2·3	
M_1	M–D	24	2·3±0·03	0·17	2·0–2·7	
	Tri B	23	1·9±0·03	0·15	1·6–2·4	
	Tal B	23	1·9±0·03	0·14	1·7–2·5	
M_2	M–D	23	2·3±0·03	0·13	2·1–2·7	
	Tri B	23	2·0±0·03	0·14	1·8–2·5	
	Tal B	22	1·9±0·06	0·26	1·0–2·5	
M_3	M–D	20	2·7±0·06	0·27	2·4–3·7	
	Tri B	21	1·8±0·04	0·17	1·6–2·4	
	Tal B	20	1·5±0·05	0·21	1·3–2·3	
	Hypo-conulid					

* $P<0·05$

Table 41
Galago senegalensis female

Maxillary teeth		n	$\bar{X}\pm$s.e.	s.d.	Range
I^1	M–D				
	B–L				
I^2	M–D				
	B–L				
C	M–D	13	2·2±0·04	0·16	1·9–2·4
	B–L	13	1·4±0·03	0·12	1·2–1·5
P^2	M–D	18	2·0±0·04	0·15	1·8–2·3
	B–L	18	1·2±0·04	0·16	0·9–1·5
P^3	M–D	18	1·9±0·04	0·15	1·6–2·2
	B–L	18	1·6±0·05	0·20	1·2–2·0
P^4	M–D	18	2·2±0·05	0·20	1·7–2·6
	B–L	18	2·5±0·03	0·13	2·3–2·8
M^1	M–D	18	2·5±0·03	0·12	2·3–2·8
	Ant B	17	3·0±0·06	0·24	2·6–3·5
	Post B	18	3·3±0·05	0·22	3·0–3·7
M^2	M–D	19	2·3±0·03	0·12	2·2–2·6
	Ant B	19	3·2±0·06	0·24	2·9–3·8
	Post B	19	3·4±0·04	0·17	3·0–3·7
M^3	M–D	19	1·8±0·03	0·14	1·5–2·2
	Ant B	18	2·9±0·05	0·22	2·3–3·1
	Post B				

Table 42
Galago senegalensis female

Mandibular teeth		n	$\bar{X}\pm$s.e.	s.d.	Range
I_1	M–D				
	B–L				
I_2	M–D				
	B–L				
C	M–D				
	B–L				
P_2	M–D	9	1·7±0·10	0·30	1·2–2·2
	B–L	9	1·3±0·05	0·14	1·0–1·4
P_3	M–D	15	2·0±0·05	0·18	1·7–2·4
	B–L	15	1·3±0·03	0·13	1·0–1·5
P_4	M–D	20	2·0±0·03	0·14	1·8–2·4
	B–L	20	1·6–0·06	0·29	1·2–2·7
M_1	M–D	20	2·2±0·03	0·13	2·0–2·5
	Tri B	20	1·9±0·03	0·12	1·8–2·2
	Tal B	20	2·0±0·03	0·13	1·7–2·3
M_2	M–D	20	2·3±0·03	0·12	2·0–2·5
	Tri B	20	2·0±0·02	0·11	1·8–2·2
	Tal B	20	2·0±0·03	0·12	1·7–2·2
M_3	M–D	20	2·7±0·03	0·13	2·4–2·9
	Tri B	20	1·8±0·02	0·10	1·7–2·8
	Tal B	20	1·5±0·02	0·11	1·3–1·7
	Hypo-conulid				

Table 43
Tarsius (*spectrum*, *bancanus*, *synichtus*, *syrichta*) male

Maxillary teeth		*n*	$\bar{X} \pm$ s.e.	s.d.	Range
I^1	M–D				
	B–L				
I^2	M–D				
	B–L				
C	M–D	5	1·66±0·07	0·15	1·50–1·90
	B–L	5	1·64±0·04	0·09	1·50–1·80
P^2	M–D	6	1·40±0·05	0·11	1·30–1·60
	B–L	6	1·37±0·06	0·14	1·20–1·60
P^3	M–D	5	1·74±0·07	0·15	1·50–1·90
	B–L	6	2·32±0·10	0·25	2·00–2·60
P^4	M–D	4	2·08±0·05	0·10	2·00–2·20
	B–L	6	3·03±0·15	0·37	2·50–3·50
M^1	M–D	6	2·88±0·12	0·30	2·50–3·30
	Ant B	6	3·95±0·17	0·40	3·40–4·50
	Post B				
M^2	M–D	6	2·88±0·11	0·26	2·60–3·30
	Ant B	6	4·12±0·14	0·33	3·70–4·50
	Post B				
M^3	M–D	5	2·50±0·06	0·14	2·40–2·70
	Ant B	5	3·82±0·14	0·32	3·50–4·10
	Post B				

Table 44
Tarsius (*spectrum*, *bancanus*, *synichtus*, *syrichta*) male

Mandibular teeth		*n*	$\bar{X} \pm$ s.e.	s.d.	Range	*P*
I_1	M–D					
	B–L					
I_2	M–D					
	B–L					
C	M–D	5	1·70±0·08	0·17	1·5–2·0	
	B–L	5	1·78±0·07	0·16	1·6–2·0	
P_2	M–D	4	1·48±0·09	0·17	1·3–1·7	
	B–L	5	1·46±0·07	0·15	1·3–1·7	
P_3	M–D	4	1·43±0·03	0·05	1·3–1·5	
	B–L	5	1·58±0·07	0·16	1·4–1·8	
P_4	M–D	4	1·83±0·06	0·13	1·7–2·0	
	B–L	5	1·68±0·25	0·55	1·1–2·3	
M_1	M–D	6	2·97±0·08	0·21	2·7–3·2	
	Tri B	6	2·35±0·06	0·14	2·1–2·5	
	Tal B	6	2·42±0·10	0·23	2·0–2·7	
M_2	M–D	6	2·88±0·08	0·19	2·7–3·2	
	Tri B	6	2·48±0·04	0·10	2·4–2·6	
	Tal B	6	2·55±0·05	0·12	2·4–2·7	
M_3	M–D	6	3·57±0·09	0·23	3·4–4·0	
	Tri B	6	2·45±0·05	0·12	2·3–2·6	*
	Tal B	6	2·37±0·09	0·23	2·2–2·7	
	Hypoconulid	6	1·33±0·03	0·08	1·2–1·4	

* $P<0·05$

Table 45

Tarsius (*spectrum*, *bancanus*, *synichtus*, *syrichta*) female

Maxillary teeth		n	$\bar{X} \pm$ s.e.	s.d.	Range
I^1	M–D				
	B–L				
I^2	M–D				
	B–L				
C	M–D	8	1·58 ± 0·06	0·17	1·3–1·8
	B–L	7	1·56 ± 0·06	0·15	1·4–1·8
P^2	M–D	8	1·39 ± 0·02	0·06	1·3–1·5
	B–L	8	1·38 ± 0·04	0·12	1·3–1·6
P^3	M–D	9	1·74 ± 0·06	0·17	1·5–2·0
	B–L	9	2·17 ± 0·08	0·25	1·8–2·5
P^4	M–D	9	2·10 ± 0·10	0·31	1·7–2·5
	B–L	9	2·86 ± 0·09	0·27	2·5–3·2
M^1	M–D	9	2·80 ± 0·08	0·24	2·5–3·1
	Ant B	9	3·86 ± 0·12	0·36	3·4–4·5
	Post B				
M^2	M–D	9	2·70 ± 0·05	0·16	2·4–2·9
	Ant B	9	4·07 ± 0·08	0·25	3·6–4·5
	Post B				
M^3	M–D	7	2·50 ± 0·08	0·20	2·2–2·7
	Ant B	9	3·79 ± 0·09	0·28	3·4–4·1
	Post B				

Table 46

Tarsius (*spectrum*, *bancanus*, *synichtus*, *syrichta*) female

Mandibular teeth		n	$\bar{X} \pm$ s.e.	s.d.	Range
I_1	M–D				
	B–L				
I_2	M–D				
	B–L				
C	M–D	4	1·75 ± 0·10	0·19	1·5–2·0
	B–L	6	1·67 ± 0·08	0·19	1·2–1·9
P_2	M–D	8	1·36 ± 0·05	0·13	1·1–1·5
	B–L	8	1·35 ± 0·03	0·09	1·2–1·5
P_3	M–D	9	1·50 ± 0·04	0·12	1·4–1·8
	B–L	9	1·49 ± 0·04	0·11	1·4–1·7
P_4	M–D	9	1·70 ± 0·04	0·12	1·5–1·9
	B–L	9	1·70 ± 0·06	0·18	1·5–2·0
M_1	M–D	9	2·81 ± 0·06	0·19	2·6–3·2
	Tri B	9	2·18 ± 0·09	0·27	1·7–2·6
	Tal B	9	2·38 ± 0·08	0·24	2·0–2·8
M_2	M–D	9	2·71 ± 0·05	0·16	2·5–2·9
	Tri B	9	2·40 ± 0·04	0·12	2·2–2·6
	Tal B	9	2·48 ± 0·04	0·12	2·3–2·7
M_3	M–D	9	3·39 ± 0·09	0·26	3·0–3·8
	Tri B	9	2·26 ± 0·05	0·14	2·0–2·4
	Tal B	9	2·26 ± 0·03	0·09	2·1–2·4
	Hypoconulid	9	1·28 ± 0·05	0·14	1·0–1·5

Table 47
Saguinus geoffroyi male

Maxillary teeth		n	$\bar{X} \pm$ s.e.	s.d.	Range	P
I^1	M–D	10	2·4±0·05	0·15	2·2–2·7	
	B L	11	2·0±0·04	0·13	1·8–2·3	
I^2	M–D	6	1·9±0·07	0·18	1·7–2·2	
	B–L	6	1·7±0·03	0·06	1·6–1·8	
C	M–D	10	2·8±0·04	0·13	2·6–3·1	
	B–L	9	2·3±0·03	0·09	2·2–2·5	
P^2	M–D	10	2·1±0·04	0·12	1·9–2·2	
	B–L	10	2·7±0·04	0·13	2·6–3·0	
P^3	M–D	10	1·9±0·03	0·11	1·7–2·0	
	B–L	10	3·0±0·05	0·14	2·9–3·4	
P^4	M–D	10	2·0±0·04	0·14	1·9–2·3	
	B–L	10	3·4±0·05	0·15	3·1–3·6	
M^1	M–D	11	2·8±0·03	0·11	2·6–3·0	*
	Ant B	11	3·6±0·05	0·18	3·3–4·0	
	Post B					
M^2	M–D	10	1·6±0·02	0·07	1·4–1·7	
	Ant B	10	2·5±0·04	0·13	2·3–2·7	†
	Post B					
M^3	M–D					
	Ant B					
	Post B					

* $P<0{\cdot}05$
† $P<0{\cdot}01$

Table 48
Saguinus geoffroyi male

Mandibular teeth		n	$\bar{X} \pm$ s.e.	s.d.	Range	P
I_1	M–D	8	1·6±0·03	0·09	1·5–1·8	
	B–L	8	1·8±0·03	0·09	1·7–2·0	
I_2	M–D	8	1·7±0·06	0·18	1·5–2·0	
	B–L	8	1·9±0·03	0·07	1·8–2·0	
C	M–D	9	2·6±0·07	0·20	2·4–3·1	
	B–L	9	2·7±0·04	0·13	2·5–2·8	
P_2	M–D	10	2·4±0·06	0·18	2·1–2·6	
	B–L	10	2·2±0·04	0·11	2·0–2·4	
P_3	M–D	10	2·2±0·06	0·18	1·9–2·5	
	B–L	10	2·2±0·03	0·08	2·1–2·3	
P_4	M–D	10	2·3±0·02	0·17	2·1–2·3	
	B–L	10	2·3±0·03	0·10	2·2–2·5	
M_1	M–D	11	2·8±0·03	0·11	2·7–3·0	*
	Tri B	11	2·3±0·04	0·12	2·1–2·5	
	Tal B	10	2·3±0·04	0·12	2·1–2·5	
M_2	M–D	9	2·4±0·04	0·11	2·2–2·5	
	Tri B	9	1·9±0·11	0·32	1·6–2·7	
	Tal B	9	1·9±0·14	0·41	1·2–2·8	
M_3	M–D					
	Tri B					
	Tal B					
	Hypo-conulid					

* $P<0{\cdot}05$

Table 49
Saguinus geoffroyi female

Maxillary teeth		n	$\bar{X} \pm$ s.e.	s.d.	Range
I^1	M–D	14	2·4±0·06	0·20	2·0–2·6
	B–L	12	1·9±0·02	0·08	1·8–2·0
I^2	M–D	12	1·9±0·04	0·13	1·7–2·2
	B–L	12	1·7±0·03	0·11	1·4–1·8
C	M–D	13	2·9±0·06	0·20	2·6–3·2
	B–L	14	2·4±0·03	0·10	2·2–2·5
P^2	M–D	14	2·1±0·05	0·19	1·9–2·4
	B–L	14	2·8±0·05	0·20	2·4–3·1
P^3	M–D	14	2·0±0·04	0·15	1·8–2·3
	B–L	13	3·1±0·05	0·19	3·0–3·5
P^4	M–D	14	2·1±0·04	0·14	1·9–2·3
	B–L	14	3·4±0·07	0·26	3·0–3·9
M^1	M–D	14	2·9±0·03	0·12	2·7–3·1
	Ant B	14	3·9±0·06	0·21	3·4–4·1
	Post B				
M^2	M–D	14	1·6±0·03	0·10	1·4–1·8
	Ant B	14	2·7±0·03	0·12	2·4–2·9
	Post B				
M^3	M–D				
	Ant B				
	Post B				

Table 50
Saguinus geoffroyi female

Mandibular teeth		n	$\bar{X} \pm$ s.e.	s.d.	Range
I_1	M–D	12	1·7±0·03	0·11	1·5–1·8
	B–L	12	1·8±0·04	0·12	1·6–2·0
I_2	M–D	12	1·8±0·04	0·14	1·5–2·0
	B–L	12	1·9±0·03	0·10	1·8–2·0
C	M–D	12	2·6±0·03	0·10	2·4–2·8
	B–L	10	2·7±0·03	0·11	2·5–2·8
P_2	M–D	14	2·5±0·05	0·19	2·0–2·8
	B–L	14	2·3±0·03	0·13	2·0–2·4
P_3	M–D	14	2·3±0·05	0·18	2·0–2·7
	B–L	14	2·2±0·04	0·14	2·0–2·5
P_4	M–D	14	2·3±0·03	0·11	2·2–2·6
	B–L	14	2·4±0·02	0·09	2·3–2·6
M_1	M–D	15	2·9±0·03	0·12	2·7–3·1
	Tri B	15	2·3±0·02	0·08	2·2–2·5
	Tal B	15	2·3±0·02	0·09	2·2–2·5
M_2	M–D	12	2·4±0·04	0·15	2·2–2·6
	Tri B	11	2·0±0·09	0·29	1·7–2·8
	Tal B	10	2·0±0·11	0·34	1·7–2·9
M_3	M–D				
	Tri B				
	Tal B				
	Hypo-conulid				

Table 51
Aotus trivirgatus male

Maxillary teeth		n	$\bar{X} \pm$ s.e.	s.d.	Range	P
I^1	M–D	7	3·6±0·40	0·11	3·4–3·7	
	B L	7	2·7±0·02	0·05	2·6–2·7	
I^2	M–D	9	2·5±0·12	0·37	2·2–3·4	
	B–L	8	2·3±0·05	0·15	2·0–2·5	
C	M–D	7	3·1±0·05	0·14	2·9–3·2	
	B–L	7	2·9±0·05	0·14	2·7–3·1	†
P^2	M–D	9	2·2±0·05	0·16	2·0–2·5	
	B–L	8	3·2±0·08	0·22	3·0–3·7	
P^3	M–D	9	2·2±0·04	0·11	2·0–2·3	
	B–L	9	3·5±0·09	0·26	3·1–4·0	
P^4	M–D	9	2·4±0·05	0·14	2·2–2·6	
	B–L	9	3·8±0·08	0·22	3·6–4·3	
M^1	M–D	9	3·3±0·06	0·19	3·0–3·7	
	Ant B	9	4·1±0·10	0·30	3·7–4·6	
	Post B	9	4·0±0·07	0·20	3·8–4·4	
M^2	M–D	9	3·1±0·05	0·14	2·9–3·4	
	Ant B	9	3·9±0·09	0·26	3·7–4·5	*
	Post B	7	3·7±0·04	0·10	3·5–3·8	*
M^3	M–D	7	2·3±0·12	0·32	2·0–2·8	
	Ant B	7	3·4±0·12	0·32	2·9–3·9	†
	Post B					

* $P<0·05$
† $P<0·01$

Table 52
Aotus trivirgatus male

Mandibular teeth		n	$\bar{X} \pm$ s.e.	s.d.	Range	P
I_1	M–D	6	2·3±0·05	0·12	2·2–2·5	
	B–L	6	2·4±0·06	0·16	2·2–2·5	
I_2	M–D	7	2·3±0·11	0·28	1·9–2·6	
	B–L	7	2·6±0·07	0·19	2·3–2·7	
C	M–D	4	2·8±0·06	0·13	2·7–3·0	
	B–L	4	2·7±0·18	0·37	2·3–3·1	
P_2	M–D	8	2·4±0·07	0·20	2·2–2·8	
	B–L	8	2·4±0·08	0·23	2·1–2·8	
P_3	M–D	8	2·3±0·05	0·14	2·0–2·4	
	B–L	7	2·3±0·06	0·15	2·1–2·5	
P_4	M–D	9	2·5±0·05	0·15	2·2–2·7	
	B–L	9	2·7±0·04	0·12	2·5–2·8	
M_1	M–D	9	3·3±0·07	0·20	3·0–3·6	
	Tri B	9	2·9±0·05	0·15	2·6–3·0	
	Tal B	8	3·0±0·06	0·17	2·7–3·3	
M_2	M–D	9	3·3±0·07	0·19	3·0–3·5	
	Tri B	7	3·0±0·06	0·15	2·8–3·2	
	Tal B	9	2·9±0·04	0·12	2·8–3·2	
M_3	M–D	8	3·3±0·13	0·38	2·8–3·9	*
	Tri B	6	2·7±0·07	0·17	2·6–3·0	
	Tal B	5	2·7±0·10	0·22	2·4–3·0	
	Hypoconulid					

* $P<0·05$

Table 53
Aotus trivirgatus female

Maxillary teeth		n	$\bar{X}\pm$s.e.	s.d.	Range
I^1	M–D	11	3·4±0·11	0·36	2·6–3·9
	B–L	11	2·7±0·03	0·10	2·5–2·8
I^2	M–D	9	2·9±0·08	0·23	1·8–2·5
	B–L	9	2·2±0·06	0·18	1·9–2·5
C	M–D	11	3·0±0·05	0·18	2·8–3·4
	B–L	12	2·6±0·03	0·11	2·4–2·7
P^2	M–D	10	2·0±0·07	0·22	1·7–2·4
	B–L	11	3·1±0·05	0·17	2·8–3·4
P^3	M–D	12	2·1±0·05	0·16	1·8–2·3
	B–L	13	3·4±0·05	0·18	3·0–3·6
P^4	M–D	13	2·3±0·06	0·23	1·9–2·6
	B–L	12	3·6±0·05	0·16	3·4–3·9
M^1	M–D	15	3·3±0·05	0·18	3·0–3·7
	Ant B	15	3·9±0·06	0·21	3·7–4·4
	Post B	15	3·9±0·06	0·22	3·6–4·3
M^2	M–D	14	3·0±0·04	0·15	2·8–3·3
	Ant B	13	3·7±0·03	0·12	3·6–4·1
	Post B	12	3·5±0·04	0·13	3·2–3·6
M^3	M–D	13	2·3±0·14	0·51	1·7–3·7
	Ant B	13	2·9±0·07	0·23	2·5–3·2
	Post B				

Table 54
Aotus trivirgatus female

Mandibular teeth		n	$\bar{X}\pm$s.e.	s.d.	Range
I_1	M–D	10	2·3±0·11	0·35	1·9–3·2
	B–L	10	2·4±0·05	0·17	2·2–2·6
I_2	M–D	10	2·3±0·13	0·41	1·9–3·4
	B–L	10	2·4±0·05	0·16	2·2–2·7
C	M–D	10	2·7±0·05	0·16	2·4–2·9
	B–L	9	2·4±0·08	0·25	2·0–2·8
P_2	M–D	11	2·4±0·04	0·14	2·1–2·6
	B–L	11	2·3±0·06	0·20	2·0–2·6
P_3	M–D	13	2·2±0·04	0·16	1·9–2·4
	B–L	12	2·3±0·05	0·18	2·0–2·6
P_4	M–D	12	2·4±0·05	0·17	2·1–2·6
	B–L	13	2·7±0·05	0·15	2·4–2·9
M_1	M–D	15	3·3±0·04	0·14	3·0–3·5
	Tri B	15	2·9±0·04	0·14	2·7–3·2
	Tal B	15	3·0±0·03	0·13	2·8–3·2
M_2	M–D	14	3·2±0·05	0·20	2·9–3·5
	Tri B	13	2·9±0·03	0·09	2·8–3·1
	Tal B	13	2·9±0·03	0·10	2·7–3·0
M_3	M–D	11	2·9±0·05	0·17	2·7–3·2
	Tri B	11	2·6±0·05	0·16	2·3–2·9
	Tal B	10	2·5±0·05	0·17	2·3–2·8
	Hypoconulid				

Table 55
Ateles geoffroyi male

Maxillary teeth		n	$\bar{X} \pm$ s.e.	s.d.	Range
I^1	M–D	10	4·7±0·12	0·36	4·2–5·3
	B–L	12	4·2±0·10	0·35	3·7–4·8
I^2	M–D	11	3·5±0·11	0·36	3·0–4·3
	B–L	11	4·1±0·08	0·26	3·7–4·5
C	M–D	3	6·1±0·37	0·64	5·7–6·8
	B–L	3	4·9±0·26	0·45	4·4–5·3
P^2	M–D	12	3·5±0·09	0·31	3·0–4·1
	B–L	12	4·5±0·07	0·23	4·2–4·9
P^3	M–D	12	3·5±0·08	0·26	3·1–4·0
	B–L	12	5·3±0·09	0·32	4·9–5·8
P^4	M–D	12	3·5±0·05	0·17	3·2–3·8
	B–L	12	5·6±0·12	0·42	5·0–6·5
M^1	M–D	14	4·9±0·06	0·22	4·5–5·2
	Ant B	14	5·5±0·08	0·30	4·8–5·9
	Post B	12	5·6±0·11	0·39	5·0–6·2
M^2	M–D	11	4·7±0·09	0·30	4·2–5·2
	Ant B	11	5·6±0·08	0·27	5·1–6·0
	Post B	8	5·8±0·11	0·31	5·2–6·2
M^3	M–D	7	4·0±0·19	0·51	3·2–4·8
	Ant B	7	5·3±0·19	0·51	4·5–5·9
	Post B	2	5·7±0·30	0·42	5·4–6·0

Table 56
Ateles geoffroyi male

Mandibular teeth		n	$\bar{X} \pm$ s.e.	s.d.	Range
I_1	M–D	12	3·0±0·07	0·23	2·6–3·5
	B–L	11	3·6±0·08	0·27	3·0–4·0
I_2	M–D	12	3·4±0·08	0·29	2·9–4·1
	B–L	12	4·1±0·08	0·26	3·7–4·7
C	M–D	5	5·8±0·42	0·93	5·1–7·4
	B–L	6	4·8±0·09	0·21	4·5–5·1
P_2	M–D	11	4·3±0·11	0·38	3·8–5·2
	B–L	11	4·5±0·08	0·27	4·0–4·8
P_3	M–D	12	3·6±0·09	0·31	3·2–4·4
	B–L	9	4·2±0·08	0·25	3·9–4·6
P_4	M–D	12	3·6±0·07	0·24	3·2–3·9
	B–L	11	4·5±0·06	0·21	4·2–4·9
M_1	M–D	14	5·2±0·08	0·28	4·7–5·6
	Tri B	11	5·0±0·05	0·16	4·7–5·2
	Tal B	14	4·9±0·05	0·17	4·6–5·2
M_2	M–D	13	5·0±0·07	0·25	4·6–5·5
	Tri B	13	5·1±0·06	0·22	4·7–5·5
	Tal B	13	5·0±0·06	0·20	4·7–5·4
M_3	M–D	8	4·7±0·13	0·35	4·1–5·2
	Tri B	7	4·5±0·12	0·30	4·0–4·8
	Tal B	7	4·6±0·10	0·26	4·2–4·9
	Hypoconulid				

Table 57
Ateles geoffroyi female

Maxillary teeth		n	$\bar{X}\pm$s.e.	s.d.	Range
I^1	M–D	14	4·7±0·09	0·34	4·0–5·2
	B–L	14	4·4±0·09	0·33	3·8–5·1
I^2	M–D	12	3·5±0·10	0·36	3·0–4·2
	B–L	12	4·1±0·07	0·24	3·7–4·6
C	M–D	13	5·5±0·15	0·53	4·7–6·6
	B–L	14	5·1±0·13	0·47	4·2–5·8
P^2	M–D	13	3·6±0·06	0·21	3·3–4·0
	B–L	13	4·7±0·06	0·21	4·4–5·2
P^3	M–D	14	3·6±0·05	0·19	3·3–3·9
	B–L	13	5·5±0·06	0·23	5·2–5·9
P^4	M–D	14	3·6±0·05	0·18	3·4–3·9
	B–L	13	5·8±0·06	0·20	5·5–6·2
M^1	M–D	15	5·3±0·05	0·20	4·7–5·5
	Ant B	15	5·7±0·09	0·32	5·2–6·5
	Post B	14	5·8±0·09	0·32	5·4–6·4
M^2	M–D	13	5·0±0·08	0·27	4·5–5·5
	Ant B	14	5·8±0·10	0·38	5·2–6·6
	Post B	13	5·7±0·10	0·36	5·2–6·4
M^3	M–D	14	3·9±0·17	0·64	2·8–4·9
	Ant B	14	5·3±0·13	0·50	4·4–6·0
	Post B	3	5·6±0·09	0·15	5·4–5·7

Table 58
Ateles geoffroyi female

Mandibular teeth		n	$\bar{X}\pm$s.e.	s.d.	Range
I_1	M–D	14	3·0±0·04	0·13	2·8–3·2
	B–L	14	3·7±0·06	0·23	3·3–4·2
I_2	M–D	15	3·6±0·06	0·25	3·2–4·0
	B–L	15	4·2±0·07	0·25	3·9–4·8
C	M–D	13	5·0±0·17	0·59	3·5–5·7
	B–L	14	4·5±0·07	0·27	4·0–5·1
P_2	M–D	15	4·0±0·07	0·29	3·5–4·6
	B–L	15	4·4±0·07	0·27	4·0–5·0
P_3	M–D	15	3·6±0·07	0·28	3·2–4·4
	B–L	15	4·5±0·06	0·25	4·2–5·0
P_4	M–D	15	3·7±0·05	0·19	3·4–4·1
	B–L	15	4·7±0·06	0·22	4·4–5·2
M_1	M–D	15	5·3±0·05	0·20	4·8–5·6
	Tri B	13	5·1±0·07	0·25	4·8–5·6
	Tal B	12	5·0±0·06	0·22	4·7–5·4
M_2	M–D	14	5·2±0·06	0·24	4·9–5·7
	Tri B	14	5·2±0·09	0·33	4·3–5·7
	Tal B	14	5·1±0·05	0·18	4·9–5·5
M_3	M–D	14	5·0±0·08	0·32	4·3–5·5
	Tri B	14	4·8±0·08	0·28	4·2–5·2
	Tal B	14	4·7±0·06	0·21	4·2–5·0
	Hypoconulid				

Table 59
Brachyteles arachnoides (sexes pooled)

Maxillary teeth		n	$\bar{X}\pm$s.e.	s.d.	Range
I^1	M–D		4·1		3·8–4·6
	B–L		3·9		3·6–4·5
I^2	M–D		3·8		3·6–4·1
	B–L		3·8		3·5–4·2
C	M–D		6·5		5·9–7·0
	B–L		5·5		5·0–6·3
P^2	M–D		4·7		4·5–5·7
	B–L		5·5		4·6–6·0
P^3	M–D		5·0		4·5–5·6
	B–L		6·6		6·4–7·0
P^4	M–D		5·0		4·6–5·5
	B–L		7·3		7·0–7·8
M^1	M–D		6·9		6·2–7·3
	Ant B		7·8		7·3–8·2
	Post B				
M^2	M–D		6·4		5·8–6·8
	Ant B		7·6		7·3–8·2
	Post B				
M^3	M–D		5·4		5·2–5·6
	Ant B		6·4		5·9–7·3
	Post B				

Zingeser, 1974

Table 60
Brachyteles arachnoides (sexes pooled)

Mandibular teeth		n	$\bar{X}\pm$s.e.	s.d.	Range
I_1	M–D		2·8		2·3–3·2
	B–L		3·7		3·2–4·5
I_2	M–D		3·3		2·8–3·5
	B–L		4·3		3·9–4·6
C	M–D		4·5		4·1–4·9
	B–L		5·8		5·5–6·2
P_2	M–D		4·4		4·1–4·9
	B–L		5·3		5·0–5·8
P_3	M–D		4·6		4·0–5·0
	B–L		5·0		4·6–5·5
P_4	M–D		4·9		4·6–5·5
	B–L		5·0		5·0–5·2
M_1	M–D		7·2		6·9–7·4
	Tri B		5·4		5·0–5·7
	Tal B		5·6		5·0–6·0
M_2	M–D		7·1		6·8–7·6
	Tri B		5·5		5·0–6·0
	Tal B		5·7		5·2–6·0
M_3	M–D		6·2		5·2–6·9
	Tri B		5·5		5·1–6·3
	Tal B		4·8		4·4–5·3
	Hypo-conulid				

Zingeser, 1974

Table 61
Cebus apella male

Maxillary teeth		n	$\bar{X} \pm$ s.e.	s.d.	Range	P
I^1	M–D	34	4·56±0·04	0·23	3·9–5·0	
	B–L	33	4·44±0·06	0·31	3·9–5·1	*
I^2	M–D	32	3·90±0·05	0·27	3·6–4·4	†
	B–L	33	4·66±0·05	0·30	4·1–5·4	*
C	M–D	28	7·36±0·15	0·80	4·8–8·3	†
	B–L	22	6·25±0·16	0·76	3·8–7·3	†
P^2	M–D	31	3·82±0·05	0·27	3·2–4·5	
	B–L	31	6·66±0·10	0·55	5·0–7·8	
P^3	M–D	30	3·59±0·03	0·15	3·2–3·9	
	B–L	30	6·99±0·09	0·50	5·4–7·9	*
P^4	M–D	29	3·57±0·02	0·13	3·4–3·8	
	B–L	30	6·94±0·06	0·30	6·3–7·7	*
M^1	M–D	34	4·69±0·03	0·20	4·3–5·0	
	Ant B	35	6·31±0·06	0·32	5·6–6·7	†
	Post B	35	5·99±0·05	0·30	5·5–6·7	†
M^2	M–D	33	4·13±0·04	0·25	3·7–4·7	
	Ant B	34	5·85±0·06	0·33	5·3–6·8	†
	Post B	33	5·27±0·05	0·30	4·7–5·7	
M^3	M–D	21	2·76±0·10	0·45	2·3–3·3	
	Ant B	22	4·47±0·11	0·51	3·5–5·8	
	Post B	2	3·95±1·05	1·49	2·5–5·5	

* $P<0·05$
† $P<0·01$

Table 62
Cebus apella male

Mandibular teeth		n	$\bar{X} \pm$ s.e.	s.d.	Range	P
I_1	M–D	34	2·76±0·03	0·20	2·4–3·2	
	B–L	36	4·02±0·03	0·20	3·7–4·4	†
I_2	M–D	34	3·41±0·06	0·32	2·8–4·0	
	B–L	35	4·53±0·05	0·27	4·2–5·3	†
C	M–D	26	6·09±0·11	0·57	4·7–7·2	†
	B–L	22	7·01±0·15	0·71	4·6–7·9	†
P_2	M–D	29	6·00±0·13	0·67	4·6–7·0	†
	B–L	29	5·09±0·10	0·56	3·6–6·0	†
P_3	M–D	28	3·89±0·04	0·22	3·5–4·5	†
	B–L	29	5·01±0·06	0·31	4·2–5·7	†
P_4	M–D	26	3·79±0·03	0·15	3·5–4·1	
	B–L	27	5·26±0·05	0·27	4·5–5·6	†
M_1	M–D	35	5·03±0·04	0·24	4·6–5·8	*
	Tri B	35	5·09±0·05	0·31	4·2–5·8	†
	Tal B	35	4·93±0·04	0·25	4·5–5·5	*
M_2	M–D	34	4·72±0·04	0·21	4·2–5·1	*
	Tri B	33	4·97±0·04	0·25	4·5–5·7	*
	Tal B	34	4·54±0·06	0·34	3·8–5·0	
M_3	M–D	20	3·77±0·07	0·30	3·0–4·5	
	Tri B	19	4·21±0·08	0·36	3·1–5·0	
	Tal B	14	3·76±0·08	0·31	3·5–4·6	
	Hypo-conulid					

* $P<0·05$
† $P<0·01$

Table 63
Cebus apella female

Maxillary teeth		n	$\bar{X}\pm$s.e.	s.d.	Range
I^1	M–D	19	4·47±0·05	0·21	4·1–5·0
	B–L	18	4·25±0·08	0·32	3·4–4·6
I^2	M–D	21	3·69±0·04	0·20	3·4–4·3
	B–L	20	4·46±0·05	0·22	4·0–4·9
C	M–D	18	6·21±0·13	0·56	4·6–7·0
	B–L	18	5·50±0·13	0·55	4·0–6·1
P^2	M–D	18	3·71±0·16	0·67	3·0–4·4
	B–L	19	6·36±0·11	0·49	4·7–6·9
P^3	M–D	19	3·55±0·04	0·17	3·2–3·8
	B–L	20	6·64±0·11	0·50	5·0–7·1
P^4	M–D	19	3·51±0·04	0·18	3·0–3·7
	B–L	19	6·60±0·08	0·33	5·8–7·2
M^1	M–D	21	4·56±0·04	0·17	4·0–4·8
	Ant B	21	6·04±0·07	0·31	5·5–6·5
	Post B	21	5·74±0·06	0·28	5·3–6·2
M^2	M–D	19	4·04±0·05	0·20	3·7–4·3
	Ant B	18	5·66±0·06	0·26	5·4–6·1
	Post B	18	5·09±0·09	0·36	4·6–5·5
M^3	M–D	12	2·73±0·16	0·56	2·1–4·2
	Ant B	11	4·37±0·14	0·47	3·8–5·5
	Post B				

Table 64
Cebus apella female

Mandibular teeth		n	$\bar{X}\pm$s.e.	s.d.	Range
I_1	M–D	20	2·73±0·04	0·17	2·5–3·0
	B–L	18	3·78±0·06	0·24	3·4–4·2
I_2	M–D	20	3·35±0·06	0·26	2·8–3·8
	B–L	20	4·33±0·06	0·27	3·9–4·7
C	M–D	19	5·53±0·07	0·29	4·8–6·0
	B–L	16	6·09±0·12	0·49	5·6–7·0
P_2	M–D	19	4·78±0·06	0·24	4·2–5·2
	B–L	18	4·65±0·10	0·43	3·6–5·6
P_3	M–D	18	3·83±0·05	0·21	3·3–4·2
	B–L	18	4·80±0·08	0·32	3·9–5·5
P_4	M–D	18	3·83±0·04	0·18	3·5–4·0
	B–L	18	5·03±0·07	0·28	4·5–5·3
M_1	M–D	20	4·86±0·04	0·17	4·5–5·2
	Tri B	20	4·80±0·08	0·37	3·7–6·3
	Tal B	20	4·75±0·05	0·22	4·3–5·2
M_2	M–D	18	4·58±0·04	0·18	4·3–5·0
	Tri B	19	4·80±0·05	0·21	4·0–5·5
	Tal B	18	4·53±0·08	0·35	4·0–5·5
M_3	M–D	17	3·71±0·09	0·36	3·0–4·3
	Tri B	17	4·09±0·08	0·32	3·5–4·8
	Tal B	14	3·77±0·09	0·32	3·0–4·2
	Hypoconulid				

Table 65
Saimiri oerstedii male

Maxillary teeth		n	$\bar{X}$±s.e.	s.d.	Range
I^1	M–D	10	2·8±0·05	0·16	2·5–3·0
	B–L	10	2·6±0·05	0·17	2·3–2·8
I^2	M–D	10	2·3±0·07	0·22	2·0–2·6
	B–L	10	2·5±0·07	0·21	2·2–2·8
C	M–D	6	3·9±0·10	0·25	3·4–4·1
	B–L	7	3·5±0·04	0·10	3·3–3·6
P^2	M–D	9	2·4±0·06	0·18	2·2–2·8
	B–L	9	3·7±0·09	0·26	3·1–4·0
P^3	M–D	9	2·0±0·04	0·11	1·9–2·2
	B–L	8	3·9±0·04	0·10	3·8–4·1
P^4	M–D	9	1·9±0·04	0·12	1·8–2·2
	B–L	9	3·9±0·04	0·11	3·7–4·0
M^1	M–D	11	2·8±0·05	0·15	2·6–3·1
	Ant B	11	3·9±0·08	0·26	3·6–4·5
	Post B	11	3·8±0·07	0·24	3·5–4·2
M^2	M–D	11	2·6±0·04	0·13	2·4–2·8
	Ant B	11	3·6±0·05	0·17	3·3–3·9
	Post B	10	3·4±0·06	0·17	3·2–3·7
M^3	M–D	9	1·8±0·08	0·22	1·5–2·3
	Ant B	9	2·6±0·08	0·23	2·3–3·0
	Post B				

Table 66
Saimiri oerstedii male

Mandibular teeth		n	$\bar{X}$±s.e.	s.d.	Range
I_1	M–D	7	1·7±0·14	0·38	1·3–2·5
	B–L	7	2·4±0·11	0·28	1·8–2·6
I_2	M–D	7	2·1±0·09	0·25	1·7–2·4
	B–L	8	2·6±0·03	0·09	2·4–2·7
C	M–D	8	3·5±0·10	0·29	3·0–4·0
	B–L	5	3·4±0·09	0·20	3·2–3·7
P_2	M–D	10	3·4±0·14	0·44	2·5–3·8
	B–L	10	3·0±0·04	0·13	2·9–3·3
P_3	M–D	10	2·3±0·07	0·22	2·0–2·6
	B–L	10	2·7±0·05	0·15	2·5–3·0
P_4	M–D	10	2·2±0·05	0·15	2·0–2·5
	B–L	10	2·8±0·04	0·12	2·6–3·0
M_1	M–D	11	3·0±0·08	0·26	2·6–3·4
	Tri B	11	2·8±0·06	0·20	2·4–3·0
	Tal B	11	2·7±0·06	0·21	2·4–3·0
M_2	M–D	10	2·7±0·04	0·14	2·5–3·0
	Tri B	11	2·8±0·07	0·22	2·2–2·9
	Tal B	11	2·5±0·06	0·21	2·2–2·8
M_3	M–D	7	2·2±0·03	0·08	2·2–2·4
	Tri B	8	2·1±0·09	0·24	1·7–2·5
	Tal B	6	2·0±0·10	0·24	1·6–2·3
	Hypoconulid				

Table 67
Saimiri oerstedii female

Maxillary teeth		n	$\bar{X} \pm$ s.e.	s.d.	Range
I^1	M–D	6	2·7 ± 0·04	0·10	2·5–2·8
	B–L	6	2·4 ± 0·13	0·31	2·2–3·0
I^2	M–D	6	2·3 ± 0·05	0·12	2·2–2·5
	B–L	6	2·2 ± 0·12	0·29	2·0–2·8
C	M–D	4	2·7 ± 0·05	0·10	2·6–2·8
	B–L	4	3·0 ± 0·06	0·13	2·9–3·2
P^2	M–D	4	2·1 ± 0·03	0·05	2·0–2·1
	B–L	4	3·4 ± 0·12	0·24	3·2–3·7
P^3	M–D	4	2·1 ± 0·07	0·13	1·9–2·2
	B–L	4	3·7 ± 0·10	0·20	3·6–4·0
P^4	M–D	4	1·9 ± 0·07	0·14	1·8–2·1
	B–L	4	3·7 ± 0·11	0·22	3·5–4·0
M^1	M–D	6	2·7 ± 0·02	0·05	2·7–2·8
	Ant B	6	3·7 ± 0·07	0·16	3·6–4·0
	Post B	6	3·7 ± 0·08	0·19	3·5–4·0
M^2	M–D	6	2·5 ± 0·05	0·13	2·3–2·6
	Ant B	6	3·1 ± 0·05	0·13	3·2–3·5
	Post B	6	3·3 ± 0·09	0·22	3·2–3·7
M^3	M–D	4	1·7 ± 0·07	0·14	1·6–1·9
	Ant B	4	2·3 ± 0·13	0·26	2·1–2·7
	Post B				

Table 68
Saimiri oerstedii female

Mandibular teeth		n	$\bar{X} \pm$ s.e.	s.d.	Range
I_1	M–D	5	1·4 ± 0·05	0·11	1·2–1·5
	B–L	5	1·9 ± 0·04	0·09	1·7–1·9
I_2	M–D	5	1·8 ± 0·05	0·11	1·7–2·0
	B–L	5	2·3 ± 0·02	0·06	2·3–2·4
C	M–D	3	3·0 ± 0·12	0·21	2·8–3·2
	B–L	3	2·5 ± 0·19	0·32	2·1–2·7
P_2	M–D	2	2·4 ± 0·20	0·28	2·2–2·6
	B–L	2	2·7 ± 0·00	0·00	2·7–2·7
P_3	M–D	3	2·2 ± 0·00	0·00	2·2–2·2
	B–L	3	2·6 ± 0·03	0·06	2·5–2·6
P_4	M–D	3	2·2 ± 0·00	0·00	2·2–2·2
	B–L	3	2·6 ± 0·00	0·00	2·6–2·6
M_1	M–D	5	2·7 ± 0·05	0·11	2·6–2·9
	Tri B	5	2·6 ± 0·02	0·05	2·5–2·6
	Tal B	5	2·5 ± 0·04	0·09	2·4–2·6
M_2	M–D	5	2·6 ± 0·04	0·09	2·5–2·7
	Tri B	5	2·4 ± 0·04	0·08	2·3–2·5
	Tal B	5	2·3 ± 0·02	0·06	2·3–2·4
M_3	M–D	3	2·3 ± 0·03	0·06	2·2–2·3
	Tri B	3	1·9 ± 0·06	0·10	1·8–2·0
	Tal B	3	1·8 ± 0·07	0·12	1·7–1·9
	Hypo-conulid				

Table 69
Saimiri sciureus male

Maxillary teeth		n	$\bar{X} \pm$ s.e.	s.d.	Range	P
I^1	M–D	14	2·7±0·10	0·36	1·7–3·0	
	B–L	12	2·7±0·08	0·28	2·3–3·2	
I^2	M–D	12	2·2±0·05	0·17	2·0–2·5	
	B–L	10	2·4±0·09	0·28	1·8–2·8	
C	M–D	11	3·5±0·14	0·48	2·6–4·0	†
	B–L	9	3·1±0·14	0·43	2·1–3·6	
P^2	M–D	17	2·2±0·06	0·24	1·5–2·5	
	B–L	16	3·6±0·04	0·17	3·2–3·9	
P^3	M–D	17	2·0±0·04	0·17	1·7–2·0	
	B–L	17	3·9±0·04	0·16	3·7–4·2	*
P^4	M–D	18	1·9±0·04	0·15	1·6–2·2	
	B–L	18	3·9±0·04	0·16	3·6–4·2	
M^1	M–D	20	2·8±0·03	0·14	2·5–3·0	
	Ant B	20	4·1±0·04	0·19	3·7–4·5	
	Post B	20	4·0±0·05	0·21	3·5–4·4	
M^2	M–D	20	2·5±0·03	0·14	2·1–2·7	
	Ant B	19	3·6±0·08	0·35	2·8–4·4	
	Post B	19	3·5±0·09	0·40	2·6–4·5	
M^3	M–D	11	1·7±0·04	0·13	1·5–1·9	
	Ant B	12	2·9±0·04	0·14	2·5–3·0	
	Post B					

* $P<0·05$
† $P<0·01$

Table 70
Saimiri sciureus male

Mandibular teeth		n	$\bar{X} \pm$ s.e.	s.d.	Range	P
I_1	M–D	13	1·6±0·05	0·18	1·3–1·9	
	B–L	14	2·3±0·05	0·20	1·9–2·5	
I_2	M–D	11	2·0±0·06	0·20	1·8–2·3	
	B–L	11	2·7±0·06	0·20	2·3–2·9	
C	M–D	11	3·4±0·06	0·18	3·1–3·6	†
	B–L	10	3·1±0·14	0·45	2·0–3·6	
P_2	M–D	17	3·1±0·07	0·29	2·4–3·7	†
	B–L	17	3·0±0·07	0·30	2·4–3·5	†
P_3	M–D	19	2·2±0·05	0·21	1·9–2·5	
	B–L	19	2·7±0·05	0·21	2·3–3·0	
P_4	M–D	19	2·1±0·04	0·15	1·9–2·4	
	B–L	19	2·7±0·05	0·21	2·4–3·2	
M_1	M–D	21	2·9±0·04	0·19	2·5–3·3	
	Tri B	20	2·8±0·03	0·12	2·6–3·0	
	Tal B	21	2·8±0·03	0·15	2·6–3·2	
M_2	M–D	19	2·7±0·03	0·12	2·4–2·9	
	Tri B	18	2·7±0·04	0·16	2·4–3·0	
	Tal B	17	2·6±0·03	0·11	2·4–2·8	
M_3	M–D	12	2·2±0·04	0·14	2·0–2·4	
	Tri B	13	2·2±0·04	0·15	1·9–2·5	
	Tal B	9	2·0±0·07	0·21	1·7–2·4	
	Hypoconulid					

† $P<0·01$

Table 71
Saimiri sciureus female

Maxillary teeth		n	$\bar{X}\pm$s.e.	s.d.	Range
I^1	M–D	6	2·8±0·12	0·29	2·4–3·2
	B–L	6	2·8±0·07	0·17	2·6–3·0
I^2	M–D	4	2·2±0·04	0·08	2·1–2·3
	B–L	5	2·6±0·06	0·13	2·6–3·0
C	M–D	4	2·7±0·06	0·13	2·5–2·8
	B–L	5	2·9±0·12	0·26	2·5–3·2
P^2	M–D	6	2·0±0·06	0·14	1·8–2·2
	B–L	6	3·5±0·08	0·19	3·2–3·7
P^3	M–D	6	2·1±0·08	0·18	1·9–2·3
	B–L	6	3·8±0·04	0·11	3·6–3·9
P^4	M–D	6	1·9±0·05	0·11	1·7–2·0
	B–L	5	3·8±0·06	0·13	3·6–3·9
M^1	M–D	5	2·8±0·04	0·09	2·7–2·9
	Ant B	5	4·0±0·08	0·19	3·7–4·2
	Post B	5	3·9±0·07	0·16	3·7–4·1
M^2	M–D	5	2·5±0·04	0·09	2·4–2·6
	Ant B	5	3·7±0·07	0·15	3·5–3·8
	Post B	5	3·5±0·11	0·26	3·2–3·9
M^3	M–D	5	1·8±0·19	0·43	1·4–2·5
	Ant B	4	2·7±0·19	0·37	2·2–3·0
	Post B				

Table 72
Saimiri sciureus female

Mandibular teeth		n	$\bar{X}\pm$s.e.	s.d.	Range
I_1	M–D	4	1·5±0·03	0·05	1·4–1·5
	B–L	4	2·3±0·10	0·19	2·1–2·5
I_2	M–D	4	2·0±0·23	0·45	1·8–2·7
	B–L	4	2·5±0·08	0·16	2·3–2·7
C	M–D	4	2·9±0·18	0·35	2·5–3·3
	B–L	4	2·7±0·14	0·27	2·5–3·1
P_2	M–D	6	2·4±0·10	0·23	2·1–2·7
	B–L	6	2·5±0·08	0·19	2·2–2·7
P_3	M–D	6	2·3±0·08	0·20	2·0–2·5
	B–L	6	2·7±0·08	0·20	2·4–2·9
P_4	M–D	6	2·3±0·04	0·10	2·1–2·4
	B–L	6	2·7±0·12	0·29	2·4–3·0
M_1	M–D	6	2·9±0·04	0·10	2·8–3·0
	Tri B	6	2·8±0·08	0·16	2·6–3·0
	Tal B	6	2·7±0·06	0·15	2·6–2·9
M_2	M–D	6	2·6±0·07	0·16	2·4–2·8
	Tri B	5	2·5±0·08	0·18	2·3–2·8
	Tal B	5	2·5±0·08	0·17	2·4–2·8
M_3	M–D	6	2·3±0·07	0·18	2·1–2·5
	Tri B	5	2·1±0·07	0·16	1·8–2·2
	Tal B	5	2·0±0·10	0·28	1·7–2·3
	Hypo-conulid				

Table 73
Alouatta seniculus male

Maxillary teeth		n	$\bar{X} \pm$ s.e.	s.d.	Range	P
I^1	M–D	17	4·21 ± 0·10	0·41	3·3–4·7	†
	B–L	16	3·67 ± 0·08	0·31	3·1–4·4	†
I^2	M–D	18	4·04 ± 0·07	0·29	3·4–4·4	†
	B–L	18	3·64 ± 0·07	0·29	3·2–4·3	*
C	M–D	14	8·09 ± 0·16	0·61	6·9–9·2	†
	B–L	11	6·20 ± 0·14	0·48	5·6–6·8	†
P^2	M–D	18	5·41 ± 0·08	0·33	4·8–6·4	†
	B–L	18	6·13 ± 0·08	0·35	5·5–6·5	†
P^3	M–D	18	5·18 ± 0·06	0·26	4·8–5·5	†
	B–L	18	6·86 ± 0·08	0·33	6·3–7·5	†
P^4	M–D	18	5·12 ± 0·04	0·17	4·8–5·4	†
	B–L	18	7·37 ± 0·11	0·48	6·6–8·4	†
M^1	M–D	19	7·47 ± 0·06	0·26	7·0–7·9	†
	Ant B	18	7·99 ± 0·08	0·32	7·5–8·7	†
	Post B	18	8·16 ± 0·09	0·39	7·7–9·0	†
M^2	M–D	19	8·01 ± 0·10	0·41	7·2–8·8	†
	Ant B	18	8·48 ± 0·08	0·34	7·8–9·1	†
	Post B	18	8·59 ± 0·08	0·36	8·1–9·4	†
M^3	M–D	15	6·17 ± 0·12	0·47	5·5–7·3	
	Ant B	15	7·65 ± 0·10	0·38	7·2–8·5	†
	Post B					

* $P < 0·05$
† $P < 0·01$

Table 74
Alouatta seniculus male

Mandibular teeth		n	$\bar{X} \pm$ s.e.	s.d.	Range	P
I_1	M–D	16	2·94 ± 0·07	0·26	2·4–3·3	*
	B–L	16	3·36 ± 0·07	0·26	2·9–3·7	†
I_2	M–D	17	3·48 ± 0·06	0·24	2·9–3·8	†
	B–L	18	4·14 ± 0·06	0·26	3·8–4·5	†
C	M–D	13	6·85 ± 0·10	0·63	5·6–7·8	†
	B–L	14	6·52 ± 0·11	0·42	5·8–7·4	†
P_2	M–D	19	6·11 ± 0·06	0·28	5·7–6·7	†
	B–L	19	6·46 ± 0·07	0·32	5·8–7·4	†
P_3	M–D	19	5·11 ± 0·05	0·22	4·8–5·5	*
	B–L	18	5·63 ± 0·07	0·28	5·4–6·0	†
P_4	M–D	19	5·61 ± 0·08	0·35	4·9–6·2	†
	B–L	18	5·81 ± 0·08	0·34	5·4–6·5	†
M_1	M–D	20	7·58 ± 0·07	0·32	6·9–8·0	*
	Tri B	16	5·38 ± 0·07	0·28	4·9–5·9	†
	Tal B	16	5·93 ± 0·08	0·31	5·6–6·5	†
M_2	M–D	19	8·11 ± 0·09	0·38	7·3–8·8	†
	Tri B	18	5·89 ± 0·07	0·30	5·3–6·3	†
	Tal B	17	6·42 ± 0·05	0·22	6·2–6·7	†
M_3	M–D	14	8·94 ± 0·09	0·34	8·5–9·7	†
	Tri B	14	5·91 ± 0·09	0·33	5·2–6·4	†
	Tal B	12	5·85 ± 0·10	0·35	5·4–6·5	
	Hypoconulid					

* $P < 0·05$
† $P < 0·01$

Table 75
Alouatta seniculus female

Maxillary teeth		n	$\bar{X}$ ± s.e.	s.d.	Range
I^1	M–D	17	3·8 ± 0·09	0·36	3·3–4·5
	B–L	15	3·2 ± 0·07	0·27	2·9–3·8
I^2	M–D	14	3·7 ± 0·10	0·37	3·2–4·3
	B–L	13	3·3 ± 0·12	0·43	2·5–3·9
C	M–D	17	6·3 ± 0·17	0·71	5·4–7·7
	B–L	16	4·7 ± 0·17	0·69	3·9–6·2
P^2	M–D	19	4·8 ± 0·05	0·22	4·3–5·0
	B–L	19	5·6 ± 0·07	0·32	5·1–6·1
P^3	M–D	19	4·9 ± 0·05	0·23	4·3–5·4
	B–L	19	6·5 ± 0·10	0·42	5·9–7·1
P^4	M–D	20	4·9 ± 0·05	0·24	4·5–5·5
	B–L	20	6·9 ± 0·10	0·44	6·2–7·9
M^1	M–D	19	7·1 ± 0·08	0·36	6·5–8·0
	Ant B	17	7·3 ± 0·07	0·30	6·8–7·7
	Post B	17	7·4 ± 0·09	0·37	6·7–8·1
M^2	M–D	20	7·5 ± 0·11	0·49	6·5–8·6
	Ant B	19	7·7 ± 0·09	0·40	7·0–8·5
	Post B	19	7·8 ± 0·10	0·43	7·2–8·7
M^3	M–D	12	5·9 ± 0·14	0·48	5·1–6·6
	Ant B	14	7·1 ± 0·11	0·40	6·7–7·9
	Post B				

Table 76
Alouatta seniculus female

Mandibular teeth		n	$\bar{X}$ ± s.e.	s.d.	Range
I_1	M–D	13	2·7 ± 0·09	0·33	2·2–3·4
	B–L	12	3·1 ± 0·10	0·36	2·5–3·8
I_2	M–D	14	3·1 ± 0·05	0·19	2·8–3·5
	B–L	13	3·6 ± 0·11	0·41	2·7–4·2
C	M–D	16	5·6 ± 0·17	0·69	4·4–7·0
	B–L	18	5·1 ± 0·15	0·62	4·2–6·3
P_2	M–D	18	5·0 ± 0·07	0·28	4·5–5·5
	B–L	19	5·2 ± 0·09	0·40	4·5–6·1
P_3	M–D	17	4·9 ± 0·05	0·22	4·6–5·4
	B–L	18	5·1 ± 0·09	0·36	4·7–6·1
P_4	M–D	19	5·2 ± 0·07	0·29	4·8–5·8
	B–L	18	5·1 ± 0·09	0·39	4·7–6·0
M_1	M–D	20	7·2 ± 0·09	0·39	6·5–8·2
	Tri B	17	5·1 ± 0·06	0·25	4·7–5·5
	Tal B	19	5·5 ± 0·07	0·32	4·9–6·1
M_2	M–D	19	7·5 ± 0·10	0·43	6·5–8·2
	Tri B	18	5·5 ± 0·05	0·21	5·2–5·9
	Tal B	20	6·0 ± 0·08	0·36	5·4–6·7
M_3	M–D	16	8·3 ± 0·15	0·58	7·5–9·5
	Tri B	19	5·7 ± 0·06	0·27	5·2–6·1
	Tal B	15	5·6 ± 0·09	0·36	5·1–6·4
	Hypo-conulid				

Table 77
Alouatta villosa male

Maxillary teeth		n	$\bar{X}\pm$s.e.	s.d.	Range	P
I^1	M–D	11	3·7±0·09	0·28	3·3–4·2	
	B–L	11	3·7±0·07	0·24	3·4–4·1	
I^2	M–D	12	3·8±0·06	0·22	3·4–4·1	
	B–L	12	3·8±0·05	0·18	3·4–4·0	*
C	M–D	12	7·6±0·14	0·48	6·6–8·1	†
	B–L	12	5·7±0·18	0·64	4·7–6·9	†
P^2	M–D	12	4·8±0·11	0·38	4·2–5·3	
	B–L	12	6·5±0·08	0·29	6·0–6·9	†
P^3	M–D	12	4·8±0·06	0·22	4·5–5·2	
	B–L	11	7·1±0·06	0·19	6·8–7·4	†
P^4	M–D	13	4·8±0·10	0·37	4·3–5·5	
	B–L	12	7·6±0·10	0·34	6·9–8·0	
M^1	M–D	13	7·1±0·08	0·28	6·4–7·4	
	Ant B	12	8·1±0·12	0·41	7·3–8·8	†
	Post B	12	8·0±0·08	0·28	7·5–8·4	*
M^2	M–D	12	7·3±0·12	0·43	6·5–8·0	
	Ant B	11	8·2±0·16	0·54	7·3–9·0	
	Post B	11	8·0±0·12	0·41	7·4–8·8	
M^3	M–D	13	5·4±0·14	0·49	4·5–6·0	
	Ant B	12	7·7±0·15	0·51	6·5–8·5	
	Post B					

* $P<0·05$
† $P<0·01$

Table 78
Alouatta villosa male

Mandibular teeth		n	$\bar{X}\pm$s.e.	s.d.	Range	P
I_1	M–D	8	2·5±0·06	0·18	2·3–2·8	
	B–L	8	3·2±0·06	0·18	2·9–3·4	*
I_2	M–D	11	3·1±0·08	0·27	2·8–3·8	
	B–L	10	3·9±0·07	0·22	3·6–4·4	
C	M–D	12	5·2±0·16	0·55	4·9–6·5	†
	B–L	12	6·9±0·13	0·44	6·3–7·5	†
P_2	M–D	13	5·5±0·13	0·46	4·3–6·2	†
	B–L	13	6·4±0·07	0·26	6·0–6·8	†
P_3	M–D	13	4·8±0·11	0·39	3·9–5·3	
	B–L	12	5·6±0·12	0·41	4·8–6·0	†
P_4	M–D	12	5·0±0·08	0·28	4·5–5·4	
	B–L	13	6·1±0·08	0·29	5·7–6·6	†
M_1	M–D	15	7·1±0·06	0·22	6·7–7·4	†
	Tri B	14	5·5±0·07	0·27	5·0–5·9	*
	Tal B	14	6·0±0·05	0·17	5·7–6·3	†
M_2	M–D	15	7·8±0·08	0·32	7·3–8·2	†
	Tri B	14	5·9±0·08	0·30	5·3–6·4	†
	Tal B	14	6·7±0·08	0·31	6·3–7·1	†
M_3	M–D	13	7·8±0·14	0·51	6·9–8·7	
	Tri B	12	6·0±0·10	0·35	5·1–6·3	*
	Tal B	12	5·4±0·12	0·41	5·0–6·0	
	Hypoconulid					

* $P<0·05$
† $P<0·01$

Table 79
Alouatta villosa female

Maxillary teeth		n	$\bar{X}\pm$s.e.	s.d.	Range
I^1	M–D	34	3·7±0·13	0·76	2·9–4·3
	B–L	33	3·7±0·13	0·73	3·0–6·3
I^2	M–D	32	3·7±0·14	0·79	3·0–4·7
	B–L	32	3·5±0·07	0·40	3·1–4·9
C	M–D	26	6·2±0·18	0·93	3·7–8·5
	B–L	27	4·7±0·14	0·73	3·8–7·2
P^2	M–D	34	4·5±0·08	0·44	3·0–5·1
	B–L	34	6·1±0·08	0·48	5·3–6·9
P^3	M–D	35	4·7±0·06	0·33	4·0–5·4
	B–L	36	6·8±0·07	0·40	6·2–7·7
P^4	M–D	34	4·6±0·05	0·31	3·9–5·0
	B–L	36	7·3±0·10	0·59	5·6–8·1
M^1	M–D	36	6·9±0·06	0·35	6·1–7·6
	Ant B	37	7·3±0·13	0·81	5·2–8·7
	Post B	37	7·2±0·17	1·02	6·0–8·7
M^2	M–D	36	7·2±0·07	0·42	6·7–7·8
	Ant B	36	7·6±0·20	1·13	6·0–8·6
	Post B	36	7·6±0·11	0·65	6·7–8·6
M^3	M–D	29	5·4±0·12	0·65	4·2–6·4
	Ant B	27	7·3±0·15	0·78	5·8–8·6
	Post B	6	6·2±0·53	1·30	4·7–8·0

Table 80
Alouatta villosa female

Mandibular teeth		n	$\bar{X}\pm$s.e.	s.d.	Range
I_1	M–D	31	2·6±0·18	1·00	2·0–2·9
	B–L	29	3·0±0·05	0·26	2·6–3·6
I_2	M–D	32	2·9±0·05	0·25	2·3–3·5
	B–L	28	3·7±0·06	0·09	3·2–4·5
C	M–D	30	4·6±0·10	0·53	3·6–6·0
	B–L	28	5·5±0·12	0·66	4·3±8·0
P_2	M–D	35	4·6±0·07	0·38	3·5–5·6
	B–L	34	5·5±0·08	0·45	4·8–6·9
P_3	M–D	35	4·7±0·07	0·38	4·0–6·3
	B–L	36	5·7±0·07	0·44	4·6–6·0
P_4	M–D	37	4·9±0·08	0·47	4·3–7·2
	B–L	36	5·7±0·07	0·44	5·0–6·9
M_1	M–D	37	6·8±0·07	0·39	6·0–8·0
	Tri B	36	5·2±0·05	0·30	4·6–5·8
	Tal B	37	5·7±0·05	0·32	5·1–6·4
M_2	M–D	37	7·3±0·06	0·35	6·6–8·0
	Tri B	36	5·6±0·06	0·34	5·0–6·3
	Tal B	36	6·3±0·06	0·33	5·8–7·0
M_3	M–D	37	7·6±0·10	0·63	6·0–9·3
	Tri B	36	5·7±0·06	0·35	4·8–6·7
	Tal B	36	5·4±0·05	0·33	4·9–6·1
	Hypo-conulid				

Table 81
Cercopithecus cephus male

Maxillary teeth		n	$\bar{X} \pm$ s.e.	s.d.	Range	P
I^1	M–D	8	6·2±0·08	0·22	5·8–6·5	
	B–L	8	4·9±0·09	0·26	4·4–5·1	
I^2	M–D	7	3·3±0·08	0·22	3·0–3·6	
	B–L	8	3·6±0·07	0·21	3·2–3·9	
C	M–D					
	B–L	3	4·6±0·15	0·25	4·5–4·9	
P^3	M–D	7	4·0±0·09	0·24	3·7–4·3	
	B–L	9	3·9±0·04	0·13	3·7–4·1	
P^4	M–D	6	4·3±0·11	0·27	4·0–4·7	*
	B–L	9	4·5±0·07	0·19	4·2–4·8	
M^1	M–D	8	5·7±0·07	0·21	5·4–6·0	*
	Ant B	9	5·3±0·06	0·19	5·0–5·6	
	Post B	9	5·0±0·48	0·14	4·8–5·2	
M^2	M–D	9	6·1±0·07	0·22	5·8–6·5	†
	Ant B	9	6·0±0·10	0·30	5·5–6·5	
	Post B	9	5·4±0·11	0·32	4·8–5·8	
M^3	M–D	7	5·4±0·12	0·33	5·0–5·9	†
	Ant B	7	5·2±0·09	0·25	4·9–5·5	
	Post B	6	4·1±0·21	0·52	3·5–4·9	

* $P<0{\cdot}05$
† $P<0{\cdot}01$

Table 82
Cercopithecus cephus male

Mandibular teeth		n	$\bar{X} \pm$ s.e.	s.d.	Range	P
I_1	M–D	6	3·8±0·12	0·30	3·2–4·1	
	B–L	6	4·2±0·06	0·14	4·0–4·3	
I_2	M–D	4	3·5±0·16	0·31	3·1–3·8	*
	B–L	6	4·0±0·04	0·09	3·9–4·1	
C	M–D	4	4·2±0·06	0·13	4·0–4·3	
	B–L	6	6·0±0·12	0·29	5·6–6·5	†
P_3	M–D	5	8·0±0·20	0·45	7·6–8·6	†
	B–L	7	3·3±0·11	0·30	2·8–3·7	
P_4	M–D	8	5·0±0·17	0·47	4·4–5·9	†
	B–L	8	3·5±0·07	0·19	3·1–3·7	
M_1	M–D	8	5·6±0·09	0·25	5·3–6·1	
	Tri B	8	4·1±0·06	0·18	3·7–4·3	
	Tal B	8	4·2±0·07	0·19	3·9–4·5	
M_2	M–D	8	6·1±0·05	0·15	5·9–6·4	†
	Tri B	8	4·8±0·08	0·23	4·5–5·1	*
	Tal B	7	4·9±0·12	0·32	4·5–5·3	
M_3	M–D	6	6·0±0·08	0·20	5·6–6·1	†
	Tri B	6	4·7±0·11	0·26	4·2–4·9	*
	Tal B	6	4·3±0·13	0·33	3·8–4·8	
	Hypo-conulid					

* $P<0{\cdot}05$
† $P<0{\cdot}01$

Table 83
Cercopithecus cephus female

Maxillary teeth		n	$\bar{X}\pm$s.e.	s.d.	Range
I[1]	M–D	12	5·9±0·16	0·56	5·0–6·7
	B–L	11	4·6±0·11	0·38	3·7–5·1
I[2]	M–D	10	3·4±0·09	0·30	2·9–3·8
	B–L	11	3·4±0·09	0·29	2·9–3·8
C	M–D	6	5·2±0·10	0·25	5·0–5·6
	B–L	8	4·1±0·18	0·50	3·0–4·6
P[3]	M–D	9	3·8±0·08	0·24	3·4–4·1
	B–L	9	3·8±0·10	0·30	3·2–4·1
P[4]	M–D	9	4·0±0·09	0·28	3·6–4·5
	B–L	10	4·6±0·08	0·25	4·2–5·1
M[1]	M–D	11	5·4±0·11	0·36	4·7–5·9
	Ant B	11	5·1±0·10	0·34	4·4–5·5
	Post B	12	4·9±0·08	0·27	4·3–5·2
M[2]	M–D	11	5·6±0·12	0·39	5·1–6·2
	Ant B	11	5·8±0·10	0·34	5·0–6·1
	Post B	12	5·2±0·11	0·38	4·4–5·7
M[3]	M–D	7	4·8±0·10	0·27	4·5–5·2
	Ant B	8	5·0±0·15	0·42	4·3–5·6
	Post B	7	4·1±0·16	0·42	3·6–4·7

Table 84
Cercopithecus cephus female

Mandibular teeth		n	$\bar{X}\pm$s.e.	s.d.	Range
I_1	M–D	10	3·6±0·11	0·35	3·0–4·0
	B–L	9	4·0±0·09	0·28	3·5–4·3
I_2	M–D	9	3·1±0·09	0·28	2·8–3·6
	B–L	9	3·8±0·10	0·31	3·4–4·4
C	M–D	9	3·7±0·26	0·78	3·0–5·2
	B–L	9	4·5±0·21	0·63	3·2–5·3
P_3	M–D	7	6·1±0·15	0·39	5·5–6·5
	B–L	9	3·0±0·09	0·28	2·7–3·6
P_4	M–D	12	4·6±0·07	0·25	4·0–4·9
	B–L	11	3.4±0·08	0·28	3·0–3·8
M_1	M–D	12	5·4±0·09	0·32	4·8–5·9
	Tri B	10	3·8±0·10	0·32	3·1–4·1
	Tal B	9	4·0±0·11	0·32	3·3–4·4
M_2	M–D	12	5·7±0·09	0·32	5·3–6·2
	Tri B	10	4·6±0·08	0·25	4·1–4·9
	Tal B	10	4·6±0·09	0·29	4·0–5·0
M_3	M–D	8	5·5±0·10	0·29	5·1–5·8
	Tri B	7	4·5±0·13	0·33	3·9–4·9
	Tal B	6	3·9±0·11	0·26	3·5–4·2
	Hypo-conulid				

Table 85
Cercopithecus nictitans male

Maxillary teeth		n	$\bar{X} \pm$ s.e.	s.d.	Range	P
I[1]	M–D	11	6·0±0·19	0·62	4·8–6·8	
	B–L	11	4·7±0·11	0·36	4·1–5·0	
I[2]	M–D	11	3·4±0·16	0·54	2·7–4·6	
	B–L	11	3·7±0·15	0·48	3·0–4·3	
C	M–D	8	7·5±0·37	1·05	5·9–8·8	†
	B–L	6	4·8±0·24	0·58	3·9–5·3	*
P[3]	M–D	11	4·0±0·17	0·58	3·0–4·9	
	B–L	11	3·9±0·10	0·32	3·2–4·5	
P[4]	M–D	11	4·1±0·19	0·63	3·4–5·4	
	B–L	11	4·7±0·11	0·36	3·8–5·0	
M[1]	M–D	11	5·6±0·15	0·51	4·8–6·3	
	Ant B	11	5·1±0·12	0·41	4·2–5·7	
	Post B	10	4·8±0·12	0·39	4·0–5·2	
M[2]	M–D	11	6·2±0·18	0·60	5·3–7·2	
	Ant B	11	5·9±0·15	0·49	4·9–6·4	
	Post B	11	5·5±0·13	0·43	4·9–6·3	
M[3]	M–D	10	5·1±0·16	0·51	4·3–5·8	
	Ant B	10	5·3±0·12	0·39	4·5–5·7	
	Post B	9	4·3±0·19	0·58	3·3–5·2	

* $P<0{\cdot}05$
† $P<0{\cdot}01$

Table 86
Cercopithecus nictitans male

Mandibular teeth		n	$\bar{X} \pm$ s.e.	s.d.	Range	P
I_1	M–D	11	3·8±0·13	0·44	3·3–4·6	
	B–L	11	4·2±0·12	0·39	3·7–4·9	
I_2	M–D	11	3·3±0·11	0·36	2·9–4·0	
	B–L	11	4·1±0·14	0·47	3·5–4·9	
C	M–D	9	6·6±0·23	0·68	5·8–8·0	†
	B–L	7	5·2±0·35	0·93	4·2–6·8	*
P_3	M–D	8	8·2±0·27	0·77	7·2–9·2	†
	B–L	10	3·7±0·12	0·36	3·2–4·3	
P_4	M–D	10	4·9±0·21	0·67	4·1–5·6	
	B–L	9	3·6±0·09	0·27	3·1–3·9	
M_1	M–D	11	5·7±0·17	0·57	4·9–6·7	
	Tri B	11	4·0±0·12	0·40	3·0–4·5	
	Tal B	10	4·2±0·06	0·20	3·8–4·5	
M_2	M–D	10	6·2±0·14	0·45	5·7–7·0	
	Tri B	10	5·0±0·05	0·15	4·7–5·2	
	Tal B	9	5·0±0·07	0·22	4·7–5·3	
M_3	M–D	8	5·9±0·17	0·49	5·4–6·9	
	Tri B	8	4·8±0·12	0·33	4·5–5·5	
	Tal B	8	4·3±0·13	0·35	3·8–4·8	
	Hypo-conulid					

* $P<0{\cdot}05$
† $P<0{\cdot}01$

Table 87
Cercopithecus nictitans female

Maxillary teeth		n	$\bar{X} \pm$ s.e.	s.d.	Range
I^1	M–D	6	6·1 ± 0·14	0·33	5·7–6·5
	B–L	4	4·5 ± 0·11	0·22	4·2–4·7
I^2	M–D	5	3·3 ± 0·06	0·14	3·1–3·4
	B–L	5	3·6 ± 0·30	0·68	3·2–4·8
C	M–D	4	4·9 ± 0·20	0·41	4·6–5·5
	B–L	4	4·1 ± 0·13	0·25	3·8–4·4
P^3	M–D	5	3·7 ± 0·18	0·40	3·4–4·4
	B–L	5	3·9 ± 0·14	0·31	3·6–4·4
P^4	M–D	6	3·8 ± 0·15	0·36	3·5–4·5
	B–L	6	4·6 ± 0·10	0·25	4·2–4·9
M^1	M–D	7	5·5 ± 0·16	0·43	4·9–6·2
	Ant B	7	5·1 ± 0·12	0·32	4·8–5·6
	Post B	6	4·9 ± 0·14	0·34	4·5–5·4
M^2	M–D	6	5·8 ± 0·17	0·41	5·2–6·4
	Ant B	6	5·8 ± 0·08	0·19	5·6–6·1
	Post B	6	5·4 ± 0·16	0·39	5·0–6·0
M^3	M–D	5	5·3 ± 0·23	0·51	4·6–6·0
	Ant B	5	5·2 ± 0·23	0·51	4·7–6·0
	Post B	5	4·2 ± 0·33	0·73	3·2–5·0

Table 88
Cercopithecus nictitans female

Mandibular teeth		n	$\bar{X} \pm$ s.e.	s.d.	Range
I_1	M–D	7	3·6 ± 0·20	0·52	2·5–4·1
	B–L	7	4·1 ± 0·16	0·43	3·5–4·9
I_2	M–D	6	3·3 ± 0·16	0·39	2·6–3·8
	B–L	6	3·9 ± 0·21	0·52	3·5–4·9
C	M–D	2	4·5 ± 0·00	0·00	4·5–4·5
	B–L	2	3·3 ± 0·20	0·28	3·1–3·5
P_3	M–D	5	5·8 ± 0·23	0·51	5·5–6·7
	B–L	6	3·3 ± 0·24	0·59	2·8–4·5
P_4	M–D	4	4·6 ± 0·16	0·32	4·3–5·0
	B–L	5	3·6 ± 0·05	0·11	3·5–3·7
M_1	M–D	7	5·5 ± 0·17	0·46	4·7–6·2
	Tri B	6	3·9 ± 0·11	0·27	3·7–4·3
	Tal B	7	4·2 ± 0·10	0·25	3·8–4·5
M_2	M–D	7	5·8 ± 0·12	0·31	5·5–6·4
	Tri B	6	4·8 ± 0·09	0·21	4·5–5·1
	Tal B	6	4·8 ± 0·10	0·25	4·5–5·2
M_3	M–D	5	6·0 ± 0·14	0·31	5·7–6·5
	Tri B	4	4·8 ± 0·14	0·28	4·5–5·1
	Tal B	5	4·3 ± 0·17	0·39	3·8–4·8
	Hypo-conulid				

Table 89
Cercopithecus mona male

Maxillary teeth		n	$\bar{X} \pm$ s.e.	s.d.	Range	P
I^1	M–D	18	5·5±0·09	0·39	5·0–6·2	
	B–L	14	4·5±0·06	0·24	4·0–4·8	
I^2	M–D	20	3·3±0·06	0·27	3·0–4·1	
	B–L	20	3·4±0·08	0·35	2·8–4·5	
C	M–D	16	7·2±0·29	1·18	6·4–9·3	†
	B–L	14	5·2±0·22	0·81	4·1–7·5	†
P^3	M–D	23	3·9±0·07	0·35	3·3–3·9	
	B–L	23	3·8±0·07	0·35	3·3–4·9	
P^4	M–D	25	4·0±0·07	0·32	3·5–4·9	
	B–L	25	4·6±0·07	0·37	4·0–5·5	
M^1	M–D	25	5·3±0·09	0·46	4·7–7·0	
	Ant B	23	5·2±0·07	0·34	4·4–5·8	
	Post B	23	5·0±0·06	0·30	4·3–5·5	
M^2	M–D	26	5·8±0·08	0·43	5·2–7·4	
	Ant B	26	6·1±0·08	0·41	5·3–6·7	*
	Post B	26	5·6±0·08	0·40	4·7–6·6	*
M^3	M–D	23	5·1±0·11	0·54	4·6–7·1	
	Ant B	23	5·4±0·08	0·36	4·7–6·3	†
	Post B	22	4·5±0·08	0·39	3·8–5·5	

* $P < 0·05$
† $P < 0·01$

Table 90
Cercopithecus mona male

Mandibular teeth		n	$\bar{X} \pm$ s.e.	s.d.	Range	P
I_1	M–D	21	3·5±0·10	0·45	2·8–4·6	
	B–L	22	3·9±0·08	0·37	3·4–5·2	
I_2	M–D	19	3·1±0·06	0·28	2·7–3·8	†
	B–L	22	3·9±0·08	0·38	3·5–5·4	
C	M–D	11	4·3±0·25	0·83	3·4–6·0	*
	B–L	16	5·5±0·28	1·13	3·2–8·8	†
P_3	M–D	22	7·8±0·18	0·85	6·3–10·2	†
	B–L	24	3·1±0·07	0·35	2·5–3·8	
P_4	M–D	24	4·5±0·07	0·37	4·0–5·8	
	B–L	26	3·5±0·05	0·26	2·9–3·8	
M_1	M–D	27	5·3±0·08	0·43	4·8–7·2	
	Tri B	21	3·9±0·04	0·18	3·5–4·2	
	Tal B	25	4·0±0·06	0·32	3·6–5·2	
M_2	M–D	25	5·9±0·09	0·43	5·4–7·5	
	Tri B	23	5·0±0·05	0·26	4·5–5·5	†
	Tal B	25	4·9±0·06	0·29	4·5–5·8	
M_3	M–D	22	5·7±0·17	0·79	5·2–9·0	
	Tri B	23	4·8±0·05	0·25	4·5–5·6	
	Tal B	23	4·4±0·08	0·38	3·9–5·9	
	Hypo-conulid					

* $P < 0·05$
† $P < 0·01$

Table 91
Cercopithecus mona female

Maxillary teeth		n	$\bar{X} \pm$ s.e.	s.d.	Range
I^1	M–D	12	5·3 ± 0·14	0·49	4·6–6·1
	B–L	11	4·4 ± 0·11	0·37	3·8–5·1
I^2	M–D	13	3·1 ± 0·08	0·28	2·7–3·8
	B–L	13	3·3 ± 0·11	0·40	2·8–4·3
C	M–D	8	4·9 ± 0·16	0·46	4·3–5·8
	B–L	11	4·0 ± 0·14	0·47	3·3–4·8
P^3	M–D	11	3·7 ± 0·11	0·37	3·1–4·2
	B–L	12	3·6 ± 0·15	0·51	3·2–4·9
P^4	M–D	11	4·0 ± 0·10	0·34	3·5–4·5
	B–L	15	4·4 ± 0·10	0·39	3·9–5·4
M^1	M–D	13	5·3 ± 0·11	0·41	4·7–6·1
	Ant B	16	5·0 ± 0·10	0·40	4·4–5·7
	Post B	15	4·8 ± 0·10	0·40	4·3–5·6
M^2	M–D	15	5·7 ± 0·10	0·39	5·1–6·6
	Ant B	14	5·8 ± 0·11	0·43	5·2–6·7
	Post B	14	5·2 ± 0·11	0·42	4·6–6·2
M^3	M–D	9	4·9 ± 0·13	0·39	4·6–5·7
	Ant B	11	5·1 ± 0·15	0·51	4·6–6·4
	Post B	10	4·0 ± 0·11	0·34	3·4–4·5

Table 92
Cercopithecus mona female

Mandibular teeth		n	$\bar{X} \pm$ s.e.	s.d.	Range
I_1	M–D	9	3·4 ± 0·14	0·40	2·6–3·9
	B–L	7	3·8 ± 0·20	0·54	3·1–4·7
I_2	M–D	11	2·8 ± 0·07	0·23	2·5–3·2
	B–L	12	3·8 ± 0·12	0·43	3·3–4·7
C	M–D	4	3·1 ± 0·11	0·22	2·9–3·4
	B–L	9	4·3 ± 0·18	0·54	3·6–5·5
P_3	M–D	8	6·2 ± 0·49	1·37	4·5–8·4
	B–L	11	2·9 ± 0·12	0·41	2·3–3·6
P_4	M–D	14	4·3 ± 0·08	0·30	3·8–4·9
	B–L	14	3·3 ± 0·09	0·32	2·9–3·9
M_1	M–D	15	5·2 ± 0·09	0·36	4·6–5·8
	Tri B	11	3·7 ± 0·11	0·36	3·3–4·4
	Tal B	13	3·9 ± 0·10	0·36	3·5–4·6
M_2	M–D	16	5·6 ± 0·09	0·36	5·2–6·5
	Tri B	16	4·7 ± 0·10	0·38	4·1–5·5
	Tal B	15	4·7 ± 0·09	0·33	4·1–5·5
M_3	M–D	10	5·4 ± 0·11	0·34	5·0–6·1
	Tri B	11	4·5 ± 0·12	0·41	4·0–5·5
	Tal B	10	4·1 ± 0·10	0·31	3·8–4·8
	Hypoconulid				

Table 93
Cercopithecus mitis male

Maxillary teeth		n	$\bar{X} \pm$s.e.	s.d.	Range	P
I¹	M–D	30	6·0±0·08	0·42	5·2–6·8	
	B–L	25	4·9±0·18	0·92	4·2–9·0	
I²	M–D	25	3·6±0·08	0·40	2·9–4·6	
	B–L	25	3·9±0·07	0·32	3·4–4·5	
C	M–D	11	7·9±0·48	1·59	3·4–9·2	†
	B–L	10	5·7±0·24	0·76	4·2–6·8	†
P³	M–D	27	4·3±0·08	0·43	3·5–5·0	
	B–L	32	4·3±0·07	0·34	3·7–4·9	
P⁴	M–D	32	4·5±0·06	0·33	3·7–5·2	*
	B–L	32	5·1±0·07	0·38	4·4–5·7	
M¹	M–D	32	6·1±0·05	0·31	5·5–6·8	
	Ant B	28	5·5±0·07	0·35	4·8–6·1	
	Post B	27	5·3±0·07	0·37	4·5–6·2	
M²	M–D	31	6·6±0·06	0·34	5·8–7·4	*
	Ant B	30	6·3±0·08	0·43	5·5–7·0	*
	Post B	30	5·8±0·09	0·47	4·7-6·6	*
M³	M–D	27	5·8±0·08	0·43	4·9–6·7	
	Ant B	26	5·7±0·09	0·45	4·6–6·6	*
	Post B	24	4·8±0·09	0·41	4·0–5·9	*

* $P<0·05$
† $P<0·01$

Table 94
Cercopithecus mitis male

Mandibular teeth		n	$\bar{X} \pm$s.e.	s.d.	Range	P
I₁	M–D	27	3·8±0·06	0·34	3·0–4·6	
	B–L	28	4·4±0·07	0·39	3·6–5·5	*
I₂	M–D	29	3·5±0·08	0·41	2·8–4·6	
	B–L	28	4·3±0·07	0·39	3·6–5·1	
C	M–D	24	7·3±0·14	0·67	5·6–8·3	†
	B–L	22	5·3±0·09	0·44	4·4–5·8	†
P₃	M–D	18	9·2±0·34	1·44	4·6–11·2	†
	B–L	29	4·0±0·08	0·42	3·1–4·7	†
P₄	M–D	29	5·2±0·09	0·46	4·5–6·3	
	B–L	25	3·8±0·06	0·31	3·2–4·5	
M₁	M–D	30	6·2±0·06	0·34	5·5–6·8	
	Tri B	24	4·3±0·06	0·30	3·9–5·0	
	Tal B	25	4·6±0·06	0·31	4·2–5·2	
M₂	M–D	31	6·7±0·06	0·31	6·1–7·5	*
	Tri B	26	5·3±0·09	0·44	4·6–6·3	
	Tal B	27	5·4±0·08	0·40	4·6–6·1	
M₃	M–D	25	6·6±0·08	0·40	5·9–7·7	*
	Tri B	23	5·2±0·10	0·46	4·4–6·2	
	Tal B	23	4·7±0·08	0·36	3·9–5·7	
	Hypoconulid					

* $P<0·05$
† $P<0·01$

Table 95
Cercopithecus mitis female

Maxillary teeth		n	$\bar{X} \pm$s.e.	s.d.	Range
I^1	M–D	18	5·9±0·10	0·42	5·3–6·6
	B–L	13	4·8±0·16	0·59	4·2–6·4
I^2	M–D	18	3·6±0·07	0·28	3·1–4·0
	B–L	17	3·9±0·09	0·36	3·5–4·5
C	M–D	14	5·5±0·12	0·45	5·0–6·5
	B–L	13	4·4±0·08	0·30	3·7–4·7
P^3	M–D	19	4·0±0·09	0·41	3·4–4·9
	B–L	19	4·2±0·11	0·49	3·6–5·7
P^4	M–D	20	4·3±0·07	0·29	3·8–4·8
	B–L	20	5·0±0·07	0·32	4·6–5·9
M^1	M–D	21	5·9±0·08	0·37	5·5–6·8
	Ant B	21	5·4±0·08	0·36	4·9–6·2
	Post B	19	5·2±0·08	0·36	4·7–6·0
M^2	M–D	20	6·3±0·08	0·36	5·8–7·0
	Ant B	20	6·0±0·08	0·36	5·5–7·0
	Post B	20	5·5±0·07	0·33	5·0–6·3
M^3	M–D	14	5·6±0·16	0·60	4·8–7·3
	Ant B	15	5·4±0·09	0·33	5·0–5·9
	Post B	14	4·4±0·16	0·60	3·5–6·0

Table 96
Cercopithecus mitis female

Mandibular teeth		n	$\bar{X} \pm$s.e.	s.d.	Range
I_1	M–D	19	3·7±0·07	0·31	3·2–4·4
	B–L	20	4·2±0·08	0·36	3·2–4·7
I_2	M–D	18	3·3±0·08	0·35	2·7–4·0
	B–L	18	4·2±0·06	0·24	3·8–4·7
C	M–D	15	5·4±0·14	0·55	4·7–6·8
	B–L	13	3·7±0·16	0·56	3·0–5·3
P_3	M–D	17	6·7±0·18	0·75	5·6–8·5
	B–L	15	3·4±0·09	0·35	2·8–4·0
P_4	M–D	16	4·9±0·09	0·35	4·5–5·7
	B–L	15	3·7±0·05	0·21	3·4–4·0
M_1	M–D	21	6·0±0·06	0·28	5·7–6·6
	Tri B	18	4·3±0·06	0·26	3·9–4·9
	Tal B	17	4·5±0·05	0·21	4·2–5·0
M_2	M–D	21	6·5±0·08	0·36	5·6–7·2
	Tri B	17	5·2±0·09	0·38	4·4–5·8
	Tal B	18	5·2±0·07	0·31	4·7–5·8
M_3	M–D	18	6·3±0·07	0·31	5·7–6·8
	Tri B	16	5·4±0·10	0·41	4·2–5·6
	Tal B	15	4·5±0·08	0·29	4·0–5·2
	Hypoconulid				

Table 97
Cercopithecus neglectus male

Maxillary teeth		n	$\bar{X}\pm$s.e.	s.d.	Range	P
I^1	M–D	12	5·7±0·09	0·32	5·4–6·3	*
	B–L	8	4·5±0·11	0·31	4·1–4·8	
I^2	M–D	9	3·5±0·07	0·20	3·2–3·8	
	B–L	8	3·7±0·10	0·28	3·2–4·1	
C	M–D	7	8·3±0·52	1·38	5·5–9·4	†
	B–L	6	6·1±0·23	0·56	5·3–6·8	†
P^3	M–D	10	4·6±0·09	0·28	4·2–5·0	*
	B–L	9	4·3±0·05	0·14	4·1–4·4	†
P^4	M–D	10	4·9±0·06	0·19	4·7–5·1	†
	B–L	7	5·0±0·10	0·25	4·7–5·5	†
M^1	M–D	13	6·3±0·05	0·19	5·9–6·6	†
	Ant B	11	5·5±0·05	0·17	5·3–5·8	*
	Post B	11	5·3±0·06	0·21	5·0–5·6	†
M^2	M–D	12	6·9±0·07	0·23	6·5–7·2	†
	Ant B	11	6·5±0·08	0·28	6·1–7·1	†
	Post B	12	6·0±0·07	0·23	5·6–6·3	†
M^3	M–D	11	5·9±0·09	0·30	5·5–6·5	
	Ant B	10	5·8±0·08	0·26	5·3–6·1	†
	Post B	9	4·8±0·11	0·34	4·3–5·3	†

* $P<0·05$
† $P<0·01$

Table 98
Cercopithecus neglectus male

Mandibular teeth		n	$\bar{X}\pm$s.e.	s.d.	Range	P
I_1	M–D	12	3·6±0·06	0·20	3·2–3·9	*
	B–L	11	4·2±0·07	0·24	3·7–4·5	*
I_2	M–D	10	3·4±0·10	0·32	2·8–3·9	
	B–L	11	4·1±0·08	0·27	3·6–4·3	
C	M–D	6	4·5±0·16	0·39	3·9–5·1	†
	B–L	8	6·9±0·24	0·67	5·5–7·5	†
P_3	M–D	7	9·4±0·44	1·16	7·0–10·3	†
	B–L	10	3·7±0·05	0·14	3·4–3·8	†
P_4	M–D	9	5·5±0·08	0·23	5·2–5·9	
	B–L	9	3·8±0·10	0·29	3·4–4·4	
M_1	M–D	11	6·3±0·08	0·26	5·7–6·6	†
	Tri B	9	4·4±0·05	0·14	4·1–4·6	
	Tal B	10	4·6±0·05	0·17	4·2–4·8	
M_2	M–D	12	6·7±0·10	0·33	6·1–7·4	†
	Tri B	11	5·4±0·05	0·16	5·2–5·9	*
	Tal B	11	5·4±0·04	0·13	5·2–5·6	†
M_3	M–D	9	6·8±0·11	0·33	6·3–7·2	
	Tri B	9	5·3±0·07	0·21	5·0–6·0	
	Tal B	8	4·8±0·09	0·26	4·5–5·2	
	Hypoconulid					

* $P<0·05$
† $P<0·01$

Table 99
Cercopithecus neglectus female

Maxillary teeth		n	$\bar{X}\pm$s.e.	s.d.	Range
I^1	M–D	6	5·3±0·21	0·52	4·5–5·9
	B–L	3	4·3±0·10	0·17	4·2–4·5
I^2	M–D	6	3·3±0·09	0·23	3·1–3·7
	B–L	6	3·4±0·06	0·15	3·2–3·6
C	M–D	5	5·6±0·20	0·44	5·1–6·1
	B–L	6	4·0±0·07	0·17	3·8–4·3
P^3	M–D	7	4·3±0·08	0·22	4·0–4·7
	B–L	7	3·9±0·11	0·29	3·6–4·3
P^4	M–D	6	4·4±0·04	0·10	4·3–4·6
	B–L	6	4·7±0·10	0·23	4·4–4·9
M^1	M–D	7	5·9±0·11	0·28	5·6–6·3
	Ant B	6	5·2±0·10	0·25	4·9–5·6
	Post B	6	5·0±0·07	0·16	4·8–5·2
M^2	M–D	6	6·2±0·17	0·41	5·8–6·8
	Ant B	6	6·0±0·12	0·30	5·6–6·4
	Post B	7	5·4±0·11	0·30	5·0–5·9
M^3	M–D	6	5·6±0·15	0·36	5·3–6·3
	Ant B	6	5·2±0·07	0·17	5·0–5·5
	Post B	5	4·2±0·10	0·23	3·9–4·5

Table 100
Cercopithecus neglectus female

Mandibular teeth		n	$\bar{X}\pm$s.e.	s.d.	Range
I_1	M–D	6	3·2±0·17	0·42	2·6–3·8
	B–L	5	3·8±0·09	0·21	3·6–4·1
I_2	M–D	6	3·1±0·08	0·19	2·7–3·2
	B–L	5	3·9±0·15	0·33	3·4–4·2
C	M–D	6	3·4±0·09	0·23	3·1–3·6
	B–L	6	4·7±0·24	0·58	3·7–5·4
P_3	M–D	7	6·4±0·23	0·60	5·6–7·3
	B–L	7	3·2±0·08	0·22	2·8–3·4
P_4	M–D	7	5·2±0·15	0·40	4·7–6·0
	B–L	7	3·7±0·09	0·24	3·4–4·0
M_1	M–D	7	5·9±0·10	0·27	5·5–6·4
	Tri B	5	4·4±0·20	0·44	4·0–5·1
	Tal B	5	4·5±0·17	0·39	4·1–5·1
M_2	M–D	7	6·3±0·13	0·35	5·9–7·0
	Tri B	5	5·2±0·08	0·18	5·1–5·5
	Tal B	6	5·2±0·12	0·29	4·7–5·5
M_3	M–D	5	6·4±0·16	0·35	6·0–6·7
	Tri B	6	5·1±0·04	0·11	4·9–5·2
	Tal B	6	4·6±0·07	0·17	4·4–4·9
	Hypo-conulid				

Table 101
Cercopithecus ascanius male

Maxillary teeth		n	$\bar{X} \pm$ s.e.	s.d.	Range
I^1	M–D	17	5·42±0·11	0·45	4·6–6·2
	B–L	17	4·19±0·06	0·26	3·7–4·5
I^2	M–D	15	2·86±0·06	0·24	2·5–3·4
	B–L	17	3·13±0·06	0·23	2·7–3·4
C	M–D	10	6·13±0·19	0·60	5·3–6·8
	B–L	11	4·24±0·08	0·26	3·9–4·7
P^3	M–D	9	3·37±0·10	0·40	3·0–3·4
	B–L	18	3·60±0·06	0·27	3·0–3·9
P^4	M–D	16	3·81±0·04	0·17	3·6–4·2
	B–L	18	4·24±0·06	0·23	3·7–4·5
M^1	M–D	17	5·09±0·05	0·21	4·7–5·5
	Ant B	18	4·78±0·09	0·38	3·7–5·2
	Post B	18	4·49±0·07	0·28	4·1–4·9
M^2	M–D	18	5·54±0·06	0·24	5·2–6·0
	Ant B	18	5·46±0·07	0·29	5·0–6·1
	Post B	18	4·94±0·07	0·28	4·7–5·6
M^3	M–D	17	5·37±0·10	0·49	4·6–6·3
	Ant B	17	4·74±0·07	0·30	4·3–5·5
	Post B	10	3·68±0·17	0·55	2·9–4·5

J. E. Sirianni, 1974

Table 102
Cercopithecus ascanius male

Mandibular teeth		n	$\bar{X} \pm$ s.e.	s.d.	Range
I_1	M–D	18	3·28±0·05	0·22	2·9–3·7
	B–L	15	3·59±0·05	0·18	3·2–3·9
I_2	M–D	15	2·97±0·07	0·27	2·5–3·5
	B–L	15	3·45±0·08	0·30	2·8–3·9
C	M–D	13	4·11±0·17	0·61	3·3–5·7
	B–L	16	5·13±0·10	0·38	4·5–5·6
P_3	M–D	16	6·78±0·23	0·56	6·0–7·6
	B–L	16	3·07±0·07	0·27	2·7–3·5
P_4	M–D	15	4·23±0·09	0·34	3·7–4·9
	B–L	16	3·34±0·05	0·20	3·1–3·8
M_1	M–D	15	5·20±0·07	0·27	4·6–5·6
	Tri B	16	3·79±0·05	0·19	3·3–4·2
	Tal B	16	3·98±0·04	0·16	3·7–4·3
M_2	M–D	19	5·54±0·05	0·22	5·1–6·0
	Tri B	17	4·59±0·05	0·22	4·3–5·0
	Tal B	17	4·56±0·05	0·22	4·3–5·0
M_3	M–D	17	5·32±0·09	0·39	4·5–5·9
	Tri B	15	4·27±0·05	0·20	4·1–4·7
	Tal B	15	3·80±0·06	0·24	3·5–4·3
	Hypo-conulid				

J. E. Sirianni, 1974

Table 103
Cercopithecus ascanius female

Maxillary teeth		n	$\bar{X}\pm$s.e.	s.d.	Range
I^1	M–D	12	5·20±0·11	0·37	4·6–5·8
	B–L	12	4·06±0·08	0·27	3·6–4·6
I^2	M–D	12	2·92±0·06	0·20	2·7–3·3
	B–L	13	3·05±0·06	0·20	2·8–3·5
C	M–D	10	4·40±0·12	0·37	3·6–5·1
	B–L	9	3·56±0·09	0·28	3·1–4·0
P^3	M–D	8	3·45±0·09	0·27	2·9–3·7
	B–L	13	3·55±0·06	0·23	3·1–4·0
P^4	M–D	12	3·72±0·04	0·15	3·5–4·1
	B–L	13	4·22±0·07	0·25	3·8–4·7
M^1	M–D	12	4·91±0·09	0·31	4·6–5·8
	Ant B	13	4·76±0·08	0·30	4·3–5·4
	Post B	13	4·51±0·08	0·28	4·2–5·1
M^2	M–D	13	5·29±0·08	0·29	4·9–5·9
	Ant B	13	5·32±0·10	0·39	4·8–6·1
	Post B	13	4·89±0·09	0·33	4·5–5·5
M^3	M–D	10	4·27±0·10	0·31	3·8–4·9
	Ant B	7	4·56±0·09	0·23	4·4–5·0
	Post B	6	3·48±0·14	0·33	3·1–3·9

J. E. Sirianni, 1974

Table 104
Cercopithecus ascanius female

Mandibular teeth		n	$\bar{X}\pm$s.e.	s.d.	Range
I_1	M–D	11	3·27±0·06	0·20	3·0–3·7
	B–L	10	3·52±0·07	0·21	3·3–3·9
I_2	M–D	8	2·98±0·10	0·29	2·6–3·4
	B–L	11	3·39±0·04	0·12	3·2–3·6
C	M–D	9	3·78±0·05	0·16	2·5–3·0
	B–L	10	3·96±0·07	0·23	3·4–4·2
P_3	M–D	5	5·44±0·26	0·59	4·6–6·2
	B–L	10	2·65±0·06	0·18	2·4–3·0
P_4	M–D	11	4·24±0·08	0·27	3·9–4·8
	B–L	13	3·14±0·06	0·22	2·8–3·5
M_1	M–D	11	5·01±0·08	0·26	4·8–5·7
	Tri B	13	3·62±0·05	0·16	3·5–3·9
	Tal B	13	3·77±0·06	0·20	3·5–4·3
M_2	M–D	11	5·33±0·07	0·24	5·0–6·0
	Tri B	13	4·40±0·06	0·22	4·1–4·6
	Tal B	12	4·40±0·07	0·25	4·0–4·8
M_3	M–D	10	5·14±0·09	0·27	4·7–5·7
	Tri B	11	4·25±0·06	0·21	3·8–4·5
	Tal B	11	3·69±0·06	0·21	3·3–4·0
	Hypo-conulid				

J. E. Sirianni, 1974

Table 105
Cercopithecus aethiops male

Maxillary teeth		n	$\bar{X}\pm$s.e.	s.d.	Range
I^1	M–D	19	5·44±0·08	0·35	5·0–6·2
	B–L	20	4·61±0·07	0·33	3·9–5·1
I^2	M–D	24	3·47±0·08	0·40	2·9–4·2
	B–L	23	3·90±0·07	0·33	3·3–4·6
C	M–D	17	7·42±0·17	0·73	6·0–9·2
	B–L	17	5·24±0·14	0·57	4·4–6·9
P^3	M–D	21	4·02±0·08	0·35	3·2–4·7
	B–L	25	4·25±0·11	0·55	3·1–5·6
P^4	M–D	27	4·25±0·06	0·29	3·6±4·9
	B–L	27	5·03±0·07	0·36	4·3–5·6
M^1	M–D	28	5·74±0·06	0·31	5·3–6·6
	Ant B	27	5·59±0·07	0·34	4·9–6·4
	Post B	26	5·12±0·06	0·32	4·5–5·7
M^2	M–D	29	6·45±0·06	0·33	5·9–7·0
	Ant B	28	6·47±0·08	0·42	5·8–7·4
	Post B	27	5·82±0·08	0·39	5·1–6·9
M^3	M–D	22	5·37±0·10	0·49	4·6–6·3
	Ant B	23	5·72±0·13	0·63	3·9–7·2
	Post B	21	4·51±0·14	0·65	3·6–6·4

J. E. Sirianni, 1974

Table 106
Cercopithecus aethiops male

Mandibular teeth		n	$\bar{X}\pm$s.e.	s.d.	Range
I_1	M–D	26	3·64±0·06	0·31	3·1–4·3
	B–L	18	4·24±0·09	0·38	3·6–5·1
I_2	M–D	26	3·24±0·05	0·27	2·7–3·7
	B–L	23	4·29±0·11	0·54	3·6–5·7
C	M–D	19	4·34±0·13	0·58	3·3–5·3
	B–L	24	5·65±0·17	0·69	5·7–8·2
P_3	M–D	20	8·88±0·18	0·82	7·7–10·3
	B–L	24	3·64±0·07	0·34	3·0–4·5
P_4	M–D	24	4·73±0·07	0·36	4·1–5·5
	B–L	23	3·72±0·05	0·26	3·3–4·3
M_1	M–D	26	5·88±0·05	0·26	5·4–6·5
	Tri B	26	4·40±0·05	0·28	3·9–5·0
	Tal B	25	4·49±0·06	0·31	3·6–5·1
M_2	M–D	25	6·52±0·07	0·34	5·8–7·4
	Tri B	26	5·31±0·06	0·32	4·8–6·1
	Tal B	25	5·16±0·08	0·42	4·1–6·3
M_3	M–D	21	6·29±0·08	0·35	5·6–6·7
	Tri B	20	5·25±0·09	0·41	4·7–6·3
	Tal B	18	4·53±0·11	0·47	3·7–5·7
	Hypo-conulid				

J. E. Sirianni, 1974

Table 107
Cercopithecus aethiops female

Maxillary teeth		n	$\bar{X}$ ± s.e.	s.d.	Range
I^1	M–D	20	5·08 ± 0·08	0·34	4·5–5·8
	B–L	20	4·38 ± 0·08	0·36	3·7–5·0
I^2	M–D	18	3·28 ± 0·07	0·30	2·8–3·8
	B–L	20	3·73 ± 0·07	0·32	3·2–4·5
C	M–D	18	5·16 ± 0·10	0·43	4·5–5·8
	B–L	18	3·92 ± 0·09	0·40	3·5–5·0
P^3	M–D	11	3·61 ± 0·13	0·43	3·1–4·3
	B–L	20	3·92 ± 0·10	0·45	3·1–5·2
P^4	M–D	19	3·91 ± 0·08	0·37	3·4–4·8
	B–L	20	4·66 ± 0·08	0·34	4·3–5·5
M^1	M–D	20	5·36 ± 0·06	0·26	5·1–5·9
	Ant B	19	5·23 ± 0·08	0·36	4·7–6·0
	Post B	20	4·83 ± 0·07	0·30	4·4–5·5
M^2	M–D	20	5·96 ± 0·08	0·36	5·4–6·9
	Ant B	20	5·94 ± 0·08	0·37	5·2–6·8
	Post B	20	5·35 ± 0·05	0·24	4·9–5·8
M^3	M–D	11	4·95 ± 0·14	0·46	4·5–5·8
	Ant B	13	5·20 ± 0·12	0·42	4·5–6·2
	Post B	9	4·09 ± 0·15	0·44	3·3–4·7

J. E. Sirianni, 1974

Table 108
Cercopithecus aethiops female

Mandibular teeth		n	$\bar{X}$ ± s.e.	s.d.	Range
I_1	M–D	19	3·33 ± 0·06	0·26	3·0–4·0
	B–L	18	3·87 ± 0·10	0·42	3·1–4·8
I_2	M–D	14	3·09 ± 0·06	0·22	2·7–3·3
	B–L	17	3·92 ± 0·09	0·35	3·4–4·6
C	M–D	14	3·23 ± 0·12	0·46	2·3–4·2
	B–L	18	4·9 ± 0·14	0·61	4·1–6·6
P_3	M–D	16	6·71 ± 0·22	0·91	5·5–9·5
	B–L	17	3·02 ± 0·08	0·33	2·6–4·0
P_4	M–D	16	4·47 ± 0·08	0·34	4·1–5·2
	B–L	18	3·44 ± 0·05	0·23	3·0–4·0
M_1	M–D	17	5·59 ± 0·40	0·17	5·4–5·9
	Tri B	17	4·18 ± 0·05	0·21	3·8–4·7
	Tal B	18	4·19 ± 0·05	0·22	3·9–4·6
M_2	M–D	19	6·02 ± 0·06	0·26	5·7–6·8
	Tri B	18	4·97 ± 0·06	0·26	4·6–5·5
	Tal B	18	4·89 ± 0·05	0·21	4·6–5·3
M_3	M–D	11	5·76 ± 0·13	0·42	5·0–6·3
	Tri B	12	4·79 ± 0·09	0·33	4·4–5·3
	Tal B	13	4·25 ± 0·12	0·42	6·7–5·0
	Hypo-conulid				

J. E. Sirianni, 1974

Table 109
Cercocebus albigena male

Maxillary teeth		n	$\bar{X}\pm$s.e.	s.d.	Range	P
I[1]	M–D	28	8·20±0·07	0·37	7·5–8·8	†
	B–L	27	6·72±0·09	0·44	5·9–8·0	†
I[2]	M–D	23	5·87±0·08	0·36	5·4–6·3	†
	B–L	25	5·34±0·06	0·30	4·8–6·2	†
C	M–D	22	7·71±0·13	0·61	6·8–9·3	†
	B–L	24	5·28±0·16	0·77	3·9–6·6	†
P[3]	M–D	30	5·38±0·09	0·50	4·3–6·4	†
	B–L	29	5·81±0·05	0·29	5·4–6·3	†
P[4]	M–D	30	4·74±0·06	0·31	4·2–5·5	†
	B–L	30	5·90±0·05	0·25	5·4–6·4	†
M[1]	M–D	31	7·00±0·05	0·25	6·5–7·4	†
	Ant B	31	6·73±0·05	0·28	6·1–7·4	†
	Post B	30	6·57±0·05	0·27	6·1–7·2	
M[2]	M–D	29	7·58±0·04	0·22	7·1–8·2	†
	Ant B	29	7·36±0·05	0·26	7·0–8·0	†
	Post B	29	6·79±0·06	0·34	6·2–7·4	†
M[3]	M–D	24	6·81±0·07	0·33	6·4–7·5	†
	Ant B	26	6·88±0·06	0·29	6·5–7·5	†
	Post B	18	5·63±0·11	0·48	4·5–6·2	†

† $P<0·01$

Table 110
Cercocebus albigena male

Mandibular teeth		n	$\bar{X}\pm$s.e.	s.d.	Range	P
I_1	M–D	27	5·79±0·10	0·51	4·5–6·6	†
	B–L	21	6·28±0·09	0·40	5·6–7·3	†
I_2	M–D	25	4·81±0·11	0·54	3·7–6·4	†
	B–L	24	5·60±0·07	0·36	4·7–6·4	†
C	M–D	14	5·07±0·21	0·79	4·3–7·3	†
	B–L	20	7·40±0·07	0·29	6·7–7·9	†
P_3	M–D	22	7·42±0·28	1·32	4·7–10·7	†
	B–L	28	4·67±0·09	0·46	3·9–5·6	†
P_4	M–D	30	5·32±0·07	0·35	4·8–6·5	*
	B–L	29	4·93±0·05	0·26	4·5–5·5	†
M_1	M–D	30	6·88±0·04	0·20	6·5–7·2	†
	Tri B	30	5·57±0·04	0·21	5·1–6·0	†
	Tal B	30	5·67±0·05	0·28	5·3–6·6	†
M_2	M–D	31	7·62±0·05	0·28	6·8–8·2	†
	Tri B	31	6·56±0·05	0·27	6·0–7·1	†
	Tal B	31	6·18±0·05	0·28	5·7–6·8	†
M_3	M–D	27	8·32±0·09	0·45	7·3–9·1	†
	Tri B	27	6·39±0·06	0·31	5·9–7·1	†
	Tal B	25	5·41±0·06	0·31	4·9–6·1	†
	Hypoconulid	16	2·88±0·11	0·45	1·7–3·3	†

* $P<0·05$
† $P<0·01$

Table 111
Cercocebus albigena female

Maxillary teeth		n	$\bar{X} \pm$s.e.	s.d.	Range
I^1	M–D	29	7·50±0·07	0·35	6·6–8·2
	B–L	27	6·29±0·08	0·42	5·3–6·9
I^2	M–D	29	5·28±0·08	0·41	4·4–6·7
	B–L	28	4·95±0·06	0·34	4·4–5·7
C	M–D	28	6·08±0·08	0·40	5·2–6·7
	B–L	28	4·60±0·03	0·16	4·4–5·1
P^3	M–D	28	4·93±0·08	0·41	4·1–5·6
	B–L	29	5·42±0·06	0·33	4·2–5·9
P^4	M–D	29	4·53±0·05	0·26	3·9–5·0
	B–L	29	5·52±0·04	0·21	5·0–5·9
M^1	M–D	29	6·56±0·04	0·21	6·1–6·9
	Ant B	28	6·22±0·04	0·22	5·8–6·7
	Post B	28	6·03±0·04	0·22	5·5–6·5
M^2	M–D	27	7·10±0·07	0·34	6·4–7·7
	Ant B	27	6·74±0·05	0·27	6·2–7·2
	Post B	25	6·20±0·05	0·23	5·7–6·7
M^3	M–D	21	6·46±0·07	0·33	6·0-7·3
	Ant B	21	6·27±0·06	0·26	5·8–6·7
	Post B	18	5·14±0·09	0·39	4·6–6·0

Table 112
Cercocebus albigena female

Mandibular teeth		n	$\bar{X} \pm$s.e.	s.d.	Range
I_1	M–D	29	5·37±0·06	0·32	4·4–5·8
	B–L	26	5·91±0·07	0·35	5·4-6·8
I_2	M–D	26	4·37±0·08	0·42	3·3–5·4
	B–L	26	5·25±0·05	0·27	4·6–5·8
C	M–D	24	4·05±0·07	0·32	3·6–4·7
	B–L	25	5·53±0·10	0·49	3·6–6·4
P_3	M–D	25	5·91±0·22	1·12	4·1–9·0
	B–L	26	4·02±0·07	0·37	3·1–4·8
P_4	M–D	29	5·09±0·06	0·31	4·6–5·6
	B–L	28	4·63±0·06	0·30	4·1–5·5
M_1	M–D	29	6·49±0·03	0·15	6·2–6·8
	Tri B	29	5·18±0·04	0·21	4·8–5·6
	Tal B	29	5·33±0·04	0·20	4·7–5·6
M_2	M–D	29	7·12±0·06	0·31	6·5–7·7
	Tri B	29	6·06±0·05	0·28	5·4–6·4
	Tal B	29	5·84±0·04	0·23	5·2–6·2
M_3	M–D	22	7·65±0·11	0·53	6·7–8·7
	Tri B	24	5·86±0·06	0·30	5·2–6·5
	Tal B	20	5·08±0·05	0·24	4·6–5·7
	Hypo-conulid	16	2·39±0·13	0·53	1·4–3·4

Table 113
Cercocebus torquatus male

Maxillary teeth		n	$\bar{X}\pm$s.e.	s.d.	Range	P
I^1	M–D	12	7·4±0·18	0·64	6·0-8·4	
	B–L	11	6·5±0·30	0·98	4·4–8·0	
I^2	M–D	11	4·7±0·18	0·60	3·9–5·7	
	B–L	11	5·2±0·21	0·69	4·2–6·1	
C	M–D	8	8·5±0·65	1·84	6·2–10·6	*
	B–L	6	6·2±0·47	1·16	5·1–7·7	*
P^3	M–D	8	5·7±0·13	0·35	5·0–6·0	
	B–L	9	6·1±0·11	0·32	5·3–6·3	
P^4	M–D	9	5·7±0·26	0·78	5·1–7·7	
	B–L	9	7·1±0·19	0·57	5·8–7·7	
M^1	M–D	14	7·5±0·15	0·55	6·3–8·5	
	Ant B	14	7·8±0·22	0·81	6·1–8·7	
	Post B	14	6·9±0·18	0·67	6·5–7·8	
M^2	M–D	10	8·0±0·19	0·59	6·7–8·7	
	Ant B	10	8·4±0·27	0·84	6·8–9·5	
	Post B	9	7·7±0·30	0·91	6·2–8·7	
M^3	M–D	6	7·7±0·35	0·85	6·2–8·3	
	Ant B	6	8·0±0·34	0·82	6·5–8·7	
	Post B	6	6·6±0·38	0·92	4·8–7·3	

* $P<0·05$

Table 114
Cercocebus torquatus male

Mandibular teeth		n	$\bar{X}\pm$s.e.	s.d.	Range	P
I_1	M–D	12	5·1±0·15	0·52	3·7–5·5	
	B–L	12	5·9±0·23	0·81	4·1–7·2	
I_2	M–D	11	4·2±0·15	0·49	3·4–5·2	
	B–L	13	5·6±0·24	0·85	4·3–7·3	
C	M–D	5	5·1±0·30	0·68	4·3–5·7	†
	B–L	6	7·5±0·65	1·58	5·6–9·1	†
P_3	M–D	5	12·1±1·30	2·90	8·1–14·5	†
	B–L	9	4·6±0·17	0·51	3·7–5·4	
P_4	M–D	8	6·1±0·14	0·40	5·4–6·7	
	B–L	8	5·7±0·36	1·01	3·5–6·6	
M_1	M–D	13	7·3±0·16	0·57	6·1–8·2	
	Tri B	14	6·1±0·17	0·65	4·6–6·9	
	Tal B	13	5·8±0·14	0·52	4·6–6·4	
M_2	M–D	10	8·2±0·18	0·57	7·2-8·9	
	Tri B	10	7·3±0·28	0·88	5·6–8·2	
	Tal B	9	6·8±0·28	0·84	5·3–7·7	
M_3	M–D	6	9·7±1·40	3·43	9·1–10·2	
	Tri B	6	7·2±0·31	0·75	5·7–7·7	
	Tal B	6	6·3±0·29	0·70	5·0–6·8	*
	Hypoconulid	4	4·2±0·16	0·31	3·7–4·4	

* $P<0·05$
† $P<0·01$

Table 115
Cercocebus torquatus female

Maxillary teeth		n	$\bar{X} \pm$ s.e.	s.d.	Range
I^1	M–D	10	7·4±0·16	0·48	6·4–8·1
	B–L	9	6·4±0·11	0·33	5·9–6·9
I^2	M–D	9	4·6±0·14	0·43	3·9–5·1
	B–L	10	4·9±0·11	0·34	4·3–5·5
C	M–D	7	6·5±0·04	0·11	6·4–6·7
	B–L	7	5·0±0·11	0·29	4·7–5·4
P^3	M–D	8	5·4±0·07	0·20	5·1–5·7
	B–L	8	5·9±0·07	0·21	5·6–6·3
P^4	M–D	8	5·6±0·06	0·18	5·4–5·9
	B–L	8	6·9±0·10	0·27	6·6–7·3
M^1	M–D	10	7·2±0·22	0·71	5·3–7·8
	Ant B	10	7·6±0·08	0·25	7·2–7·9
	Post B	10	6·9±0·07	0·22	6·5–7·2
M^2	M–D	8	8·1±0·09	0·26	7·7–8·5
	Ant B	8	8·4±0·08	0·23	8·0–8·7
	Post B	8	7·8±0·14	0·39	7·4–8·6
M^3	M–D	2	8·5±0·20	0·28	8·3–8·7
	Ant B	2	8·4±0·15	0·21	8·2–8·5
	Post B	2	7·2±0·15	0·21	7·0–7·3

Table 116
Cercocebus torquatus female

Mandibular teeth		n	$\bar{X} \pm$ s.e.	s.d.	Range
I_1	M–D	10	5·0±0·08	0·24	4·5–5·3
	B–L	9	5·6±0·16	0·48	4·8–6·2
I_2	M–D	9	4·2±0·08	0·24	3·8–4·6
	B–L	10	5·3±0·15	0·47	4·2–5·8
C	M–D	8	4·1±0·10	0·28	3·5–4·5
	B–L	8	5·6±0·10	0·27	5·1–5·9
P_3	M–D	7	8·5±0·23	0·62	7·5–9·5
	B–L	8	4·3±0·07	0·21	4·0–4·6
P_4	M–D	7	5·9±0·10	0·25	5·5–6·2
	B–L	8	5·6±0·12	0·33	5·0–6·1
M_1	M–D	10	7·3±0·07	0·22	6·9–7·6
	Tri B	10	6·0±0·07	0·24	5·7–6·4
	Tal B	10	5·8±0·09	0·27	5·5–6·2
M_2	M–D	8	8·3±0·12	0·35	7·6–8·7
	Tri B	8	7·3±0·09	0·26	6·9–7·6
	Tal B	8	6·8±0·10	0·29	6·4–7·2
M_3	M–D	2	9·3±0·65	0·92	8·6–9·9
	Tri B	2	7·6±0·15	0·21	7·4–7·7
	Tal B	2	7·1±0·80	1·13	6·3–7·9
	Hypoconulid	2	3·0±0·25	0·35	2·7–3·2

Table 117
Cercocebus galeritus male

Maxillary teeth		n	$\bar{X} \pm$ s.e.	s.d.	Range	P
I[1]	M–D	7	7·9±0·23	0·60	7·0–8·5	
	B–L	8	7·1±0·14	0·39	6·2–7·5	
I[2]	M–D	10	4·8±0·14	0·45	4·2–5·4	
	B–L	8	5·2±0·10	0·29	4·9–5·7	†
C	M–D	7	9·7±0·19	0·50	8·9–10·5	†
	B–L	8	6·6±0·23	0·65	5·8–7·6	†
P[3]	M–D	10	5·5±0·17	0·55	4·7–6·2	
	B–L	10	6·5±0·16	0·51	5·5–7·3	†
P[4]	M–D	9	5·7±0·14	0·43	4·7–6·1	
	B–L	10	7·8±0·22	0·68	6·0–8·4	*
M[1]	M–D	10	7·3±0·09	0·29	6·8–7·8	
	Ant B	10	8·1±0·23	0·72	6·6–9·2	*
	Post B	10	7·4±0·14	0·43	6·6–8·1	*
M[2]	M–D	10	7·9±0·12	0·37	7·4–8·5	
	Ant B	10	8·6±0·11	0·34	8·1–9·2	*
	Post B	10	8·0±0·15	0·47	7·1–8·8	
M[3]	M–D	9	6·8±0·16	0·49	6·1–7·4	
	Ant B	9	8·0±0·15	0·44	7·4–8·5	*
	Post B	8	6·5±0·22	0·61	5·6–7·3	

* $P<0·05$
† $P<0·01$

Table 118
Cercocebus galeritus male

Mandibular teeth		n	$\bar{X} \pm$ s.e.	s.d.	Range	P
I_1	M–D	9	5·9±0·11	0·32	5·3–6·4	
	B–L	9	6·5±0·10	0·31	5·8–6·8	*
I_2	M–D	9	4·1±0·11	0·32	3·6–4·7	
	B–L	10	5·7±0·09	0·27	5·3–6·0	†
C	M–D	4	4·9±0·08	0·15	4·8–5·1	†
	B–L	7	7·9±0·55	1·45	4·7–8·7	†
P_3	M–D	5	11·9±0·18	0·39	11·5–12·4	†
	B–L	9	4·5±0·08	0·25	4·3–5·0	†
P_4	M–D	10	6·2±0·10	0·31	5·5–6·6	*
	B–L	10	5·9±0·10	0·31	5·5–6·5	*
M_1	M–D	10	7·4±0·12	0·39	6·9–8·2	*
	Tri B	10	6·4±0·10	0·32	5·9–6·7	*
	Tal B	10	6·2±0·12	0·39	5·6–6·7	
M_2	M–D	10	8·1±0·12	0·39	7·5–8·9	
	Tri B	8	7·5±0·10	0·27	7·2–7·9	
	Tal B	9	7·2±0·11	0·34	6·7–7·8	
M_3	M–D	9	8·5±0·21	0·62	7·5–9·5	
	Tri B	9	7·3±0·09	0·28	6·9–7·7	*
	Tal B	9	6·6±0·19	0·56	5·8–7·8	*
	Hypo-conulid	6	3·5±0·17	0·41	2·7–3·8	

* $P<0·05$
† $P<0·01$

Table 119
Cercocebus galeritus female

Maxillary teeth		n	$\bar{X} \pm$ s.e.	s.d.	Range
I^1	M–D	10	7·6±0·22	0·70	6·2–8·7
	B–L	10	6·7±0·13	0·40	6·2–7·4
I^2	M–D	9	4·6±0·11	0·32	4·1–5·1
	B–L	9	4·7±0·11	0·33	4·2–5·4
C	M–D	6	6·0±0·16	0·39	5·4–6·4
	B–L	6	4·8±0·08	0·18	4·6–5·0
P^3	M–D	10	5·1±0·08	0·26	4·5–5·4
	B–L	10	6·0±0·12	0·39	5·2–6·5
P^4	M–D	9	5·6±0·07	0·22	5·1–5·8
	B–L	10	7·2±0·15	0·47	6·3–8·0
M^1	M–D	10	7·1±0·10	0·32	6·4–7·5
	Ant B	10	7·4±0·14	0·44	6·3–8·0
	Post B	10	6·9±0·13	0·40	6·0–7·6
M^2	M–D	10	7·8±0·12	0·37	6·9–8·3
	Ant B	10	8·0±0·18	0·58	7·0–8·7
	Post B	10	7·6±0·17	0·54	6·4–8·1
M^3	M–D	7	6·8±0·10	0·26	6·5–7·2
	Ant B	8	7·4±0·18	0·52	6·6–8·3
	Post B	7	6·1±0·15	0·40	5·6–6·7

Table 120
Cercocebus galeritus female

Mandibular teeth		n	$\bar{X} \pm$ s.e.	s.d.	Range
I_1	M–D	10	5·5±0·17	0·52	4·6–6·2
	B–L	8	5·9±0·21	0·60	4·7–6·7
I_2	M–D	8	3·8±0·13	0·37	3·1–4·2
	B–L	10	5·1±0·12	0·39	4·5–5·7
C	M–D	9	3·8±0·07	0·19	3·5–4·1
	B–L	8	5·7±0·21	0·60	4·8–6·5
P_3	M–D	8	7·5±0·14	0·41	7·0–8·3
	B–L	9	4·0±0·09	0·25	3·6–4·3
P_4	M–D	10	5·8±0·10	0·32	5·2–6·1
	B–L	9	5·5±0·19	0·58	4·5–6·5
M_1	M–D	10	7·0±0·10	0·29	6·5–7·4
	Tri B	9	5·9±0·13	0·40	5·0–6·4
	Tal B	9	5·8±0·15	0·46	4·9–6·6
M_2	M–D	10	7·8±0·11	0·34	7·0–8·3
	Tri B	10	7·2±0·18	0·57	5·9–8·1
	Tal B	10	6·8±0·22	0·68	5·3–8·0
M_3	M–D	8	8·1±0·23	0·64	7·3–8·9
	Tri B	9	6·8±0·14	0·43	5·8–7·3
	Tal B	8	6·4±0·13	0·37	5·8–6·9
	Hypo-conulid	7	2·6±0·28	0·74	1·4–3·4

Table 121
Macaca nemestrina male

Maxillary teeth		*n*	$\bar{X}\pm$s.e.	s.d.	Range	*P*
I^1	M–D	15	7·6±0·14	0·56	6·4–8·5	
	B–L	13	6·5±0·14	0·52	5·5–7·2	
I^2	M–D	15	5·5±0·13	0·52	4·4–6·4	
	B–L	14	5·8±0·11	0·40	5·0–6·5	†
C	M–D	11	11·7±0·31	1·03	9·4–13·0	†
	B–L	9	8·0±0·19	0·57	7·2–9·1	†
P^3	M–D	16	5·9±0·11	0·45	5·0–6·7	†
	B–L	17	6·3±0·09	0·36	5·8–7·0	*
P^4	M–D	15	5·7±0·08	0·30	5·0–6·0	*
	B–L	15	6·8±0·09	0·33	6·4–7·6	*
M^1	M–D	17	7·4±0·08	0·31	6·8–7·9	
	Ant B	19	7·2±0·12	0·53	5·6–8·2	
	Post B	15	6·8±0·09	0·37	6·3–7·5	*
M^2	M–D	17	8·6±0·09	0·38	7·8–9·3	†
	Ant B	17	8·5±0·12	0·48	7·8–9·3	†
	Post B	17	7·9±0·12	0·50	7·0–8·7	†
M^3	M–D	16	8·7±0·13	0·52	7·6–8·9	†
	Ant B	16	8·5±0·12	0·48	7·6–9·3	†
	Post B	16	7·7±0·18	0·73	6·4–9·3	†

* $P<0·05$
† $P<0·01$

Table 122
Macaca nemestrina male

Mandibular teeth		*n*	$\bar{X}\pm$s.e.	s.d.	Range	*P*
I_1	M–D	17	5·3±0·10	0·41	4·6–5·9	
	B–L	17	6·1±0·10	0·41	5·3–6·9	
I_2	M–D	14	5·2±0·15	0·58	4·5–6·5	
	B–L	14	5·9±0·21	0·79	3·6–6·8	
C	M–D	15	10·9±0·25	0·95	9·1–12·7	†
	B–L	14	7·2±0·18	0·69	6·3–8·6	†
P_3	M–D	12	15·8±0·38	1·32	12·8–17·3	†
	B–L	16	5·8±0·13	0·52	4·6–6·7	†
P_4	M–D	16	6·6±0·13	0·50	5·8–7·6	*
	B–L	15	5·5±0·10	0·38	5·0–6·2	
M_1	M–D	19	7·5±0·07	0·31	6·8–8·0	
	Tri B	19	5·8±0·05	0·21	5·5–6·3	
	Tal B	18	5·8±0·05	0·20	5·4–6·2	*
M_2	M–D	17	8·6±0·12	0·50	7·8–9·5	*
	Tri B	18	7·2±0·13	0·57	6·2–8·4	*
	Tal B	18	6·8±0·17	0·46	5·9–7·8	†
M_3	M–D	15	10·9±0·22	0·87	9·6–12·5	
	Tri B	16	7·8±0·12	0·49	7·0–8·5	†
	Tal B	16	7·1±0·13	0·52	6·0–8·0	†
	Hypo-conulid	16	5·0±0·14	0·54	3·7–6·0	

* $P<0·05$
† $P<0·01$

Table 123
Macaca nemestrina female

Maxillary teeth		n	$\bar{X}\pm$s.e.	s.d.	Range
I^1	M–D	9	7·3±0·41	1·22	5·8–9·0
	B–L	7	6·1±0·20	0·53	5·5–6·8
I^2	M–D	7	5·0±0·29	0·77	3·9–5·8
	B–L	7	5·2±0·20	0·54	4·4–5·8
C	M–D	7	7·3±0·31	0·83	6·0–8·2
	B–L	7	5·5±0·25	0·67	4·4–6·2
P^3	M–D	7	5·2±0·22	0·57	4·2–5·8
	B–L	7	5·9±0·23	0·62	5·0–6·7
P^4	M–D	8	5·2±0·23	0·64	3·8–5·7
	B–L	8	6·3±0·21	0·61	5·4–7·0
M^1	M–D	11	7·1±0·21	0·69	5·7–7·8
	Ant B	10	7·1±0·15	0·48	6·0–7·6
	Post B	11	6·5±0·13	0·42	5·8–7·0
M^2	M–D	9	7·9±0·25	0·74	6·3–8·7
	Ant B	9	7·9±0·18	0·52	6·8–8·5
	Post B	9	7·2±0·13	0·40	6·5–7·7
M^3	M–D	8	7·8±0·27	0·77	6·6–9·0
	Ant B	7	7·7±0·24	0·65	6·7–8·7
	Post B	8	6·7±0·23	0·64	5·9–7·6

Table 124
Macaca nemestrina female

Mandibular teeth		n	$\bar{X}\pm$s.e.	s.d.	Range
I_1	M–D	7	5·4±0·26	0·69	4·4–6·2
	B–L	7	6·1±0·25	0·65	5·0–7·0
I_2	M–D	7	4·8±0·21	0·55	4·0–5·4
	B–L	6	5·6±0·31	0·76	4·6–6·5
C	M–D	6	7·0±0·44	1·08	4·9–8·0
	B–L	6	4·9±0·33	0·81	3·6–5·9
P_3	M–D	5	10·4±0·76	1·70	7·8–12·2
	B–L	6	5·5±1·04	2·55	3·9–10·6
P_4	M–D	9	6·0±0·28	0·85	4·4–7·0
	B–L	7	5·2±0·20	0·54	4·2–5·7
M_1	M–D	10	7·3±0·20	0·64	5·9–8·0
	Tri B	10	5·6±0·15	0·46	4·5–6·0
	Tal B	10	5·4±0·15	0·47	4·5–6·2
M_2	M–D	8	7·9±0·26	0·72	6·6–8·8
	Tri B	7	6·6±0·25	0·65	5·4–7·3
	Tal B	7	6·2±0·20	0·52	5·2–6·7
M_3	M–D	9	10·0±0·38	1·15	7·6–11·2
	Tri B	8	6·9±0·32	0·91	5·1–7·9
	Tal B	8	6·2±0·31	0·87	4·9–7·6
	Hypo-conulid	6	4·7±0·31	0·77	3·6–5·8

Table 125
Macaca mulatta male

Maxillary teeth		n	$\bar{X}\pm$s.e.	s.d.	Range	P
I^1	M–D	91	6·3±0·06	0·61	5·0–8·0	
	B–L	71	6·1±0·12	0·99	4·4–8·8	*
I^2	M–D	32	4·9±0·11	0·60	3·5–6·4	
	B–L	30	5·4±0·08	0·45	4·8–7·0	†
C	M–D	27	8·9±0·37	1·93	3·9–11·3	†
	B–L	25	6·7±0·20	1·01	4·8–8·6	†
P^3	M–D	57	5·0±0·05	0·38	4·2–6·3	*
	B–L	52	6·1±0·08	0·55	5·0–7·2	
P^4	M–D	57	5·1±0·05	0·39	4·5–6·4	
	B–L	54	6·5±0·08	0·56	5·5–7·6	
M^1	M–D	95	7·2±0·06	0·55	5·9–8·5	*
	Ant B	103	7·0±0·05	0·51	6·0–8·5	†
	Post B	101	6·5±0·06	0·58	5·6–8·0	
M^2	M–D	50	7·8±0·11	0·81	6·0–9·6	
	Ant B	56	7·7±0·11	0·82	5·5–9·6	
	Post B	53	7·1±0·12	0·88	5·0–9·0	
M^3	M–D	22	7·4±0·12	0·57	6·6–8·5	
	Ant B	26	7·4±0·16	0·82	6·5–9·0	
	Post B	20	6·5±0·18	0·79	5·5–7·9	

* $P<0{\cdot}05$
† $P<0{\cdot}01$

Table 126
Macaca mulatta male

Mandibular teeth		n	$\bar{X}\pm$s.e.	s.d.	Range	P
I_1	M–D	89	4·4±0·05	0·46	3·6–6·5	†
	B–L	82	5·4±0·10	0·92	3·5–7·4	†
I_2	M–D	75	4·2±0·04	0·37	3·4–5·2	†
	B–L	70	5·3±0·08	0·68	4·0–7·5	†
C	M–D	31	7·5±0·28	1·57	3·8–9·8	†
	B–L	29	6·3±0·27	1·43	3·0–9·0	†
P_3	M–D	17	11·2±0·62	2·54	6·8–14·0	†
	B–L	28	4·5±0·10	0·52	3·4–5·6	†
P_4	M–D	55	5·4±0·07	0·52	4·4–7·6	
	B–L	47	5·0±0·07	0·50	3·9–6·5	†
M_1	M–D	97	7·2±0·06	0·56	5·9–9·5	*
	Tri B	95	5·6±0·05	0·47	4·6–7·0	*
	Tal B	93	5·5±0·04	0·39	4·6–6·6	
M_2	M–D	60	7·8±0·10	0·81	3·8–9·0	
	Tri B	66	6·6±0·08	0·66	5·5–8·8	
	Tal B	59	6·2±0·07	0·57	5·0–7·3	
M_3	M–D	26	9·2±0·22	1·11	6·4–12·0	*
	Tri B	32	6·5±0·13	0·74	4·7–7·8	
	Tal B	27	5·9±0·12	0·64	4·5–7·1	
	Hypo-conulid	21	4·3±0·09	0·43	3·6–5·2	

* $P<0{\cdot}05$
† $P<0{\cdot}01$

Table 127
Macaca mulatta female

Maxillary teeth		n	$\bar{X} \pm$ s.e.	s.d.	Range
I^1	M–D	78	6·2±0·06	0·50	4·7–7·6
	B–L	59	5·7±0·13	1·00	3·4–9·7
I^2	M–D	33	4·7±0·12	0·62	3·7–7·2
	B–L	29	4·9±0·10	0·55	4·0–6·8
C	M–D	37	6·0±0·06	0·35	5·2–6·9
	B–L	35	5·0±0·06	0·38	4·0–5·8
P^3	M–D	55	4·9±0·06	0·41	4·0–6·0
	B–L	54	6·0±0·06	0·43	5·0–6·9
P^4	M–D	57	5·0±0·05	0·37	4·2–6·0
	B–L	56	6·3±0·06	0·46	5·4–7·2
M^1	M–D	79	7·1±0·06	0·57	5·7–8·7
	Ant B	83	6·8±0·04	0·40	5·3–8·0
	Post B	79	6·4±0·05	0·46	4·9–7·6
M^2	M–D	33	7·6±0·12	0·68	5·8–8·7
	Ant B	52	7·6±0·09	0·67	5·9–10·1
	Post B	39	7·0±0·10	0·61	5·8–8·4
M^3	M–D	16	7·3±0·18	0·71	6·0–8·7
	Ant B	15	7·2±0·22	0·84	6·4–8·9
	Post B	13	6·5±0·21	0·74	5·5–7·5

Table 128
Macaca mulatta female

Mandibular teeth		n	$\bar{X} \pm$ s.e.	s.d.	Range
I_1	M–D	81	4·2±0·04	0·39	3·5–5·6
	B–L	75	5·0±0·10	0·84	3·4–7·5
I_2	M–D	78	4·0±0·04	0·36	3·2–5·2
	B–L	75	4·8±0·07	0·63	3·4–6·8
C	M–D	29	5·3±0·12	0·67	4·1–6·9
	B–L	23	4·2±0·11	0·52	3·5–6·0
P_3	M–D	20	7·5±0·30	1·34	4·8–11·8
	B–L	31	4·0±0·08	0·45	3·2–5·2
P_4	M–D	50	5·3±0·05	0·36	4·5–6·0
	B–L	47	4·8±0·06	0·42	4·0–5·7
M_1	M–D	87	7·0±0·05	0·47	6·0–8·3
	Tri B	82	5·5±0·05	0·41	4·5–6·6
	Tal B	80	5·4±0·06	0·54	4·2–8·6
M_2	M–D	49	7·8±0·08	0·59	5·2–8·9
	Tri B	59	6·4±0·07	0·54	5·0–7·5
	Tal B	51	6·1±0·09	0·62	3·9–7·2
M_3	M–D	11	8·3±0·43	1·42	5·7–11·2
	Tri B	17	6·1±0·19	0·80	4·6–8·0
	Tal B	10	5·7±0·24	0·76	4·7–7·0
	Hypoconulid	7	4·6±0·43	1·13	3·5–6·7

Table 129
Macaca fascicularis male

Maxillary teeth		n	$\bar{X}\pm$s.e.	s.d.	Range	P
I^1	M–D	51	6·6±0·09	0·67	5·0–8·2	
	B–L	46	6·2±0·11	0·76	4·5–7·8	
I^2	M–D	48	4·5±0·08	0·53	3·5–5·7	
	B–L	49	5·0±0·07	0·52	3·7–5·9	*
C	M–D	35	9·2±0·25	1·46	5·6–12·2	†
	B–L	33	6·6±0·14	0·81	4·8–8·4	†
P^3	M–D	53	4·8±0·05	0·38	4·0–5·6	†
	B–L	55	5·5±0·07	0·50	4·4–6·5	
P^4	M–D	54	4·7±0·05	0·35	4·0–5·7	
	B–L	55	5·9±0·07	0·53	4·9–7·2	
M^1	M–D	65	6·5±0·06	0·50	5·3–7·7	*
	Ant B	65	6·4±0·07	0·55	5·5–7·6	
	Post B	64	6·0±0·07	0·53	5·2–7·5	
M^2	M–D	58	7·2±0·08	0·61	6·0–9·5	
	Ant B	57	7·2±0·10	0·72	5·9–9·4	
	Post B	56	6·7±0·09	0·69	5·5–9·2	
M^3	M–D	47	7·1±0·10	0·69	6·0–9·1	*
	Ant B	44	6·9±0·12	0·81	5·4–9·5	*
	Post B	45	6·1±0·12	0·77	5·0–8·8	*

* $P<0·05$
† $P<0·01$

Table 130
Macaca fascicularis male

Mandibular teeth		n	$\bar{X}\pm$s.e.	s.d.	Range	P
I_1	M–D	55	4·6±0·06	0·43	4·0–5·7	
	B–L	53	5·9±0·10	0·70	4·6–7·5	*
I_2	M–D	51	3·9±0·05	0·36	3·0–4·6	
	B–L	49	5·2±0·09	0·66	4·0–6·6	†
C	M–D	40	8·3±0·21	1·32	5·6–12·0	†
	B–L	36	5·1±0·18	1·07	3·5–9·0	†
P_3	M–D	40	10·9±0·27	1·72	7·1–15·5	†
	B–L	51	4·5±0·08	0·54	3·5–6·4	†
P_4	M–D	53	5·1±0·05	0·38	4·2–6·6	
	B–L	53	4·7±0·06	0·45	3·9–6·2	
M_1	M–D	66	6·5±0·06	0·52	5·6–7·7	
	Tri B	58	5·1±0·06	0·44	4·5–6·2	
	Tal B	64	5·0±0·06	0·46	4·3–5·8	
M_2	M–D	58	7·4±0·07	0·55	6·2–9·0	
	Tri B	55	6·1±0·08	0·60	4·6–7·7	
	Tal B	55	5·7±0·07	0·53	4·7–6·9	
M_3	M–D	45	8·8±0·14	0·94	6·8–11·2	
	Tri B	45	6·0±0·13	0·85	5·5–7·8	
	Tal B	43	5·5±0·09	0·60	4·3–6·8	
	Hypoconulid	38	4·1±0·10	0·62	3·0–5·7	

* $P<0·05$
† $P<0·01$

Table 131
Macaca fascicularis female

Maxillary teeth		n	$\bar{X} \pm$ s.e.	s.d.	Range
I^1	M–D	46	6·5 ± 0·11	0·72	4·7–7·6
	B–L	39	6·1 ± 0·10	0·65	5·0–7·6
I^2	M–D	47	4·4 ± 0·08	0·51	3·3–5·6
	B–L	44	4·8 ± 0·06	0·43	3·7–5·9
C	M–D	40	5·8 ± 0·12	0·78	4·3–7·8
	B–L	37	5·0 ± 0·09	0·53	4·0–6·7
P^3	M–D	45	4·5 ± 0·07	0·48	3·6–5·6
	B–L	45	5·4 ± 0·07	0·48	4·6–6·9
P^4	M–D	45	4·7 ± 0·07	0·44	3·8–5·8
	B–L	45	5·8 ± 0·08	0·53	5·0–7·4
M^1	M–D	53	6·3 ± 0·07	0·54	4·3–7·6
	Ant B	51	6·2 ± 0·07	0·50	5·5–8·2
	Post B	49	5·9 ± 0·07	0·57	5·0–7·9
M^2	M–D	46	7·2 ± 0·09	0·58	5·7–8·5
	Ant B	48	7·1 ± 0·08	0·58	6·0–8·7
	Post B	47	6·5 ± 0·08	0·57	5·6–7·8
M^3	M–D	32	6·8 ± 0·11	0·62	5·8–8·5
	Ant B	31	6·6 ± 0·07	0·40	6·0–8·0
	Post B	29	5·7 ± 0·08	0·41	5·0–6·8

Table 132
Macaca fascicularis female

Mandibular teeth		n	$\bar{X} \pm$ s.e.	s.d.	Range
I_1	M–D	48	4·6 ± 0·07	0·49	3·4–5·8
	B–L	48	5·6 ± 0·10	0·71	4·0–6·9
I_2	M–D	49	3·9 ± 0·17	0·47	3·1–5·7
	B–L	47	4·9 ± 0·08	0·55	3·5–5·8
C	M–D	43	6·0 ± 0·11	0·74	5·0–8·3
	B–L	40	3·8 ± 0·11	0·70	2·8–6·0
P_3	M–D	38	7·9 ± 0·19	1·17	6·2–11·4
	B–L	44	3·9 ± 0·07	0·44	3·2–4·8
P_4	M–D	47	5·0 ± 0·07	0·48	4·2–6·0
	B–L	46	4·6 ± 0·07	0·48	3·8–5·7
M_1	M–D	56	6·4 ± 0·07	0·49	5·0–7·5
	Tri B	52	5·1 ± 0·05	0·39	4·0–5·9
	Tal B	52	4·9 ± 0·06	0·39	4·0–6·2
M_2	M–D	49	7·3 ± 0·09	0·59	6·0–8·7
	Tri B	48	6·0 ± 0·07	0·48	5·2–7·4
	Tal B	46	5·6 ± 0·07	0·47	4·7–6·9
M_3	M–D	34	8·5 ± 0·14	0·82	5·9–10·5
	Tri B	35	5·9 ± 0·09	0·51	5·0–7·6
	Tal B	33	5·2 ± 0·09	0·53	4·2–7·6
	Hypoconulid	28	3·9 ± 0·11	0·56	2·7–5·0

Table 133
Macaca speciosa male

Maxillary teeth		n	$\bar{X} \pm$ s.e.	s.d.	Range
I^1	M–D	7	6·8±0·14	0·37	6·2–7·3
	B–L	7	6·4±0·11	0·28	6·0–6·7
I^2	M–D	7	5·6±0·10	0·28	5·4–6·2
	B–L	7	6·1±0·10	0·27	5·8–6·6
C	M–D	3	10·3±0·18	0·31	10·0–10·6
	B–L	3	7·8±0·32	0·56	7·2–8·3
P^3	M–D	7	6·2±0·15	0·39	5·7–6·6
	B–L	7	6·9±0·10	0·26	6·5–7·3
P^4	M–D	6	6·3±0·08	0·20	6·1–6·6
	B–L	6	7·6±0·09	0·22	7·3–7·9
M^1	M–D	7	8·4±0·20	0·54	7·8–9·0
	Ant B	7	8·5±0·22	0·59	7·8–9·6
	Post B	7	7·9±0·20	0·52	7·3–8·9
M^2	M–D	7	9·7±0·17	0·44	8·9–10·2
	Ant B	7	9·4±0·12	0·31	8·8–9·8
	Post B	7	8·6±0·08	0·20	8·3–8·9
M^3	M–D	4	9·8±0·14	0·28	9·5–10·1
	Ant B	4	9·2±0·18	0·35	8·8–9·5
	Post B	4	8·4±0·20	0·40	7·8–8·7

Table 134
Macaca speciosa male

Mandibular teeth		n	$\bar{X} \pm$ s.e.	s.d.	Range
I_1	M–D	7	4·8±0·11	0·29	4·2–5·1
	B–L	7	6·1±0·14	0·36	5·6–6·6
I_2	M–D	6	4·5±0·11	0·28	4·0–4·7
	B–L	6	6·0±0·26	0·64	5·3–6·9
C	M–D	6	9·9±0·43	1·04	8·0–10·9
	B–L	6	7·0±0·67	1·63	4·4–9·0
P_3	M–D	5	13·1±0·64	1·43	11·7–15·3
	B–L	7	5·8±0·31	0·81	5·0–7·3
P_4	M–D	5	6·9±0·12	0·26	6·5–7·2
	B–L	5	6·2±0·10	0·23	6·0–6·5
M_1	M–D	7	8·4±0·22	0·57	7·8–9·3
	Tri B	7	6·3±0·16	0·42	5·8–6·9
	Tal B	7	6·3±0·19	0·50	5·6–7·0
M_2	M–D	6	9·7±0·18	0·44	9·3–10·3
	Tri B	7	7·8±0·13	0·35	7·0–8·0
	Tal B	7	7·3±0·21	0·54	6·6–8·0
M_3	M–D	4	12·2±0·41	0·82	11·7–13·4
	Tri B	4	7·9±0·07	0·13	7·7–8·0
	Tal B	4	7·2±0·27	0·53	6·5–7·8
	Hypoconulid	3	5·9±0·13	0·23	5·8–6·2

Table 135
Macaca speciosa female

Maxillary teeth		n	$\bar{X} \pm$ s.e.	s.d.	Range
I^1	M–D	2	6·3 ± 0·30	0·42	6·0–6·6
	B–L	2	5·9 ± 0·35	0·50	5·5–6·2
I^2	M–D	2	5·1 ± 0·45	0·64	4·6–5·5
	B–L	2	5·5 ± 0·05	0·71	5·4–5·6
C	M–D	2	7·1 ± 0·05	0·07	7·0–7·1
	B–L	2	6·4 ± 0·15	0·21	6·2–6·5
P^3	M–D	2	5·7 ± 0·15	0·21	5·5–5·8
	B–L	2	6·4 ± 0·35	0·50	6·0–6·7
P^4	M–D	2	5·8 ± 0·05	0·07	5·7–5·8
	B–L	2	7·2 ± 0·25	0·35	6·9–7·4
M^1	M–D	2	8·0 ± 0·20	0·28	7·8–8·2
	Ant B	2	7·6 ± 0·30	0·42	7·3–7·9
	Post B	1	6·9 ± 0·00	0·00	6·9–6·9
M^2	M–D	2	9·2 ± 0·40	0·57	8·8–9·6
	Ant B	2	8·8 ± 0·15	0·21	8·6–8·9
	Post B	2	8·3 ± 0·10	0·14	8·2–8·4
M^3	M–D				
	Ant B				
	Post B				

Table 136
Macaca speciosa female

Mandibular teeth		n	$\bar{X} \pm$ s.e.	s.d.	Range
I_1	M–D	2	4·6 ± 0·00	0·00	4·6–4·6
	B–L	2	5·8 ± 0·30	0·42	5·5–6·1
I_2	M–D	2	4·5 ± 0·05	0·07	4·5–4·6
	B–L	2	5·8 ± 0·25	0·35	5·5–6·0
C	M–D	2	7·1 ± 0·20	0·28	6·9–7·3
	B–L	2	4·5 ± 0·25	0·35	4·2–4·7
P_3	M–D	2	8·9 ± 0·25	0·35	8·6–9·1
	B–L	2	4·7 ± 0·10	0·14	4·6–4·8
P_4	M–D	2	6·0 ± 0·15	0·21	5·9–6·2
	B–L	2	5·7 ± 0·20	0·28	5·5–5·9
M_1	M–D	3	8·4 ± 0·18	0·31	8·1–8·7
	Tri B	2	6·2 ± 0·35	0·50	5·8–6·5
	Tal B				
M_2	M–D	2	9·2 ± 0·30	0·42	8·9–9·5
	Tri B	2	7·4 ± 0·20	0·28	7·2–7·6
	Tal B	2	6·5 ± 0·10	0·14	6·4–6·5
M_3	M–D	2	11·3 ± 0·65	0·92	10·6–11·9
	Tri B	2	7·2 ± 0·30	0·42	6·9–7·5
	Tal B	2	6·7 ± 0·05	0·07	6·6–6·7
	Hypo-conulid	2	5·2 ± 0·15	0·21	5·0–5·3

Table 137
Macaca niger male

Maxillary teeth		n	$\bar{X}\pm$s.e.	s.d.	Range	P
I^1	M–D	5	6·3±0·58	1·30	4·8–7·5	
	B–L	9	5·9±0·48	1·43	3·4–7·3	
I^2	M–D	9	3·8±0·24	0·73	2·3–4·5	
	B–L	9	4·9±0·41	1·22	2·7–6·0	
C	M–D	5	7·6±1·14	2·56	4·8–10·0	
	B–L	3	4·9±1·21	2·09	3·5–7·3	
P^3	M–D	7	4·9±0·10	0·25	4·5–5·2	
	B–L	6	6·0±0·08	0·19	5·8–6·3	†
P^4	M–D	7	5·3±0·06	0·16	5·0–5·4	
	B–L	7	6·2±0·12	0·31	5·7–6·6	*
M^1	M–D	10	7·1±0·15	0·48	6·0–7·5	*
	Ant B	10	6·5±0·30	0·94	4·6–7·2	
	Post B	10	6·0±0·25	0·79	4·6–6·7	
M^2	M–D	8	8·3±0·12	0·35	8·0–9·0	*
	Ant B	8	7·8±0·16	0·46	7·2–8·5	
	Post B	7	7·4±0·14	0·36	6·9–7·9	
M^3	M–D	3	8·4±0·35	0·61	7·7–8·9	
	Ant B	3	7·9±0·43	0·75	7·5–8·8	
	Post B	3	6·8±0·40	0·69	6·4–7·6	*

* $P<0·05$
† $P<0·01$

Table 138
Macaca niger male

Mandibular teeth		n	$\bar{X}\pm$s.e.	s.d.	Range	P
I_1	M–D	9	4·6±0·32	0·95	2·7–5·4	
	B–L	10	5·2±0·37	1·16	3·0–6·7	
I_2	M–D	8	3·1±0·30	0·85	2·0–4·8	
	B–L	9	4·9±0·36	1·09	3·0–5·8	
C	M–D	5	5·2±0·23	0·52	4·6–5·8	
	B–L	5	6·3±1·40	3·14	3·0–9·5	
P_3	M–D	3	10·6±2·48	4·30	6·4–15·0	
	B–L	4	4·8±0·27	0·53	4·0–5·2	†
P_4	M–D	7	6·0±0·13	0·36	5·4–6·4	
	B–L	7	5·1±0·03	0·09	5·0–5·2	†
M_1	M–D	9	7·1±0·08	0·24	6·6–7·3	*
	Tri B	8	5·2±0·36	1·02	3·5–6·1	
	Tal B	9	5·3±0·30	0·91	3·9–6·5	
M_2	M–D	8	7·9±0·12	0·33	7·3–8·3	
	Tri B	8	7·0±0·22	0·61	5·6–7·5	
	Tal B	8	6·5±0·20	0·56	5·3–7·3	
M_3	M–D	3	11·0±0·66	1·14	10·2–12·3	
	Tri B	4	7·1±0·30	0·61	6·6–8·0	
	Tal B	3	6·4±0·49	0·85	5·8–7·4	
	Hypoconulid					

* $P<0·05$
† $P<0·01$

Table 139
Macaca niger female

Maxillary teeth		n	$\bar{X} \pm$ s.e.	s.d.	Range
I^1	M–D	5	6·2 ± 0·31	0·69	5·1–6·9
	B–L	5	5·2 ± 0·46	1·02	3·4–5·9
I^2	M–D	6	3·2 ± 0·32	0·78	2·0–4·1
	B–L	6	4·1 ± 0·46	1·13	2·6–5·0
C	M–D	5	6·3 ± 0·33	0·74	5·0–6·8
	B–L	5	5·2 ± 0·38	0·85	3·7–5·7
P^3	M–D	4	4·5 ± 0·24	0·47	4·0–4·9
	B–L	5	5·6 ± 0·05	0·11	5·5–5·7
P^4	M–D	4	4·9 ± 0·20	0·40	4·6–5·4
	B–L	4	5·8 ± 0·10	0·21	5·6–6·0
M^1	M–D	7	6·4 ± 0·20	0·52	5·6–7·0
	Ant B	7	6·0 ± 0·39	1·03	4·3–6·8
	Post B	6	5·7 ± 0·38	0·92	3·9–6·3
M^2	M–D	6	7·5 ± 0·27	0·67	6·3–8·3
	Ant B	6	7·1 ± 0·48	1·16	4·8–8·0
	Post B	6	6·4 ± 0·39	0·95	4·6–7·2
M^3	M–D	3	7·7 ± 0·15	0·25	7·4–7·9
	Ant B	3	7·0 ± 0·12	0·20	6·8–7·2
	Post B	3	5·6 ± 0·12	0·21	5·4–5·8

Table 140
Macaca niger female

Mandibular teeth		n	$\bar{X} \pm$ s.e.	s.d.	Range
I_1	M–D	4	4·7 ± 0·13	0·25	4·4–5·0
	B–L	5	5·4 ± 0·14	0·30	5·0–5·8
I_2	M–D	5	3·2 ± 0·43	0·95	2·0–4·3
	B–L	6	4·6 ± 0·37	0·90	2·8–5·2
C	M–D	6	4·9 ± 0·38	0·93	4·0–6·0
	B–L	6	4·5 ± 0·49	1·19	2·9–5·8
P_3	M–D	2	9·9 ± 0·50	0·71	9·4–10·4
	B–L	4	3·7 ± 0·10	0·21	3·4–3·9
P_4	M–D	4	5·7 ± 0·04	0·08	5·6–5·8
	B–L	4	4·7 ± 0·10	0·21	4·5–4·9
M_1	M–D	7	6·6 ± 0·15	0·39	5·9–7·2
	Tri B	5	4·9 ± 0·48	1·07	3·6–5·8
	Tal B	7	4·7 ± 0·27	0·71	3·7–5·6
M_2	M–D	6	7·4 ± 0·26	0·64	6·2–8·1
	Tri B	5	6·3 ± 0·44	0·97	4·6–7·0
	Tal B	6	6·0 ± 0·25	0·62	4·8–6·5
M_3	M–D	2	9·6 ± 0·60	0·85	9·0–10·2
	Tri B	3	6·5 ± 0·09	0·15	6·4–6·7
	Tal B	2	5·9 ± 0·05	0·07	5·8–5·9
	Hypoconulid				

Table 141
Papio cynocephalus male

Maxillary teeth		n	$\bar{X}\pm$s.e.	s.d.	Range	P
I^1	M–D	34	10·7±0·16	0·95	9·0–13·3	†
	B–L	31	9·3±0·16	0·88	6·0–10·7	†
I^2	M–D	29	8·1±0·14	0·75	6·6–10·0	†
	B–L	29	8·5±0·14	0·74	6·0–9·6	†
C	M–D	27	14·0±0·45	2·35	10·4–18·6	†
	B–L	19	11·1±0·39	1·68	7·6–13·9	†
P^3	M–D	31	7·3±0·10	0·56	6·3–8·4	†
	B–L	33	7·9±0·08	0·45	6·9–9·3	†
P^4	M–D	34	7·7±0·07	0·43	6·7–8·4	†
	B–L	34	8·7±0·07	0·42	8·0–10·0	†
M^1	M–D	39	10·9±0·08	0·48	10·0–12·1	†
	Ant B	39	10·1±0·09	0·56	9·1–11·6	†
	Post B	38	9·5±0·07	0·46	8·3–10·5	†
M^2	M–D	37	12·9±0·10	0·61	11·6–14·0	†
	Ant B	38	12·2±0·29	1·80	10·1–14·3	†
	Post B	36	10·8±0·11	0·65	9·0–12·0	†
M^3	M–D	27	13·4±0·14	0·71	11·9–14·5	†
	Ant B	27	12·6±0·16	0·83	11·0–14·5	†
	Post B	24	10·8±0·14	0·67	9·3–11·8	†

† $P<0{\cdot}01$

Table 142
Papio cynocephalus male

Mandibular teeth		n	$\bar{X}\pm$s.e.	s.d.	Range	P
I_1	M–D	37	7·9±0·11	0·65	6·0–9·0	†
	B–L	33	9·0±0·11	0·62	7·7–10·2	†
I_2	M–D	32	7·0±0·11	0·63	5·4–8·3	†
	B–L	29	8·6±0·13	0·71	7·1–9·9	†
C	M–D	17	8·6±0·24	0·97	5·3–10·1	†
	B–L	22	13·7±0·57	2·67	10·7–16·3	†
P_3	M–D	19	17·9±1·42	5·19	11·8–20·4	†
	B–L	29	6·8±0·12	0·64	5·7–7·8	†
P_4	M–D	32	8·4±0·10	0·56	7·5–9·8	†
	B–L	29	7·2±0·06	0·35	6·6–7·9	†
M_1	M–D	39	10·8±0·07	0·46	9·7–11·7	†
	Tri B	37	8·3±0·08	0·48	7·6–9·4	†
	Tal B	39	8·8±0·09	0·54	7·8–10·1	†
M_2	M–D	40	12·8±0·09	0·57	11·6–13·7	†
	Tri B	39	10·4±0·13	0·80	8·6–12·0	†
	Tal B	39	10·1±0·12	0·77	8·7–11·9	†
M_3	M–D	26	16·1±0·16	0·82	13·7–17·6	†
	Tri B	26	11·3±0·17	0·84	9·7–13·0	†
	Tal B	25	10·3±0·14	0·70	8·8–11·7	†
	Hypo-conulid	25	6·3±0·21	1·06	3·4–7·6	†

† $P<0{\cdot}01$

Table 143
Papio cynocephalus female

Maxillary teeth		n	$\bar{X} \pm$ s.e.	s.d.	Range
I^1	M–D	35	9·8 ± 0·15	0·89	7·4–11·7
	B–L	32	8·4 ± 0·11	0·61	7·1–9·6
I^2	M–D	35	7·1 ± 0·11	0·65	6·0–8·5
	B–L	31	7·4 ± 0·12	0·68	5·9–8·9
C	M–D	29	8·3 ± 0·25	1·33	6·7–10·7
	B–L	29	7·3 ± 0·19	1·04	5·1–10·1
P^3	M–D	31	6·5 ± 0·08	0·43	5·9–8·0
	B–L	32	7·2 ± 0·09	0·49	6·1–8·6
P^4	M–D	32	7·1 ± 0·07	0·38	6·1–7·9
	B–L	32	8·1 ± 0·09	0·50	7·2–8·9
M^1	M–D	38	10·3 ± 0·05	0·33	9·4–11·1
	Ant B	36	9·3 ± 0·06	0·37	8·6–10·0
	Post B	36	8·8 ± 0·08	0·46	8·0–10·0
M^2	M–D	31	11·7 ± 0·07	0·37	11·0–12·4
	Ant B	32	10·9 ± 0·10	0·55	9·9–11·8
	Post B	29	10·1 ± 0·13	0·70	8·5–11·9
M^3	M–D	22	11·8 ± 0·13	0·61	10·5–13·0
	Ant B	21	11·2 ± 0·23	1·04	9·0–13·0
	Post B	20	9·9 ± 0·16	0·71	9·0–11·9

Table 144
Papio cynocephalus female

Mandibular teeth		n	$\bar{X} \pm$ s.e.	s.d.	Range
I_1	M–D	37	7·2 ± 0·10	0·63	4·9–8·1
	B–L	34	8·0 ± 0·11	0·66	6·2–9·0
I_2	M–D	32	6·2 ± 0·10	0·57	4·4–7·2
	B–L	31	7·3 ± 0·11	0·63	5·5–8·4
C	M–D	28	5·9 ± 0·15	0·79	4·6–7·9
	B–L	28	8·3 ± 0·15	0·77	6·3–9·7
P_3	M–D	28	10·2 ± 0·50	1·65	7·8–14·3
	B–L	35	5·2 ± 0·11	0·64	4·4–7·2
P_4	M–D	36	7·5 ± 0·07	0·41	6·7–8·7
	B–L	33	6·6 ± 0·07	0·39	5·9–7·5
M_1	M–D	37	10·1 ± 0·06	0·39	9·5–11·1
	Tri B	37	7·7 ± 0·07	0·45	6·7–8·9
	Tal B	36	8·1 ± 0·06	0·39	7·4–9·0
M_2	M–D	37	11·6 ± 0·09	0·54	10·7–12·9
	Tri B	35	9·5 ± 0·09	0·53	8·4–10·9
	Tal B	32	9·4 ± 0·09	0·52	7·9–10·4
M_3	M–D	26	14·3 ± 0·15	0·78	12·7–16·1
	Tri B	25	10·2 ± 0·10	0·49	9·0–11·3
	Tal B	24	9·5 ± 0·15	0·71	8·1–11·0
	Hypoconulid	23	5·5 ± 0·15	0·69	3·7–6·6

Table 145
Theropithecus gelada male

Maxillary teeth		n	$\bar{X}\pm$s.e.	s.d.	Range
I[1]	M–D				
	B–L	16	6·70±0·10	0·50	5·9–7·9
I[2]	M–D				
	B–L				
C	M–D	26	14·80±0·20	1·10	13·0–17·9
	B–L	26	9·00±0·10	0·60	7·7–10·6
P[3]	M–D	25	6·80±0·10	0·70	6·0–8·6
	B–L	28	7·20±0·10	0·40	6·2–8·2
P[4]	M–D	12	7·20±0·10	0·40	6·7–8·0
	B–L	25	8·00±0·10	0·40	7·3–8·8
M[1]	M–D				
	Ant B	16	9·30±0·10	0·30	8·8–10·0
	Post B	14	8·70±0·10	0·30	8·3–9·2
M[2]	M–D	18	13·40±0·20	0·80	11·9–14·8
	Ant B	23	10·90±0·10	0·40	10·0–11·8
	Post B	22	10·00±0·10	0·40	9·3–11·0
M[3]	M–D	37	13·70±0·10	0·70	11·6–14·9
	Ant B	37	11·30±0·10	0·70	9·3–13·0
	Post B	37	9·70±0·10	0·50	9·0–10·7

G. G. Eck, unpublished data, 1975

Table 146
Theropithecus gelada male

Mandibular teeth		n	$\bar{X}\pm$s.e.	s.d.	Range
I_1	M–D				
	B–L	15	6·30±0·10	0·30	6·0–7·0
I_2	M–D				
	B–L	17	6·00±0·10	0·50	5·3–7·1
C	M–D	27	6·40±0·10	0·50	5·4–7·8
	B–L	27	11·40±0·10	0·80	9·9–13·3
P_3	M–D	29	18·10±0·20	1·10	16·6–21·5
	B–L	30	5·10±0·04	0·20	4·7–5·6
P_4	M–D	6	8·50±0·20	0·40	8·1–9·4
	B–L	18	6·50±0·10	0·30	6·0–7·0
M_1	M–D				
	Tri B	12	7·50±0·10	0·30	7·1–8·0
	Tal B	12	7·80±0·10	0·40	7·3–8·6
M_2	M–D	8	13·60±0·20	0·40	13·0–14·4
	Tri B	20	9·40±0·10	0·50	8·6–10·6
	Tal B	19	9·10±0·10	0·40	8·5–9·7
M_3	M–D	30	16·80±0·20	0·80	15·2–18·8
	Tri B	35	9·90±0·10	0·70	11·5–9·1
	Tal B	35	9·10±0·10	0·40	8·4–9·9
	Hypo-conulid				

G. G. Eck, unpublished data, 1975

Table 147
Theropithecus gelada female

Maxillary teeth		n	X̄ ± s.e.	s.d.	Range
I^1	M–D				
	B–L	16	6·00 ± 0·10	0·40	5·3–6·5
I^2	M–D				
	B–L				
C	M–D	18	7·40 ± 0·10	0·40	6·4–8·1
	B–L	18	5·40 ± 0·10	0·30	4·9–5·8
P^3	M–D	11	6·20 ± 0·10	0·40	5·5–6·8
	B–L	24	6·70 ± 0·10	0·40	5·7–7·2
P^4	M–D	8	7·00 ± 0·10	0·30	6·3–7·4
	B–L	22	7·60 ± 0·10	0·40	6·9–8·3
M^1	M–D	3	10·70 ± 0·20	0·40	10·1–11·0
	Ant B	15	8·70 ± 0·10	0·20	8·3–9·0
	Post B	15	8·00 ± 0·10	0·20	7·7–8·4
M^2	M–D	9	12·90 ± 0·20	0·70	12·1–14·3
	Ant B	20	10·50 ± 0·10	0·40	9·7–11·2
	Post B	19	9·30 ± 0·10	0·40	8·7–10·2
M^3	M–D	18	12·80 ± 0·10	0·50	11·9–13·5
	Ant B	20	10·70 ± 0·10	0·60	9·5–11·8
	Post B	20	9·10 ± 0·10	0·50	8·0–10·2

G. G. Eck, unpublished data, 1975

Table 148
Theropithecus gelada female

Mandibular teeth		n	X̄ ± s.e.	s.d.	Range
I_1	M–D				
	B–L	15	5·60 ± 0·10	0·30	5·0–6·1
I_2	M–D				
	B–L	17	5·30 ± 0·10	0·30	4·7–5·8
C	M–D	21	3·40 ± 0·10	0·40	2·1–3·9
	B–L	18	6·90 ± 0·10	0·40	6·1–7·9
P_3	M–D	17	10·40 ± 0·20	1·00	8·8–12·1
	B–L	19	4·50 ± 0·10	0·40	4·1–5·7
P_4	M–D	8	8·00 ± 0·10	0·30	7·4–8·5
	B–L	18	6·20 ± 0·10	0·20	5·7–6·6
M_1	M–D				
	Tri B	11	7·20 ± 0·10	0·20	6·8–7·5
	Tal B	11	7·50 ± 0·10	0·30	7·1–8·1
M_2	M–D	6	13·10 ± 0·30	0·70	11·9–13·8
	Tri B	18	8·90 ± 0·10	0·40	8·3–9·5
	Tal B	18	8·70 ± 0·10	0·40	7·9–9·3
M_3	M–D	14	16·20 ± 0·20	0·70	15·2–17·8
	Tri B	20	9·50 ± 0·10	0·30	8·9–10·0
	Tal B	20	8·71 ± 0·10	0·30	8·1–9·3
	Hypo-conulid				

G. G. Eck, unpublished data, 1975

Table 149
Colobus polykomos male

Maxillary teeth		n	$\bar{X} \pm$ s.e.	s.d.	Range	P
I^1	M–D	43	5·1±0·07	0·47	4·0–6·1	
	B–L	44	4·7±0·08	0·51	3·9–6·0	
I^2	M–D	42	4·6±0·06	0·42	3·7–5·6	*
	B–L	42	4·4±0·11	0·73	3·6–5·6	
C	M–D	40	9·8±0·81	0·51	4·7–10·1	†
	B–L	38	6·5±0·15	0·93	4·2–8·6	†
P^3	M–D	49	5·3±0·08	0·56	4·4–8·0	*
	B–L	48	5·6±0·07	0·45	4·6–6·4	†
P^4	M–D	47	5·2±0·05	0·37	4·3–6·1	
	B–L	48	6·7±0·08	0·55	5·8–8·0	†
M^1	M–D	49	7·0±0·06	0·42	6·0–8·5	*
	Ant B	49	6·5±0·07	0·46	5·7–8·0	†
	Post B	45	6·3±0·06	0·42	5·6–7·7	*
M^2	M–D	48	7·5±0·10	0·70	4·0–8·6	*
	Ant B	48	7·4±0·07	0·49	6·4–8·5	†
	Post B	47	6·9±0·08	0·58	5·5–8·1	†
M^3	M–D	43	7·6±0·07	0·46	6·8–8·7	
	Ant B	44	6·9±0·09	0·57	5·0–8·0	†
	Post B	44	6·4±0·11	0·71	4·7–7·5	†

* $P<0·05$
† $P<0·01$

Table 150
Colobus polykomos male

Mandibular teeth		n	$\bar{X} \pm$ s.e.	s.d.	Range	P
I_1	M–D	38	3·7±0·07	0·40	3·0–5·4	
	B–L	38	4·4±0·06	0·37	3·6–5·3	
I_2	M–D	37	3·8±0·10	0·63	3·0–6·7	
	B–L	38	4·8±0·07	0·44	3·7–6·1	*
C	M–D	37	6·9±0·20	1·22	4·7–9·1	
	B–L	36	6·9±0·23	1·39	4·4–9·4	†
P_3	M–D	44	9·1±0·77	1·50	4·2–10·0	
	B–L	45	5·2±0·12	0·80	3·7–6·9	*
P_4	M–D	48	6·1±0·06	0·43	5·1–7·1	
	B–L	46	4·9±0·07	0·49	3·9–6·1	†
M_1	M–D	47	7·2±0·06	0·42	6·3–8·3	†
	Tri B	45	5·5±0·06	0·40	4·5–6·2	†
	Tal B	46	5·7±0·06	0·43	4·6–6·4	†
M_2	M–D	47	7·7±0·10	0·65	5·0–8·8	*
	Tri B	45	6·3±0·08	0·52	5·0–7·4	†
	Tal B	47	6·5±0·07	0·51	5·3–7·4	†
M_3	M–D	43	9·6±0·10	0·67	8·1–11·1	
	Tri B	44	6·3±0·08	0·52	5·0–7·3	†
	Tal B	43	6·2±0·07	0·48	5·0–7·1	†
	Hypoconulid	40	4·2±0·08	0·53	3·1–5·6	

* $P<0·05$
† $P<0·01$

Table 151
Colobus polykomos female

Maxillary teeth		n	$\bar{X} \pm$ s.e.	s.d.	Range
I^1	M–D	28	4·9±0·07	0·34	4·1–5·7
	B–L	29	4·5±0·08	0·43	3·5–5·3
I^2	M–D	26	4·4±0·07	0·35	3·9–5·3
	B–L	26	4·3±0·08	0·39	3·6–5·0
C	M–D	23	7·0±0·13	0·62	5·0–8·0
	B–L	23	5·5±0·11	0·55	4·6–7·0
P^3	M–D	29	5·1±0·06	0·30	4·6–5·9
	B–L	29	5·4±0·09	0·46	4·2–6·6
P^4	M–D	28	5·1±0·05	0·27	4·5–5·5
	B–L	29	6·2±0·10	0·54	5·0–7·5
M^1	M–D	30	6·8±0·07	0·37	6·1–7·5
	Ant B	29	6·2±0·08	0·42	5·5–6·9
	Post B	29	6·1±0·08	0·41	5·3–6·8
M^2	M–D	29	7·2±0·11	0·57	5·0–8·0
	Ant B	28	6·9±0·09	0·47	6·0–8·0
	Post B	28	6·5±0·13	0·68	4·1–7·7
M^3	M–D	27	7·3±0·15	0·78	4·9–9·0
	Ant B	27	6·5±0·07	0·39	5·8–7·3
	Post B	27	5·9±0·11	0·57	4·7–7·0

Table 152
Colobus polykomos female

Mandibular teeth		n	$\bar{X} \pm$ s.e.	s.d.	Range
I_1	M–D	27	3·7±0·08	0·43	3·1–5·2
	B–L	26	4·2±0·06	0·30	3·8–4·8
I_2	M–D	23	3·7±0·07	0·33	3·2–4·5
	B–L	24	4·7±0·07	0·36	4·1–5·6
C	M–D	26	5·6±0·22	1·11	3·9–8·5
	B–L	27	5·3±0·18	0·53	3·9–7·1
P_3	M–D	28	7·1±0·19	1·01	4·7–8·8
	B–L	29	4·9±0·13	0·70	3·6–6·5
P_4	M–D	30	5·9±0·07	0·37	5·0–6·7
	B–L	30	4·7±0·07	0·38	3·9–5·4
M_1	M–D	30	7·0±0·07	0·36	6·4–7·7
	Tri B	31	5·2±0·07	0·40	4·4–5·8
	Tal B	31	5·5±0·07	0·39	4·7–6·2
M_2	M–D	30	7·4±0·07	0·40	6·6–8·1
	Tri B	30	5·9±0·09	0·48	5·0–6·9
	Tal B	30	6·2±0·09	0·50	5·2–7·6
M_3	M–D	26	9·4±0·13	0·67	8·2–10·5
	Tri B	24	6·0±0·09	0·44	5·3–7·0
	Tal B	25	5·9±0·07	0·34	5·3–6·8
	Hypo-conulid	25	4·1±0·13	0·65	3·1–5·6

Table 153
Colobus badius male

Maxillary teeth		n	$\bar{X}\pm$s.e.	s.d.	Range	P
I^1	M–D	23	5·5±0·09	0·42	4·8–6·3	
	B–L	22	4·9±0·06	0·28	4·2–5·4	
I^2	M–D	18	4·5±0·09	0·39	3·7–5·1	
	B–L	19	4·7±0·07	0·32	4·1–5·4	
C	M–D	17	9·8±0·23	0·94	8·3–11·3	†
	B–L	14	7·2±0·16	0·60	6·2–8·0	†
P^3	M–D	24	5·3±0·09	0·44	4·3–6·0	†
	B–L	22	5·3±0·08	0·39	4·6–6·3	
P^4	M–D	22	5·0±0·09	0·41	4·2–5·8	
	B–L	22	5·8±0·10	0·46	4·9–6·6	
M^1	M–D	26	7·0±0·08	0·39	6·3–7·9	
	Ant B	24	6·0±0·07	0·36	5·4–6·7	
	Post B	23	5·8±0·07	0·36	5·1–6·6	
M^2	M–D	24	7·3±0·08	0·41	6·5–8·1	
	Ant B	22	6·5±0·09	0·43	5·7–7·5	
	Post B	22	6·1±0·09	0·40	5·2–7·1	
M^3	M–D	22	7·0±0·09	0·40	6·5–7·8	
	Ant B	20	6·1±0·10	0·43	5·5–7·2	
	Post B	17	5·7±0·10	0·43	4·9–6·6	

† $P<0·01$

Table 154
Colobus badius male

Mandibular teeth		n	$\bar{X}\pm$s.e.	s.d.	Range	P
I_1	M–D	21	3·9±0·06	0·27	3·4–4·5	
	B–L	22	4·7±0·07	0·33	4·1–5·2	
I_2	M–D	21	3·7±0·04	0·20	3·4–4·1	
	B–L	19	5·1±0·09	0·39	4·3–5·7	
C	M–D	9	6·9±0·47	1·41	5·6–9·3	†
	B–L	16	8·2±0·23	0·91	6·1–9·3	†
P_3	M–D	20	9·2±0·33	1·46	6·2–11·3	
	B–L	24	5·2±0·16	0·77	3·6±6·6	†
P_4	M–D	19	5·7±0·09	0·39	5·0–6·4	
	B–L	24	4·6±0·09	0·46	3·7–5·5	
M_1	M–D	23	7·2±0·08	0·39	6·5–7·9	
	Tri B	19	5·0±0·08	0·35	4·5–5·7	
	Tal B	20	5·2±0·08	0·38	4·7–6·0	
M_2	M–D	24	7·4±0·10	0·47	6·1–8·2	
	Tri B	25	5·6±0·09	0·46	4·7–6·4	
	Tal B	26	5·8±0·10	0·49	5·0–6·8	
M_3	M–D	22	9·1±0·11	0·51	7·9–10·0	
	Tri B	23	5·7±0·08	0·39	5·0–6·4	
	Tal B	21	5·7±0·10	0·46	4·9–6·9	
	Hypo-conulid	19	3·8±0·11	0·49	3·1–4·9	

† $P<0·01$

Table 155
Colobus badius female

Maxillary teeth		n	$\bar{X} \pm$ s.e.	s.d.	Range
I^1	M–D	26	5·5 ± 0·10	0·50	4·6–6·4
	B–L	20	4·8 ± 0·07	0·31	4·3–5·4
I^2	M–D	25	4·4 ± 0·08	0·41	3·8–5·3
	B–L	25	4·5 ± 0·07	0·34	3·8–5·0
C	M–D	17	6·9 ± 0·27	1·11	5·8–10·8
	B–L	19	5·5 ± 0·14	0·60	4·6–6·8
P^3	M–D	23	4·8 ± 0·07	0·34	4·4–5·6
	B–L	23	5·3 ± 0·10	0·50	4·4–6·2
P^4	M–D	21	5·0 ± 0·05	0·25	4·6–5·5
	B–L	25	5·8 ± 0·08	0·40	5·2–6·6
M^1	M–D	26	6·9 ± 0·08	0·40	6·3–8·1
	Ant B	25	6·0 ± 0·07	0·33	5·5–6·8
	Post B	26	5·8 ± 0·07	0·33	5·3–6·4
M^2	M–D	27	7·2 ± 0·08	0·42	6·4–8·6
	Ant B	26	6·6 ± 0·08	0·39	5·9–7·3
	Post B	23	6·2 ± 0·09	0·42	5·5–7·3
M^3	M–D	25	7·0 ± 0·09	0·46	6·0–8·2
	Ant B	19	6·6 ± 0·13	0·57	5·9–7·7
	Post B	20	5·9 ± 0·10	0·46	5·2–6·9

Table 156
Colobus badius female

Mandibular teeth		n	$\bar{X} \pm$ s.e.	s.d.	Range
I_1	M–D	21	4·0 ± 0·07	0·32	3·3–4·4
	B–L	20	4·7 ± 0·09	0·39	3·9–5·4
I_2	M–D	21	3·8 ± 0·05	0·23	3·4–4·2
	B–L	23	5·1 ± 0·09	0·45	4·3–6·1
C	M–D	16	4·4 ± 0·15	0·58	3·7–6·4
	B–L	21	6·2 ± 0·12	0·56	5·0–7·5
P_3	M–D	20	8·5 ± 0·22	0·99	6·1–10·5
	B–L	22	4·2 ± 0·11	0·52	3·6–5·4
P_4	M–D	20	5·6 ± 0·08	0·35	4·8–6·2
	B–L	24	4·5 ± 0·10	0·46	4·0–5·8
M_1	M–D	26	7·1 ± 0·09	0·43	6·4–8·2
	Tri B	21	5·0 ± 0·07	0·32	4·6–5·8
	Tal B	22	5·2 ± 0·07	0·30	4·7–5·8
M_2	M–D	23	7·3 ± 0·09	0·43	6·6–8·5
	Tri B	22	5·6 ± 0·07	0·32	5·1–6·2
	Tal B	23	5·9 ± 0·08	0·41	5·1–6·8
M_3	M–D	20	8·9 ± 0·14	0·61	7·8–10·1
	Tri B	19	5·8 ± 0·09	0·37	5·1–6·6
	Tal B	17	5·7 ± 0·09	0·39	5·0–6·5
	Hypoconulid	17	3·7 ± 0·08	0·33	3·1–4·4

Table 157
Nasalis larvatus male

Maxillary teeth		n	$\bar{X}\pm$s.e.	s.d.	Range	P
I^1	M–D	21	5·7±0·16	0·72	4·4–8·0	
	B–L	20	4·9±0·07	0·33	4·3–6·0	†
I^2	M–D	20	4·3±0·07	0·31	3·8–5·0	
	B–L	20	4·9±0·10	0·47	4·5–6·6	†
C	M–D	18	8·2±0·28	1·18	4·1–9·4	†
	B–L	18	6·4±0·12	0·50	5·4–7·1	†
P^3	M–D	21	5·2±0·08	0·35	4·5–6·0	†
	B–L	21	5·9±0·12	0·54	5·0–7·8	†
P^4	M–D	21	5·3±0·33	0·52	4·6–6·9	
	B–L	21	6·1±0·07	0·31	5·5–7·0	†
M^1	M–D	20	7·0±0·07	0·32	6·0–7·4	
	Ant B	21	6·3±0·04	0·18	5·9–6·7	†
	Post B	21	6·3±0·06	0·27	5·4–6·6	*
M^2	M–D	21	7·9±0·07	0·34	7·4–8·9	*
	Ant B	21	7·3±0·10	0·46	5·7–8·0	†
	Post B	21	6·9±0·99	0·45	5·9–7·9	*
M^3	M–D	20	7·8±0·10	0·46	6·9–8·5	
	Ant B	20	7·1±0·01	0·42	5·6–7·7	*
	Post B	20	6·2±0·13	0·57	4·5–7·0	

* $P<0·05$
† $P<0·01$

Table 158
Nasalis larvatus male

Mandibular teeth		n	$\bar{X}\pm$s.e.	s.d.	Range	P
I_1	M–D	19	4·1±0·36	0·58	3·2–4·5	
	B–L	19	4·9±0·18	0·78	4·0–7·9	†
I_2	M–D	21	4·1±0·12	0·56	3·4–5·7	
	B–L	21	5·0±0·06	0·29	4·6–5·7	†
C	M–D	19	5·6±0·07	0·31	5·1–6·1	*
	B–L	20	6·9±0·17	0·75	4·5–7·9	†
P_3	M–D	21	10·0±0·21	0·95	7·7–11·5	†
	B–L	20	5·4±0·13	0·60	4·2–6·3	†
P_4	M–D	21	5·3±0·09	0·39	4·6–6·2	
	B–L	21	4·8±0·10	0·43	4·1–5·6	
M_1	M–D	20	7·3±0·08	0·34	6·7–7·9	†
	Tri B	21	5·3±0·09	0·39	4·9–6·5	*
	Tal B	21	5·7±0·09	0·40	5·1–6·6	†
M_2	M–D	21	8·0±0·11	0·51	6·6–9·1	*
	Tri B	20	6·3±0·08	0·36	5·5–7·0	*
	Tal B	20	6·5±0·06	0·25	6·1–7·0	
M_3	M–D	18	10·1±0·12	0·49	9·1–11·4	†
	Tri B	18	6·6±0·08	0·33	6·0–7·3	
	Tal B	18	6·4±0·07	0·31	5·7–7·0	
	Hypo-conulid	18	4·7±0·09	0·38	4·0–5·0	

* $P<0·05$
† $P<0·01$

Table 159
Nasalis larvatus female

Maxillary teeth		n	$\bar{X} \pm$ s.e.	s.d.	Range
I^1	M–D	13	5·3±0·11	0·41	4·9–6·3
	B–L	13	4·3±0·10	0·35	3·4–4·7
I^2	M–D	13	4·2±0·12	0·45	3·6–5·0
	B–L	13	4·2±0·08	0·28	3·6–4·6
C	M–D	13	6·1±0·08	0·28	5·5–6·5
	B–L	13	5·0±0·10	0·35	4·5–5·9
P^3	M–D	13	4·8±0·07	0·27	4·2–5·0
	B–L	13	5·4±0·08	0·29	4·9–5·8
P^4	M–D	13	4·9±0·09	0·33	4·4–5·4
	B–L	13	5·8±0·08	0·28	5·1–6·0
M^1	M–D	14	6·8±0·10	0·37	6·2–7·6
	Ant B	14	6·0±0·08	0·31	5·5–6·6
	Post B	14	6·1±0·07	0·24	5·6–6·5
M^2	M–D	13	7·5±0·10	0·38	6·7–8·0
	Ant B	13	6·8±0·09	0·34	6·0–7·3
	Post B	13	6·5±0·11	0·40	5·9–7·0
M^3	M–D	12	7·5±0·11	0·37	6·8–8·0
	Ant B	12	6·7±0·09	0·32	6·0–7·0
	Post B	12	5·9±0·08	0·27	5·4–6·4

Table 160
Nasalis larvatus female

Mandibular teeth		n	$\bar{X} \pm$ s.e.	s.d.	Range
I_1	M–D	14	3·7±0·07	0·27	3·0–4·1
	B–L	14	4·2±0·07	0·27	3·7–4·6
I_2	M–D	14	3·8±0·06	0·22	3·2–4·0
	B–L	14	4·4±0·07	0·27	4·0–4·9
C	M–D	13	5·1±0·22	0·80	3·4–5·9
	B–L	13	5·6±0·09	0·33	5·0–6·0
P_3	M–D	12	5·9±0·10	0·35	5·3–6·4
	B–L	12	4·8±0·11	0·39	4·1–5·6
P_4	M–D	13	5·1±0·07	0·26	4·6–5·6
	B–L	13	4·7±0·08	0·27	4·2–5·0
M_1	M–D	13	6·9±0·10	0·36	6·3–7·7
	Tri B	13	5·0±0·06	0·21	4·6–5·4
	Tal B	13	5·3±0·08	0·28	5·0–5·9
M_2	M–D	13	7·7±0·10	0·37	7·0–8·1
	Tri B	13	6·1±0·06	0·22	5·7–6·6
	Tal B	13	6·3±0·07	0·25	5·8–6·7
M_3	M–D	12	9·6±0·12	0·40	9·1–10·0
	Tri B	12	6·5±0·10	0·35	5·9–7·0
	Tal B	12	6·2±0·10	0·34	5·5–6·8
	Hypoconulid	12	4·6±0·09	0·33	4·1–5·0

Nasalis concolor male measurements not available

Table 161
Nasalis concolor female

Maxillary teeth		n	$\bar{X}\pm$s.e.	s.d.	Range
I^1	M–D	5	4·6±0·15	0·34	4·2–5·1
	B–L	5	4·1±0·10	0·23	3·9–4·5
I^2	M–D	5	4·0±0·09	0·21	3·8–4·2
	B–L	5	4·1±0·09	0·19	3·8–4·3
C	M–D	4	5·2±0·08	0·15	5·1–5·4
	B–L	4	4·6±0·09	0·17	4·4–4·8
P^3	M–D	5	4·5±0·11	0·25	4·2–4·8
	B–L	5	5·3±0·10	0·23	4·9–5·5
P^4	M–D	5	4·3±0·08	0·18	4·1–4·5
	B–L	5	5·6±0·08	0·18	5·3–5·8
M^1	M–D	7	6·5±0·08	0·20	6·1–6·7
	Ant B	7	5·5±0·06	0·16	5·2–5·7
	Post B	7	5·6±0·08	0·20	5·2–5·8
M^2	M–D	5	7·0±0·07	0·15	6·8–7·2
	Ant B	5	6·0±0·11	0·24	5·6–6·2
	Post B	5	5·8±0·07	0·15	5·6–6·0
M^3	M–D	3	6·9±0·23	0·40	6·5–7·3
	Ant B	3	5·7±0·15	0·25	5·5–6·0
	Post B	3	5·0±0·18	0·31	4·7–5·3

Table 162
Nasalis concolor female

Mandibular teeth		n	$\bar{X}\pm$s.e.	s.d.	Range
I_1	M–D	6	3·5±0·08	0·19	3·3–3·8
	B–L	6	3·8±0·05	0·12	3·6–3·9
I_2	M–D	5	3·4±0·08	0·19	3·2–3·6
	B–L	5	4·1±0·06	0·13	4·0–4·3
C	M–D	4	4·8±0·21	0·42	4·3–5·2
	B–L	4	4·3±0·23	0·46	3·8–4·9
P_3	M–D	4	6·4±0·14	0·29	6·0–6·7
	B–L	5	4·2±0·16	0·35	3·7–4·6
P_4	M–D	5	4·7±0·09	0·20	4·4–4·9
	B–L	5	4·4±0·10	0·22	4·1–4·6
M_1	M–D	7	6·5±0·09	0·24	6·0–6·8
	Tri B	7	4·6±0·06	0·16	4·3–4·8
	Tal B	7	4·8±0·05	0·13	4·6–5·0
M_2	M–D	5	7·0±0·12	0·27	6·7–7·4
	Tri B	5	5·5±0·10	0·23	5·2–5·7
	Tal B	5	5·4±0·12	0·27	5·1–5·8
M_3	M–D	3	8·7±0·36	0·62	8·0–9·2
	Tri B	4	5·4±0·17	0·34	5·2–5·9
	Tal B	4	5·4±0·06	0·13	5·2–5·5
	Hypo-conulid	3	3·7±0·10	0·17	3·5–3·8

Table 163
Pygathrix nemaeus male

Maxillary teeth		*n*	$\bar{X}$ ± s.e.	s.d.	Range
I^1	M–D	8	5·3 ± 0·13	0·36	4·8–5·9
	B–L	7	5·4 ± 0·07	0·19	5·2–5·6
I^2	M–D	7	4·6 ± 0·20	0·52	4·0–5·2
	B–L	7	5·2 ± 0·12	0·32	4·8–5·7
C	M–D	4	6·8 ± 0·59	1·18	5·8–8·4
	B–L	4	6·1 ± 0·24	0·48	5·5–6·5
P^3	M–D	6	4·9 ± 0·16	0·40	4·4–5·5
	B–L	6	6·3 ± 0·14	0·34	5·9–6·7
P^4	M–D	6	4·7 ± 0·05	0·13	4·6–4·9
	B–L	5	6·6 ± 0·11	0·25	6·3–6·9
M^1	M–D	7	6·9 ± 0·14	0·36	6·4–7·4
	Ant B	7	6·6 ± 0·08	0·21	6·3–6·8
	Post B	6	6·2 ± 0·24	0·58	5·1–6·7
M^2	M–D	7	7·4 ± 0·15	0·39	6·9–8·1
	Ant B	6	7·2 ± 0·17	0·40	6·6–7·7
	Post B	6	6·9 ± 0·20	0·49	6·2–7·5
M^3	M–D	6	7·5 ± 0·28	0·53	6·5–8·1
	Ant B	5	6·7 ± 0·11	0·25	6·5–7·1
	Post B	6	6·0 ± 0·15	0·38	5·7–6·7

Table 164
Pygathrix nemaeus male

Mandibular teeth		*n*	$\bar{X}$ ± s.e.	s.d.	Range
I_1	M–D	6	3·9 ± 0·07	0·18	3·6–4·0
	B–L	6	4·9 ± 0·10	0·25	4·6–5·2
I_2	M–D	7	3·9 ± 0·10	0·27	3·5–4·3
	B–L	7	5·2 ± 0·12	0·32	4·8–5·7
C	M–D	6	6·2 ± 0·29	0·70	5·5–7·5
	B–L	6	5·3 ± 0·14	0·34	4·7–5·6
P_3	M–D	6	8·4 ± 0·26	0·64	7·2–9·0
	B–L	6	5·6 ± 0·15	0·37	5·2–6·1
P_4	M–D	6	5·3 ± 0·07	0·16	5·0–5·4
	B–L	5	5·1 ± 0·18	0·41	4·7–5·6
M_1	M–D	8	7·0 ± 0·10	0·29	6·6–7·4
	Tri B	7	5·4 ± 0·08	0·22	5·1–5·6
	Tal B	6	5·3 ± 0·11	0·26	5·0–5·6
M_2	M–D	7	7·6 ± 0·21	0·56	7·0–8·8
	Tri B	6	6·1 ± 0·13	0·31	5·6–6·5
	Tal B	5	6·2 ± 0·18	0·40	5·7–6·6
M_3	M–D	6	9·2 ± 0·12	0·30	8·8–9·5
	Tri B	4	6·2 ± 0·10	0·21	6·0–6·5
	Tal B	5	6·0 ± 0·08	0·17	5·7–6·1
	Hypoconulid	5	4·4 ± 0·22	0·48	3·8–4·9

Table 165
Pygathrix nemaeus female

Maxillary teeth		n	$\bar{X} \pm$ s.e.	s.d.	Range
I^1	M–D	4	4·8±0·20	0·40	4·2–5·1
	B–L	4	4·7±0·17	0·33	4·3–5·1
I^2	M–D	4	4·2±0·21	0·42	3·7–4·7
	B–L	4	4·9±0·13	0·25	4·6–5·2
C	M–D	3	5·4±0·31	0·53	5·0–6·0
	B–L	3	5·1±0·29	0·50	4·6–5·6
P^3	M–D	3	4·8±0·22	0·38	4·5–5·2
	B–L	4	5·7±0·05	0·10	5·5–5·7
P^4	M–D	4	4·5±0·14	0·29	4·3–4·9
	B–L	4	6·2±0·12	0·24	5·9–6·5
M^1	M–D	5	6·6±0·16	0·35	6·1–7·0
	Ant B	5	6·3±0·10	0·22	6·0–6·5
	Post B	5	6·2±0·16	0·36	5·7–6·5
M^2	M–D	4	6·9±0·15	0·30	6·5–7·2
	Ant B	3	6·9±0·22	0·38	6·5–7·2
	Post B				
M^3	M–D	4	6·7±0·13	0·26	6·4–7·0
	Ant B	4	6·5±0·18	0·36	6·2–7·0
	Post B	3	6·6±0·20	0·35	6·3–7·0

Table 166
Pygathrix nemaeus female

Mandibular teeth		n	$\bar{X} \pm$ s.e.	s.d.	Range
I_1	M–D	5	3·8±0·23	0·51	3·0–4·4
	B–L	4	4·4±0·11	0·22	4·1–4·6
I_2	M–D	4	3·6±0·13	0·26	3·4–4·0
	B–L	4	4·8±0·16	0·31	4·3–5·0
C	M–D	4	6·0±0·53	1·06	4·7–7·3
	B–L	4	4·5±0·20	0·41	3·9–4·8
P_3	M–D	2	7·1±0·45	0·64	6·6–7·5
	B–L	4	5·0±0·07	0·13	4·8–5·1
P_4	M–D	4	5·1±0·17	0·34	4·7–5·5
	B–L	4	4·8±0·11	0·22	4·5–5·0
M_1	M–D	5	6·8±0·09	0·19	6·5–7·0
	Tri B	5	5·2±0·12	0·28	5·0–5·7
	Tal B	5	5·3±0·10	0·22	5·1–5·6
M_2	M–D	4	7·1±0·14	0·29	6·8–7·5
	Tri B	4	5·9±0·07	0·13	5·7–6·0
	Tal B	4	6·0±0·12	0·25	5·8–6·3
M_3	M–D	4	8·5±0·24	0·47	8·2–9·2
	Tri B	4	5·9±0·07	0·13	5·7–6·0
	Tal B	3	5·7±0·15	0·25	5·5–6·0
	Hypoconulid	3	4·2±0·29	0·50	3·7–4·7

Pygathrix (Rhinopithecus) roxellanae male measurements not available

Table 167
Pygathrix (Rhinopithecus) roxellanae
female

Maxillary teeth		n	$\bar{X} \pm$ s.e.	s.d.	Range
I^1	M–D	12	6·14±0·18	0·61	4·5–6·8
	B–L	11	5·27±0·12	0·38	4·8–6·0
I^2	M–D	12	4·89±0·08	0·27	4·5–5·3
	B–L	12	4·88±0·13	0·46	4·1–6·1
C	M–D	11	6·74±0·31	1·03	5·6–9·6
	B–L	11	5·37±0·18	0·59	4·4–6·6
P^3	M–D	6	4·65±0·17	0·41	4·1–5·3
	B–L	10	6·00±0·19	0·59	5·1–6·9
P^4	M–D	12	5·37±0·15	0·51	4·3–6·1
	B–L	13	6·94±0·12	0·43	5·7–7·4
M^1	M–D	19	8·02±0·15	0·65	6·2–9·2
	Ant B	19	7·56±0·13	0·54	6·0–8·2
	Post B	19	7·54±0·13	0·59	5·7–8·3
M^2	M–D	13	8·74±0·26	0·95	6·1–10·1
	Ant B	12	8·98±0·08	0·27	8·5–9·5
	Post B	10	7·97±0·12	0·37	7·6–8·6
M^3	M–D	12	8·29±0·31	1·08	5·4–9·4
	Ant B	11	8·62±0·08	0·27	8·3–9·1
	Post B	9	7·18±0·06	0·18	6·9–7·5

Table 168
Pygathrix (Rhinopithecus) roxellanae
female

Mandibular teeth		n	$\bar{X} \pm$ s.e.	s.d.	Range
I_1	M–D	10	4·45±0·12	0·38	3·8–5·1
	B–L	11	4·95±0·12	0·41	4·3–5·5
I_2	M–D	10	4·24±0·11	0·34	3·7–4·8
	B–L	11	5·17±0·11	0·37	4·5–5·7
C	M–D	11	4·59±0·16	0·53	3·9–5·2
	B–L	10	5·73±0·24	0·75	4·7–7·3
P_3	M–D	11	8·25±0·24	0·79	6·7–10·0
	B–L	12	4·38±0·14	0·48	3·8–5·8
P_4	M–D	11	5·43±0·15	0·49	4·6–6·2
	B–L	11	5·15±0·09	0·29	4·8–5·8
M_1	M–D	18	7·89±0·17	0·72	6·3–9·1
	Tri B	18	6·14±0·09	0·38	5·3–6·8
	Tal B	18	6·53±0·10	0·42	5·5–7·2
M_2	M–D	12	8·65±0·21	0·74	6·5–9·5
	Tri B	11	7·20±0·17	0·55	5·6–7·8
	Tal B	11	7·33±0·19	0·64	5·6–8·0
M_3	M–D	11	10·35±0·42	1·38	6·8–11·8
	Tri B	10	7·24±0·25	0·78	5·2–8·1
	Tal B	10	7·04±0·17	0·53	5·6–7·6
	Hypoconulid	10	4·60±0·36	1·15	2·0–5·8

Table 169
Presbytis phayrei male

Maxillary teeth		n	$\bar{X} \pm$ s.e.	s.d.	Range
I^1	M–D	4	5·1 ± 0·10	0·21	4·9–5·3
	B–L	3	4·8 ± 0·17	0·29	4·5–5·0
I^2	M–D	2	4·6 ± 0·15	0·21	4·4–4·7
	B–L	3	4·5 ± 0·29	0·50	4·0–5·0
C	M–D	1	6·6 ± 0·00	0·00	6·6–6·6
	B–L	1	5·1 ± 0·00	0·00	5·1–5·1
P^3	M–D	4	4·5 ± 0·09	0·17	4·3–4·6
	B–L	3	5·2 ± 0·17	0·29	5·0–5·5
P^4	M–D	4	4·2 ± 0·06	0·13	4·0–4·3
	B–L	4	6·1 ± 0·07	0·13	5·9–6·2
M^1	M–D	4	6·0 ± 0·13	0·26	5·7–6·3
	Ant B	4	6·3 ± 0·08	0·15	6·2–6·5
	Post B	3	5·9 ± 0·03	0·06	5·9–6·0
M^2	M–D	4	6·4 ± 0·18	0·36	6·1–6·9
	Ant B	4	6·7 ± 0·12	0·25	6·4–6·9
	Post B	4	6·4 ± 0·16	0·31	6·0–6·7
M^3	M–D	2	6·1 ± 0·05	0·07	6·0–6·1
	Ant B	2	6·4 ± 0·45	0·64	5·9–6·8
	Post B	2	6·0 ± 0·10	0·14	5·9–6·1

Table 170
Presbytis phayrei male

Mandibular teeth		n	$\bar{X} \pm$ s.e.	s.d.	Range
I_1	M–D	3	3·5 ± 0·06	0·10	3·4–3·6
	B–L	3	4·5 ± 0·18	0·31	4·2–4·8
I_2	M–D	4	3·5 ± 0·12	0·24	3·2–3·7
	B–L	4	4·5 ± 0·14	0·27	4·1–4·7
C	M–D	3	8·7 ± 0·56	0·96	7·6–9·4
	B–L	1	4·6 ± 0·00	0·00	4·6–4·6
P_3	M–D	3	4·6 ± 0·15	0·25	4·4–4·9
	B–L	3	4·9 ± 0·46	0·80	4·1–5·7
P_4	M–D	3	8·7 ± 0·56	0·96	7·6–9·4
	B–L	3	4·4 ± 0·07	0·12	4·3–4·5
M_1	M–D	4	6·4 ± 0·10	0·19	6·2–6·6
	Tri B	4	4·9 ± 0·13	0·25	4·6–5·2
	Tal B	4	5·1 ± 0·19	0·39	4·7–5·5
M_2	M–D	4	6·8 ± 0·13	0·27	6·4–7·0
	Tri B	3	5·3 ± 0·42	0·72	4·5–5·9
	Tal B	4	5·4 ± 0·36	0·71	4·4–6·0
M_3	M–D	1	8·0 ± 0·00	0·00	8·0–8·0
	Tri B	2	5·5 ± 0·05	0·07	5·4–5·5
	Tal B	2	5·0 ± 0·15	0·21	4·8–5·1
	Hypo-conulid	1	3·6 ± 0·00	0·00	3·6–3·6

Table 171
Presbytis phayrei female

Maxillary teeth		n	$\bar{X} \pm$ s.e.	s.d.	Range
I^1	M–D	3	4·9±0·15	0·27	4·6–5·1
	B–L	1	4·7±0·00	0·00	4·7–4·7
I^2	M–D	4	4·1±0·25	0·50	3·4–4·6
	B–L	4	4·6±0·24	0·48	4·1–5·1
C	M–D	3	6·9±0·37	0·64	6·4–7·6
	B–L	3	5·6±0·00	0·00	5·6–5·6
P^3	M–D	4	4·3±0·05	0·10	4·2–4·4
	B–L	4	5·3±0·17	0·34	4·9–5·6
P^4	M–D	4	4·2±0·16	0·31	3·8–4·5
	B–L	3	6·1±0·12	0·21	5·9–6·3
M^1	M–D	5	5·9±0·21	0·46	5·3–6·5
	Ant B	5	6·2±0·08	0·18	6·0–6·4
	Post B	4	5·9±0·13	0·25	5·6–6·2
M^2	M–D	4	6·2±0·21	0·42	5·7–6·7
	Ant B	4	6·6±0·16	0·31	6·3–7·0
	Post B	4	6·2±0·23	0·45	5·8–6·7
M^3	M–D	4	6·1±0·17	0·33	5·7–6·5
	Ant B	4	6·2±0·27	0·54	5·6–6·9
	Post B	3	5·3±0·55	0·95	4·3–6·2

Table 172
Presbytis phayrei female

Mandibular teeth		n	$\bar{X} \pm$ s.e.	s.d.	Range
I_1	M–D	3	3·7±0·36	0·62	3·2–4·4
	B–L	4	4·3±0·06	0·13	4·2–4·5
I_2	M–D	4	3·3±0·09	0·18	3·1–3·5
	B–L	4	4·5±0·21	0·42	4·2–5·1
C	M–D	4	6·3±0·23	0·45	5·9–6·9
	B–L	4	4·8±0·52	1·05	3·7–6·0
P_3	M–D	4	7·8±0·70	1·40	7·0–9·9
	B–L	4	4·5±0·13	0·25	4·2–4·8
P_4	M–D	4	4·6±0·10	0·21	4·3–4·8
	B–L	4	4·4±0·13	0·27	4·0–4·6
M_1	M–D	5	6·1±0·17	0·39	5·8–6·8
	Tri B	4	4·9±0·05	0·10	4·8–5·0
	Tal B	5	5·0±0·11	0·24	4·7–5·3
M_2	M–D	4	6·4±0·06	0·13	6·3–6·6
	Tri B	2	5·4±0·30	0·42	5·1–5·7
	Tal B	2	5·6±0·40	0·57	5·2–6·0
M_3	M–D	4	7·9±0·29	0·57	7·4–8·7
	Tri B	2	5·7±0·00	0·00	5·7–5·7
	Tal B	3	5·5±0·26	0·45	5·0–5·9
	Hypo-conulid	3	3·7±0·21	0·36	3·3–4·0

Table 173
Presbytis johnii male

Maxillary teeth		n	$\bar{X} \pm$ s.e.	s.d.	Range
I^1	M–D	7	4·8 ± 0·16	0·41	4·4–5·4
	B–L	3	4·3 ± 0·10	0·17	4·2–4·5
I^2	M–D	5	4·2 ± 0·16	0·36	3·7–4·7
	B–L	5	4·3 ± 0·17	0·38	3·8–4·8
C	M–D	4	8·3 ± 0·17	0·35	8·0–8·8
	B–L	4	6·8 ± 0·15	0·29	6·5–7·2
P^3	M–D	6	4·7 ± 0·11	0·28	4·2–4·9
	B–L	5	5·1 ± 0·34	0·76	4·2–6·0
P^4	M–D	6	4·6 ± 0·11	0·27	4·2–4·9
	B–L	4	5·6 ± 0·21	0·42	5·2–6·2
M^1	M–D	8	6·4 ± 0·17	0·49	5·7–7·1
	Ant B	6	6·1 ± 0·13	0·31	5·8–6·5
	Post B	6	5·8 ± 0·10	0·24	5·5–6·2
M^2	M–D	5	6·7 ± 0·18	0·40	6·3–7·3
	Ant B	5	6·5 ± 0·43	0·10	4·9–7·4
	Post B	5	6·1 ± 0·30	0·68	5·0–6·7
M^3	M–D	5	6·7 ± 0·18	0·41	5·9–7·0
	Ant B	5	6·7 ± 0·18	0·41	6·0–7·0
	Post B	5	5·8 ± 0·34	0·75	4·6–6·6

Table 174
Presbytis johnii male

Mandibular teeth		n	$\bar{X} \pm$ s.e.	s.d.	Range
I_1	M–D	5	3·1 ± 0·12	0·27	3·0–3·6
	B–L	4	4·1 ± 0·06	0·12	4·0–4·2
I_2	M–D	4	3·6 ± 0·07	0·13	3·4–3·7
	B–L	4	4·2 ± 0·10	0·19	4·0–4·4
C	M–D	4	6·9 ± 0·40	0·79	5·7–7·5
	B–L	4	4·9 ± 0·29	0·57	4·0–5·2
P_3	M–D	5	8·9 ± 0·65	1·45	6·5–10·2
	B–L	4	4·8 ± 0·35	0·70	4·0–5·5
P_4	M–D	5	5·1 ± 0·18	0·40	4·8–5·8
	B–L	4	4·5 ± 0·21	0·42	4·0–5·0
M_1	M–D	7	6·6 ± 0·19	0·49	5·7–7·1
	Tri B	4	5·0 ± 0·12	0·24	4·7–5·2
	Tal B	4	5·1 ± 0·13	0·26	4·8–5·4
M_2	M–D	5	6·8 ± 0·12	0·27	6·6–7·2
	Tri B	3	5·7 ± 0·15	0·27	5·4–5·9
	Tal B	4	5·8 ± 0·19	0·38	5·5–6·3
M_3	M–D	5	8·4 ± 0·53	1·19	6·4–9·6
	Tri B	3	5·9 ± 0·12	0·20	5·7–6·1
	Tal B	3	5·6 ± 0·20	0·35	5·3–6·0
	Hypoconulid	3	4·4 ± 0·29	0·49	3·8–4·7

Table 175
Presbytis johnii female

Maxillary teeth		n	$\bar{X}\pm$s.e.	s.d.	Range
I^1	M–D	3	4·6±0·34	0·59	3·9–5·0
	B–L	2	4·5±0·00	0·00	4·5–4·5
I^2	M–D	1	3·7±0·00	0·00	3·7–3·7
	B–L	1	4·3±0·00	0·00	4·3–4·3
C	M–D	2	6·4±0·30	0·42	6·1–6·7
	B–L	2	5·3±0·30	0·42	5·0–5·6
P^3	M–D	2	4·4±0·40	0·57	4·0–4·8
	B–L	1	6·0±0·00	0·00	6·0–6·0
P^4	M–D	2	4·6±0·35	0·50	4·2–4·9
	B–L	1	6·8±0·00	0·00	6·8–6·8
M^1	M–D	3	6·1±0·34	0·59	5·7–6·8
	Ant B	2	6·4±0·00	0·00	6·4–6·4
	Post B	2	6·3±0·05	0·07	6·2–6·3
M^2	M–D	2	7·0±0·20	0·28	6·8–7·2
	Ant B	1	7·6±0·00	0·00	7·6–7·6
	Post B	2	7·2±0·35	0·50	6·8–7·5
M^3	M–D	2	6·6±0·60	0·85	6·0–7·2
	Ant B	2	7·0±0·70	0·99	6·3–7·7
	Post B	2	6·7±0·95	1·34	5·7–7·6

Table 176
Presbytis johnii female

Mandibular teeth		n	$\bar{X}\pm$s.e.	s.d.	Range
I_1	M–D	2	3·0±0·00	0·00	3·0–3·0
	B–L	3	4·1±0·15	0·25	3·9–4·4
I_2	M–D	2	3·4±0·05	0·07	3·3–3·4
	B–L	2	4·4±0·20	0·28	4·2–4·6
C	M–D	2	6·1±0·25	0·35	5·8–6·3
	B–L	2	4·5±0·15	0·21	4·3–4·6
P_3	M–D	2	7·1±0·40	0·57	6·7–7·5
	B–L	2	4·6±0·45	0·64	4·1–5·0
P_4	M–D	2	4·9±0·30	0·42	4·6–5·2
	B–L	2	4·9±0·35	0·50	4·5–5·2
M_1	M–D	3	6·6±0·22	0·38	6·2–6·9
	Tri B	2	5·3±0·25	0·35	5·0–5·5
	Tal B	3	5·5±0·24	0·42	5·2–6·0
M_2	M–D	2	7·3±0·40	0·57	6·9–7·7
	Tri B				
	Tal B	1	6·1±0·00	0·00	6·1–6·1
M_3	M–D	2	8·8±0·20	0·28	8·6–9·0
	Tri B	1	6·7±0·00	0·00	6·7–6·7
	Tal B	1	6·5±0·00	0·00	6·5–6·5
	Hypo-conulid	1	4·8±0·00	0·00	4·8–4·8

Table 177
Presbytis aygula male

Maxillary teeth		n	$\bar{X} \pm$ s.e.	s.d.	Range	P
I^1	M–D	14	4·0±0·08	0·30	3·6–4·8	
	B–L	14	4·5±0·10	0·39	3·9–5·2	
I^2	M–D	15	4·2±0·08	0·30	3·8–4·6	*
	B–L	15	4·5±0·09	0·35	3·9–5·0	
C	M–D	11	6·1±0·13	0·45	5·1–6·5	†
	B–L	13	4·7±0·09	0·33	4·2–5·4	*
P^3	M–D	17	4·0±0·09	0·36	3·4–4·4	
	B–L	17	4·8±0·09	0·37	4·2–5·5	
P^4	M–D	17	3·9±0·05	0·21	3·6–4·3	
	B–L	15	5·5±0·09	0·35	4·5–6·0	
M^1	M–D	20	5·3±0·05	0·22	4·8–5·7	
	Ant B	16	5·6±0·07	0·28	5·0–6·0	*
	Post B	18	5·4±0·06	0·26	4·9–5·9	*
M^2	M–D	19	5·5±0·07	0·29	5·1–6·0	
	Ant B	17	6·0±0·08	0·33	5·2–6·6	
	Post B	18	5·6±0·07	0·28	5·2–6·2	
M^3	M–D	17	5·2±0·09	0·38	4·9–6·2	
	Ant B	16	5·8±0·10	0·41	4·9–6·4	
	Post B	17	5·0±0·09	0·38	4·2–5·7	

* $P<0·05$
† $P<0·01$

Table 178
Presbytis aygula male

Mandibular teeth		n	$\bar{X} \pm$ s.e.	s.d.	Range	P
I_1	M–D	16	3·2±0·04	0·14	2·9–3·4	
	B–L	18	4·4±0·07	0·28	3·8–4·8	
I_2	M–D	16	3·4±0·06	0·25	3·1–4·1	
	B–L	18	4·4±0·07	0·31	3·8–4·7	
$\underline{C}$	M–D	15	5·4±0·10	0·37	5·0–6·0	†
	B–L	14	4·2±0·07	0·25	3·8–4·6	†
P_3	M–D	16	6·9±0·12	0·47	5·7–7·6	†
	B–L	17	4·3±0·06	0·25	3·7–4·6	†
P_4	M–D	17	4·5±0·24	0·40	4·0–5·1	
	B–L	9	4·1±0·13	0·40	3·7–5·0	
M_1	M–D	20	5·5±0·05	0·24	5·0–5·8	
	Tri B	14	4·5±0·12	0·44	4·0–5·8	
	Tal B	16	5·0±0·14	0·57	4·6–6·8	*
M_2	M–D	19	5·7±0·05	0·22	5·4–6·0	
	Tri B	16	5·1±0·09	0·35	4·6–5·8	
	Tal B	17	5·4±0·17	0·48	4·7–6·5	
M_3	M–D	17	6·0±0·08	0·32	5·2–6·6	
	Tri B	13	4·9±0·09	0·33	4·3–5·5	
	Tal B	11	4·9±0·15	0·48	4·3–5·8	
	Hypoconulid	7	4·3±0·66	1·74	2·3–6·0	

* $P<0·05$
† $P<0·01$

Table 179
Presbytis aygula female

Maxillary teeth		n	$\bar{X}\pm$s.e.	s.d.	Range
I^1	M–D	19	4·2±0·07	0·29	3·7–4·9
	B–L	18	4·6±0·06	0·26	4·0–5·0
I^2	M–D	19	4·0±0·05	0·22	3·5–4·4
	B–L	18	4·6±0·04	0·16	4·3–4·9
C	M–D	16	5·5±0·06	0·25	5·0–6·0
	B–L	16	5·0±0·06	0·22	4·7–5·5
P^3	M–D	17	4·0±0·06	0·26	3·5–4·4
	B–L	17	4·8±0·09	0·38	4·2–5·5
P^4	M–D	16	3·9±0·06	0·22	3·4–4·2
	B–L	16	5·4±0·07	0·29	4·9–5·8
M^1	M–D	26	5·3±0·05	0·28	4·7–5·8
	Ant B	25	5·3±0·06	0·31	5·0–6·0
	Post B	24	5·2±0·05	0·25	4·8–5·8
M^2	M–D	21	5·4±0·05	0·23	4·8–5·8
	Ant B	21	5·9±0·05	0·23	5·4–6·4
	Post B	18	5·4±0·06	0·23	5·0–5·8
M^3	M–D	18	5·2±0·11	0·45	4·9–6·6
	Ant B	17	5·6±0·08	0·35	4·7–6·0
	Post B	17	4·8±0·07	0·30	4·4–5·4

Table 180
Presbytis aygula female

Mandibular teeth		n	$\bar{X}\pm$s.e.	s.d.	Range
I_1	M–D	22	3·3±0·08	0·37	2·9–4·4
	B–L	22	4·3±0·04	0·18	3·9–4·6
I_2	M–D	21	3·4±0·04	0·17	3·1–3·7
	B–L	22	4·6±0·05	0·25	4·0–4·9
C	M–D	16	4·5±0·20	0·79	3·5–5·6
	B–L	15	4·8±0·20	0·78	3·5–5·7
P_3	M–D	15	5·5±0·27	1·05	4·4–5·9
	B–L	16	3·9±0·11	0·44	3·4–4·7
P_4	M–D	16	4·3±0·05	0·20	4·0–4·5
	B–L	16	4·0±0·06	0·23	3·6–4·4
M_1	M–D	24	5·4±0·06	0·29	4·8–6·0
	Tri B	21	4·3±0·05	0·24	3·9–4·7
	Tal B	23	4·7±0·06	0·28	4·2–5·4
M_2	M–D	24	5·6±0·05	0·27	4·9–6·1
	Tri B	21	5·0±0·06	0·26	4·5–5·5
	Tal B	20	5·1±0·06	0·27	4·7–5·8
M_3	M–D	20	6·1±0·14	0·61	5·0–7·5
	Tri B	17	5·0±0·07	0·28	4·6–5·4
	Tal B	16	4·7±0·17	0·67	2·7–6·0
	Hypoconulid	11	3·5±0·39	1·28	2·0–5·9

Table 181
Presbytis cristatus male

Maxillary teeth		n	$\bar{X}\pm$s.e.	s.d.	Range	P
I^1	M–D	20	4·67±0·11	0·49	4·0–5·9	*
	B–L	19	4·08±0·10	0·44	3·6–5·3	
I^2	M–D	22	3·76±0·06	0·28	3·4–4·5	
	B–L	22	3·86±0·08	0·35	2·7–4·3	
C	M–D	20	6·23±0·20	0·87	4·0–7·8	†
	B–L	20	5·19±0·20	0·91	3·1–6·7	
P^3	M–D	18	4·34±0·08	0·33	3·9–6·6	
	B–L	18	4·96±0·12	0·49	4·0–5·7	
P^4	M–D	19	4·15±0·06	0·26	3·7–4·6	
	B–L	19	5·44±0·07	0·30	5·0–6·0	
M^1	M–D	22	5·95±0·10	0·45	5·2–6·6	
	Ant B	22	5·88±0·06	0·30	5·3–6·5	
	Post B	22	5·59±0·08	0·36	5·1–6·5	
M^2	M–D	21	6·34±0·09	0·39	5·5–7·0	
	Ant B	21	6·56±0·07	0·33	6·0–7·2	
	Post B	21	6·12±0·08	0·35	5·6–7·0	*
M^3	M–D	18	6·06±0·09	0·39	5·5–7·0	
	Ant B	17	6·31±0·07	0·28	5·7–6·8	†
	Post B	17	5·43±0·12	0·48	4·5–6·1	

* $P<0·05$
† $P<0·01$

Table 182
Presbytis cristatus male

Mandibular teeth		n	$\bar{X}\pm$s.e.	s.d.	Range	P
I_1	M–D	20	3·11±0·09	0·41	2·0–3·7	
	B–L	20	3·51±0·13	0·59	1·9–4·5	
I_2	M–D	17	3·16±0·06	0·25	2·6–3·7	
	B–L	17	3·68±0·10	0·43	2·9–4·3	
C	M–D	19	4·62±0·12	0·52	3·5–5·6	†
	B–L	19	5·67±0·20	0·89	2·5–6·7	
P_3	M–D	19	6·20±0·08	0·36	5·5–6·9	†
	B–L	19	3·99±0·06	0·26	3·4–4·7	
P_4	M–D	19	4·64±0·09	0·38	4·0–5·0	
	B–L	19	4·04±0·04	0·19	3·7–4·3	
M_1	M–D	21	6·12±0·07	0·32	5·7–6·8	
	Tri B	21	4·74±0·05	0·21	4·3–5·2	
	Tal B	21	4·86±0·05	0·21	4·5–5·3	
M_2	M–D	20	6·50±0·10	0·43	5·8–7·1	
	Tri B	20	5·51±0·07	0·30	5·0–6·0	
	Tal B	20	5·47±0·08	0·37	4·7–6·1	
M_3	M–D	18	7·90±0·14	0·61	7·0–8·9	
	Tri B	18	5·46±0·09	0·39	4·9–6·3	
	Tal B	18	5·16±0·09	0·37	4·5–6·0	
	Hypoconulid	16	3·38±0·11	0·45	2·7–4·2	

† $P<0·01$

Table 183
Presbytis cristatus female

Maxillary teeth		n	$\bar{X} \pm$ s.e.	s.d.	Range
I^1	M–D	33	4·26±0·12	0·67	2·4–5·0
	B–L	30	3·95±0·09	0·48	2·2–4·8
I^2	M–D	29	3·73±0·07	0·40	2·7–4·7
	B–L	28	3·87±0·07	0·37	2·5–4·4
C	M–D	33	5·35±0·10	0·55	3·8–6·5
	B–L	33	5·13±0·10	0·55	3·1–6·2
P^3	M–D	32	4·20±0·05	0·28	3·9–5·0
	B–L	32	4·90±0·10	0·56	3·8–6·2
P^4	M–D	32	4·10±0·05	0·26	3·5–4·6
	B–L	32	5·53±0·06	0·31	5·1–6·2
M^1	M–D	32	5·79±0·07	0·39	5·0–6·4
	Ant B	32	5·84±0·06	0·31	5·2–6·9
	Post B	32	5·62±0·06	0·35	5·0–6·5
M^2	M–D	32	6·27±0·07	0·37	5·6–7·0
	Ant B	32	6·43±0·06	0·34	5·7–7·0
	Post B	32	5·93±0·05	0·29	5·4–6·6
M^3	M–D	31	6·04±0·06	0·33	5·4–6·9
	Ant B	31	6·09±0·05	0·27	5·8–6·9
	Post B	30	5·18±0·07	0·40	4·6–6·3

Table 184
Presbytis cristatus female

Mandibular teeth		n	$\bar{X} \pm$ s.e.	s.d.	Range
I_1	M–D	28	2·95±0·06	0·33	2·0–3·4
	B–L	28	3·57±0·06	0·32	3·0–4·0
I_2	M–D	30	3·03±0·05	0·25	2·3–3·6
	B–L	29	3·78±0·05	0·26	3·1–4·5
C	M–D	33	4·07±0·10	0·55	3·0–5·0
	B–L	32	5·07±0·06	0·35	4·6–6·0
P_3	M–D	31	5·34±0·07	0·36	4·6–6·3
	B–L	32	3·88±0·05	0·29	3·4–4·6
P_4	M–D	32	4·53±0·07	0·37	3·7–5·3
	B–L	32	4·17±0·05	0·28	3·7–5·0
M_1	M–D	33	5·98±0·07	0·39	5·3–7·0
	Tri B	32	4·71±0·04	0·24	4·3–5·1
	Tal B	32	4·88±0·03	0·19	4·5–5·3
M_2	M–D	32	6·34±0·06	0·34	5·8–7·0
	Tri B	30	5·43±0·06	0·31	4·9–6·1
	Tal B	31	5·47±0·05	0·30	5·0–6·0
M_3	M–D	32	7·61±0·11	0·63	6·3–9·0
	Tri B	32	5·43±0·06	0·34	5·0–6·1
	Tal B	32	5·25±0·06	0·34	4·9–6·3
	Hypoconulid	28	3·46±0·08	0·43	2·7–4·7

Table 185
Hylobates klossii male

Maxillary teeth		n	$\bar{X}\pm$s.e.	s.d.	Range
I^1	M–D	3	4·8±0·09	0·15	4·6–4·9
	B–L	3	3·8±0·03	0·06	3·8–3·9
I^2	M–D	3	3·6±0·15	0·25	3·4–3·9
	B–L	3	3·5±0·18	0·31	3·2–3·8
C	M–D	4	7·6±0·47	0·94	6·7–8·8
	B–L	4	5·4±0·18	0·35	5·0–5·8
P^3	M–D	3	4·7±0·32	0·56	4·2–5·3
	B–L	4	4·8±0·11	0·22	4·6–5·1
P^4	M–D	4	4·1±0·13	0·25	3·7–4·3
	B–L	4	5·0±0·18	0·36	4·7–5·5
M^1	M–D	4	5·5±0·09	0·17	5·2–5·6
	Ant B	3	5·9±0·07	0·12	5·8–6·0
	Post B	4	6·1±0·09	0·18	5·9–6·3
M^2	M–D	4	5·5±0·16	0·31	5·1–5·8
	Ant B	4	6·0±0·15	0·30	5·6–6·3
	Post B	4	6·1±0·13	0·25	5·7–6·3
M^3	M–D	4	4·2±0·12	0·25	4·0–4·5
	Ant B	3	5·5±0·03	0·06	5·5–5·6
	Post B				

Table 186
Hylobates klossii male

Mandibular teeth		n	$\bar{X}\pm$s.e.	s.d.	Range	P
I_1	M–D	4	2·9±0·09	0·17	2·7–3·1	
	B–L	4	3·3±0·16	0·33	2·9–3·7	
I_2	M–D	3	3·0±0·12	0·21	2·8–3·2	
	B–L	4	3·7±0·08	0·15	3·5–3·8	
C	M–D	3	5·1±0·12	0·21	4·9–5·3	*
	B–L	4	6·3±0·17	0·33	6·0–6·7	
P_3	M–D	4	5·9±0·28	0·56	5·2–6·5	
	B–L	4	3·9±0·07	0·13	3·7–4·0	
P_4	M–D	4	4·6±0·19	0·37	4·1–5·0	
	B–L	4	4·1±0·06	0·13	3·9–4·2	
M_1	M–D	4	6·0±0·10	0·20	5·7–6·1	
	Tri B	5	4·7±0·06	0·13	4·5–5·8	
	Tal B	5	4·8±0·07	0·15	4·6–5·0	
M_2	M–D	4	6·1±0·21	0·41	5·7–6·6	
	Tri B	4	5·0±0·08	0·15	4·8–5·1	
	Tal B	4	5·1±0·05	0·10	5·0–5·2	
M_3	M–D	4	5·0±0·29	0·58	4·4–5·7	
	Tri B	4	4·4±0·15	0·29	4·1–4·7	
	Tal B	3	4·4±0·34	0·59	3·7–4·8	
	Hypo-conulid					

* $P<0.05$

Table 187
Hylobates klossii female

Maxillary teeth		n	$\bar{X} \pm$ s.e.	s.d.	Range
I^1	M–D	4	4·5 ± 0·18	0·35	4·3–5·0
	B–L	4	4·0 ± 0·07	0·13	3·8–4·1
I^2	M–D	2	3·6 ± 0·10	0·14	3·5–3·7
	B–L	3	3·7 ± 0·07	0·12	3·6–3·8
C	M–D	2	7·0 ± 0·10	0·14	6·9–7·1
	B–L	2	4·8 ± 0·35	0·50	4·4–5·1
P^3	M–D	4	4·6 ± 0·07	0·14	4·5–4·8
	B–L	4	4·7 ± 0·13	0·26	4·4–4·9
P^4	M–D	3	4·0 ± 0·13	0·23	3·9–4·3
	B–L	4	4·9 ± 0·18	0·36	4·4–5·2
M^1	M–D	4	5·2 ± 0·08	0·15	5·0–5·3
	Ant B	4	5·7 ± 0·14	0·28	5·4–6·0
	Post B	4	5·7 ± 0·14	0·29	5·4–5·9
M^2	M–D	4	5·3 ± 0·12	0·25	5·0–5·5
	Ant B	4	5·7 ± 0·16	0·32	5·4–6·1
	Post B				
M^3	M–D	2	3·8 ± 0·45	0·64	3·3–4·2
	Ant B	4	5·7 ± 0·13	0·26	5·4–6·0
	Post B	2	5·0 ± 0·65	0·92	4·3–5·6

Table 188
Hylobates klossii female

Mandibular teeth		n	$\bar{X} \pm$ s.e.	s.d.	Range
I_1	M–D	3	2·8 ± 0·03	0·06	2·8–2·9
	B–L	4	3·2 ± 0·04	0·08	3·1–3·3
I_2	M–D	4	2·9 ± 0·10	0·21	2·7–3·1
	B–L	4	3·6 ± 0·10	0·21	3·4–3·9
C	M–D	3	4·6 ± 0·09	0·15	4·4–4·7
	B–L	3	5·9 ± 0·19	0·32	5·7–6·3
P_3	M–D	4	6·2 ± 0·13	0·27	5·9–6·5
	B–L	4	3·8 ± 0·05	0·10	3·7–3·9
P_4	M–D	4	4·6 ± 0·06	0·13	4·5–4·8
	B–L	4	4·0 ± 0·05	0·10	3·9–4·1
M_1	M–D	4	5·8 ± 0·07	0·14	5·7–6·0
	Tri B	4	4·6 ± 0·13	0·25	4·2–4·8
	Tal B	4	4·7 ± 0·09	0·17	4·5–4·9
M_2	M–D	4	5·8 ± 0·13	0·26	5·5–6·1
	Tri B	4	4·9 ± 0·10	0·19	4·8–5·2
	Tal B	4	5·0 ± 0·06	0·13	4·8–5·1
M_3	M–D	3	5·4 ± 0·15	0·25	5·1–5·6
	Tri B	3	4·7 ± 0·07	0·12	4·6–4·8
	Tal B	3	4·6 ± 0·09	0·15	4·5–4·8
	Hypoconulid				

Table 189
Hylobates agilis male

Maxillary teeth		n	$\bar{X} \pm$ s.e.	s.d.	Range
I^1	M–D	5	5·2±0·16	0·35	4·7–5·6
	B–L	5	4·0±0·18	0·41	3·5–4·5
I^2	M–D	6	4·0±0·10	0·25	3·5–4·2
	B–L	7	4·0±0·10	0·25	3·6–4·3
C	M–D	8	7·2±0·13	0·35	6·8–7·8
	B–L	8	5·3±0·13	0·38	4·7–5·8
P^3	M–D	8	4·8±0·11	0·30	4·4–5·4
	B–L	8	5·0±0·14	0·39	4·4–5·6
P^4	M–D	6	4·3±0·04	0·09	4·2–4·4
	B–L	8	5·0±0·15	0·44	4·2–5·4
M^1	M–D	8	5·7±0·07	0·20	5·5–6·0
	Ant B	6	6·0±0·07	0·17	5·7–6·2
	Post B	3	5·9±0·06	0·10	5·8–6·0
M^2	M–D	8	6·1±0·08	0·23	5·7–6·4
	Ant B	6	6·5±0·06	0·14	6·3–6·7
	Post B	6	6·1±0·06	0·15	5·8–6·2
M^3	M–D	8	5·3±0·10	0·28	5·0–5·7
	Ant B	7	6·0±0·08	0·20	5·7–6·3
	Post B	3	5·3±0·22	0·38	4·9–5·6

Table 190
Hylobates agilis male

Mandibular teeth		n	$\bar{X} \pm$ s.e.	s.d.	Range	P
I_1	M–D	3	3·4±0·17	0·30	3·1–3·7	
	B–L	4	3·4±0·20	0·39	3·0–3·9	
I_2	M–D	5	3·6±0·15	0·34	3·4–4·2	
	B–L	6	4·0±0·15	0·36	3·5–4·6	
C	M–D	5	5·3±0·19	0·43	4·7–5·8	
	B–L	6	6·8±0·07	0·16	6·5–7·0	
P_3	M–D	7	6·0±0·13	0·34	5·6–6·5	
	B–L	7	4·0±0·07	0·28	3·7-4·4	
P_4	M–D	7	4·8±0·07	0·18	4·6–5·1	
	B–L	7	4·2±0·07	0·19	3·9–4·4	
M_1	M–D	6	6·0±0·09	0·22	5·7–6·3	
	Tri B	5	4·9±0·05	0·10	4·8–5·0	
	Tal B	6	5·1±0·12	0·29	4·9–5·6	
M_2	M–D	7	6·4±0·11	0·28	5·9–6·7	
	Tri B	8	5·5±0·06	0·16	5·1–5·6	
	Tal B	7	5·6±0·16	0·30	4·9–5·8	
M_3	M–D	5	6·3±0·10	0·22	6·0–6·6	†
	Tri B	7	5·5±0·07	0·18	5·3–5·8	†
	Tal B	6	5·3±0·11	0·28	4·9–5·7	*
	Hypo-conulid					

* $P < 0·05$
† $P < 0·01$

Table 191
Hylobates agilis female

Maxillary teeth		n	$\bar{X} \pm$ s.e.	s.d.	Range
I^1	M–D	6	4·9 ± 0·09	0·23	4·7–5·2
	B–L	6	4·0 ± 0·08	0·20	3·7–4·3
I^2	M–D	5	3·9 ± 0·12	0·26	3·6–4·2
	B–L	6	4·0 ± 0·14	0·34	3·8–4·7
C	M–D	7	6·9 ± 0·13	0·35	6·6–7·4
	B–L	5	4·9 ± 0·18	0·40	4·4–5·5
P^3	M–D	9	4·6 ± 0·13	0·40	4·0–5·4
	B–L	9	4·9 ± 0·13	0·38	4·3–5·3
P^4	M–D	6	4·2 ± 0·08	0·21	3·8–4·4
	B–L	9	5·0 ± 0·09	0·26	4·6–5·4
M^1	M–D	9	5·6 ± 0·12	0·37	5·2–6·1
	Ant B	4	6·0 ± 0·22	0·47	5·4–6·5
	Post B	4	5·9 ± 0·22	0·44	5·4–6·4
M^2	M–D	9	5·8 ± 0·13	0·39	5·3–6·4
	Ant B	8	6·2 ± 0·17	0·49	5·1–6·7
	Post B	7	6·0 ± 0·11	0·28	5·6–6·5
M^3	M–D	6	5·1 ± 0·20	0·48	4·4–5·7
	Ant B	2	5·9 ± 0·10	0·14	5·8–6·0
	Post B	8	3·9 ± 0·10	0·28	3·7–4·4

Table 192
Hylobates agilis female

Mandibular teeth		n	$\bar{X} \pm$ s.e.	s.d.	Range
I_1	M–D	6	3·2 ± 0·04	0·11	3·0–3·3
	B–L	7	3·5 ± 0·07	0·20	3·3–3·9
I_2	M–D	8	3·4 ± 0·08	0·22	3·0–3·7
	B–L	8	4·2 ± 0·09	0·26	3·8–4·5
C	M–D	6	4·9 ± 0·19	0·48	4·4–5·5
	B–L	9	6·7 ± 0·11	0·33	5·9–7·0
P_3	M–D	8	6·1 ± 0·16	0·44	5·3–6·6
	B–L	8	4·1 ± 0·14	0·38	3·6–4·6
P_4	M–D	8	4·7 ± 0·10	0·29	4·2–5·2
	B–L	8	4·3 ± 0·08	0·22	3·8–4·5
M_1	M–D	9	6·0 ± 0·09	0·27	5·7–6·4
	Tri B	6	4·9 ± 0·06	0·15	4·8–5·2
	Tal B	6	5·2 ± 0·11	0·26	4·8–5·6
M_2	M–D	8	6·2 ± 0·09	0·26	5·9–6·6
	Tri B	8	5·4 ± 0·05	0·15	5·1–5·6
	Tal B	7	5·5 ± 0·07	0·20	5·2–5·7
M_3	M–D	4	5·7 ± 0·13	0·26	5·5–6·1
	Tri B	4	5·1 ± 0·07	0·14	5·0–5·3
	Tal B	4	4·7 ± 0·24	0·48	4·0–5·0
	Hypo-conulid				

Table 193
Hylobates moloch male

Maxillary teeth		n	$\bar{X}\pm$s.e.	s.d.	Range
I^1	M–D	5	5·0±0·10	0·23	4·8–5·4
	B–L	4	4·1±0·14	0·29	3·9–4·5
I^2	M–D	4	4·0±0·18	0·36	3·5–4·3
	B–L	4	4·3±0·03	0·06	4·2–4·3
C	M–D	3	7·4±0·35	0·61	6·7–7·8
	B–L	3	5·5±0·17	0·30	5·2–5·8
P^3	M–D	5	4·7±0·29	0·65	3·8–5·4
	B–L	5	5·0±0·22	0·49	4·2–5·5
P^4	M–D	5	4·3±0·12	0·26	4·1–4·7
	B–L	5	5·3±0·16	0·37	4·9–5·7
M^1	M–D	5	5·9±0·11	0·25	5·5–6·1
	Ant B	3	6·5±0·12	0·21	6·3–6·7
	Post B	3	6·0±0·18	0·31	5·7–6·3
M^2	M–D	4	6·4±0·25	0·50	5·8–7·0
	Ant B	4	6·3±0·14	0·27	5·9–6·5
	Post B	4	6·5±0·17	0·33	6·2–6·9
M^3	M–D	4	5·0±0·40	0·79	4·0–5·7
	Ant B	4	6·2±0·30	0·60	5·6–7·0
	Post B	1	5·9±0·00	0·00	5·9–5·9

Table 194
Hylobates moloch male

Mandibular teeth		n	$\bar{X}\pm$s.e.	s.d.	Range
I_1	M–D	4	3·3±0·07	0·14	3·2–3·5
	B–L	4	3·6±0·09	0·18	3·4–3·8
I_2	M–D	4	3·6±0·13	0·25	3·3–3·9
	B–L	4	4·4±0·12	0·25	4·2–4·7
C	M–D	3	5·5±0·07	0·12	5·4–5·6
	B–L	3	7·2±0·20	0·35	6·9–7·6
P_3	M–D	4	6·1±0·13	0·25	5·8–6·4
	B–L	5	4·4±0·05	0·11	4·3–4·6
P_4	M–D	5	4·8±0·14	0·32	4·4–5·2
	B–L	5	4·3±0·08	0·19	4·1–4·6
M_1	M–D	5	6·4±0·10	0·23	6·0–6·6
	Tri B	3	5·2±0·20	0·35	4·8–5·4
	Tal B	3	5·2±0·21	0·36	4·8–5·5
M_2	M–D	4	6·6±0·17	0·33	6·2–7·0
	Tri B	5	5·7±0·10	0·22	5·4–6·0
	Tal B	4	5·9±0·21	0·41	5·3–6·3
M_3	M–D	3	6·2±0·31	0·53	5·8–6·8
	Tri B	2	5·5±0·15	0·21	5·3–5·6
	Tal B	2	5·3±0·30	0·42	5·0–5·6
	Hypo-conulid				

Table 195
Hylobates moloch female

Maxillary teeth		*n*	$\bar{X} \pm$ s.e.	s.d.	Range
I^1	M–D	4	5·3±0·27	0·55	4·7–5·9
	B–L	4	3·9±0·15	0·30	3·5–4·2
I^2	M–D	4	4·1±0·22	0·44	3·5–4·5
	B–L	5	4·3±0·13	0·30	4·0–4·7
C	M–D	4	7·6±0·41	0·83	7·0–8·8
	B–L	4	5·0±0·32	0·63	4·1–5·6
P^3	M–D	4	5·1±0·19	0·37	4·7–5·6
	B–L	4	5·1±0·17	0·33	4·8–5·5
P^4	M–D	5	4·5±0·14	0·32	3·9–4·7
	B–L	5	5·4±0·12	0·27	5·0–5·7
M^1	M–D	4	5·7±0·25	0·49	5·1–6·3
	Ant B	4	6·3±0·19	0·38	5·7–6·5
	Post B	4	6·3±0·19	0·39	5·8–6·7
M^2	M–D	5	5·9±0·30	0·68	5·0–6·6
	Ant B	4	6·4±0·38	0·75	5·6–7·2
	Post B	3	6·5±0·44	0·76	5·6–7·0
M^3	M–D	5	5·2±0·22	0·50	4·5–5·9
	Ant B	2	5·5±0·45	0·64	5·0–5·9
	Post B	2	5·4±1·15	1·63	4·2–6·5

Table 196
Hylobates moloch female

Mandibular teeth		*n*	$\bar{X} \pm$ s.e.	s.d.	Range
I_1	M–D	4	3·7±0·03	0·06	3·6–3·7
	B–L	5	3·7±0·13	0·30	3·3–4·0
I_2	M–D	4	3·6±0·19	0·37	3·0–3·8
	B–L	5	4·0±0·14	0·30	3·6–4·3
C	M–D	4	5·4±0·19	0·37	5·0–5·8
	B–L	4	7·0±0·42	0·84	6·0–7·9
P_3	M–D	5	6·5±0·15	0·34	6·1–6·9
	B–L	5	4·3±0·18	0·41	3·6–4·6
P_4	M–D	5	5·1±0·13	0·30	4·7–5·4
	B–L	5	4·3±0·15	0·34	3·8–4·6
M_1	M–D	5	6·2±0·20	0·44	5·5–6·7
	Tri B	3	4·9±0·19	0·32	4·5–5·1
	Tal B	2	4·9±0·35	0·50	4·5–5·2
M_2	M–D	4	6·5±0·34	0·68	5·6–7·2
	Tri B	2	5·3±0·40	0·57	4·9–5·7
	Tal B	2	5·4±0·45	0·64	4·9–5·8
M_3	M–D	4	6·3±0·50	1·01	5·2–7·3
	Tri B	3	5·3±0·46	0·79	4·4–5·9
	Tal B	3	5·2±0·46	0·79	4·3–5·8
	Hypo-conulid				

Table 197
Hylobates lar male

Maxillary teeth		n	$\bar{X}\pm$s.e.	s.d.	Range
I^1	M–D	4	5·0±0·17	0·34	4·6–5·3
	B–L	4	4·0±0·14	0·29	3·6–4·3
I^2	M–D	3	3·7±0·12	0·21	3·5–3·9
	B–L	3	3·8±0·07	0·12	3·7–3·9
C	M–D	5	7·0±0·10	0·22	6·8–7·2
	B–L	5	5·0±0·08	0·17	4·7–5·1
P^3	M–D	5	4·4±0·12	0·28	4·1–4·8
	B–L	5	4·8±0·09	0·20	4·5–5·0
P^4	M–D	3	4·0±0·15	0·25	3·8–4·3
	B–L	4	5·1±0·14	0·29	4·7–5·4
M^1	M–D	5	5·4±0·18	0·40	4·9–5·9
	Ant B	3	5·5±0·21	0·36	5·1–5·8
	Post B	4	5·9±0·31	0·61	5·1–6·6
M^2	M–D	5	5·7±0·11	0·25	5·4–6·1
	Ant B	5	6·2±0·23	0·52	5·7–6·9
	Post B	4	6·1±0·23	0·46	5·8–6·8
M^3	M–D	5	4·8±0·16	0·37	4·3–5·3
	Ant B	5	5·6±0·30	0·66	4·9–6·5
	Post B	3	5·2±0·21	0·36	4·9–5·6

Table 198
Hylobates lar male

Mandibular teeth		n	$\bar{X}\pm$s.e.	s.d.	Range
I_1	M–D	5	3·1±0·14	0·30	2·7–3·5
	B–L	5	3·3±0·13	0·29	3·0–3·7
I_2	M–D	4	3·2±0·13	0·26	3·0–3·6
	B–L	3	3·9±0·07	0·16	3·8–4·0
C	M–D	3	4·9±0·40	0·70	4·1–5·4
	B–L	4	6·3±0·15	0·29	6·0–6·7
P_3	M–D	4	5·9±0·15	0·29	5·5–6·2
	B–L	4	4·0±0·19	0·38	3·7–4·5
P_4	M–D	5	4·6±0·07	0·16	4·4–4·8
	B–L	5	4·3±0·13	0·30	4·1–4·8
M_1	M–D	5	5·6±0·23	0·51	4·9–6·2
	Tri B	2	4·4±0·15	0·21	4·2–4·5
	Tal B	4	4·6±0·22	0·43	4·2–5·2
M_2	M–D	5	5·9±0·11	0·25	5·5–6·1
	Tri B	5	5·2±0·13	0·29	4·7–5·4
	Tal B	5	5·1±0·16	0·36	4·5–5·4
M_3	M–D	5	5·4±0·10	0·21	5·1–5·7
	Tri B	4	5·0±0·14	0·29	4·7–5·4
	Tal B	3	4·8±0·23	0·40	4·4–5·2
	Hypo-conulid				

Hylobates lar female measurements not available

Table 199
Pongo pygmaeus male

Maxillary teeth		n	$\bar{X}\pm$s.e.	s.d.	Range	P
I^1	M–D	4	14·7±0·48	0·96	13·3–15·4	
	B–L	6	13·4±0·72	1·76	10·4–15·2	
I^2	M–D	4	8·8±0·36	0·73	8·0–9·7	
	B–L	4	8·4±0·14	0·28	8·1–8·7	
C	M–D	4	16·4±1·03	2·08	13·4–17·9	*
	B–L	4	13·2±0·95	1·89	10·4–14·4	*
P^3	M–D	6	10·3±0·42	1·03	9·0–11·4	
	B–L	7	13·2±0·38	1·00	12·0–14·5	*
P^4	M–D	7	10·0±0·28	0·75	9·1–11·0	
	B–L	7	13·3±0·34	0·89	12·0–14·3	*
M^1	M–D	7	12·8±0·31	0·81	11·7–13·5	*
	Ant B	7	13·8±0·31	0·82	12·5–14·6	†
	Post B	7	13·0±0·39	1·03	11·5–14·3	*
M^2	M–D	7	12·7±0·21	0·55	12·0–13·7	
	Ant B	7	14·1±0·34	0·89	13·4–15·6	†
	Post B	7	13·1±0·49	1·29	10·6–14·4	*
M^3	M–D	4	12·4±0·21	0·42	12·0–13·0	*
	Ant B	3	14·1±0·29	0·50	13·6–14·6	†
	Post B					

* $P<0·05$
† $P<0·01$

Table 200
Pongo pygmaeus male

Mandibular teeth		n	$\bar{X}\pm$s.e.	s.d.	Range	P
I_1	M–D	6	9·7±0·42	1·02	9·0–11·0	†
	B–L	7	10·1±0·52	1·36	8·6–12·0	
I_2	M–D	6	9·7±0·27	0·66	8·7–10·6	†
	B–L	7	11·0±0·52	1·37	9·2–12·8	*
C	M–D	3	12·0±0·62	1·07	11·3–13·2	
	B–L	3	14·8±0·34	0·59	14·4–15·5	†
P_3	M–D	7	15·6±0·79	2·10	13·4–18·4	
	B–L	8	11·0±0·64	1·80	8·5–14·5	*
P_4	M–D	7	11·3±0·30	0·80	10·5–12·4	
	B–L	8	11·9±0·43	1·21	10·4–13·3	*
M_1	M–D	8	13·4±0·24	0·68	12·5–14·2	
	Tri B	7	11·9±0·33	0·88	10·6–12·8	*
	Tal B	7	11·9±0·47	1·24	10·3–13·3	
M_2	M–D	8	14·2±0·21	0·59	13·3–15·0	*
	Tri B	8	13·6±0·33	0·94	12·2–14·8	†
	Tal B	8	13·0±0·35	0·99	11·5–14·5	*
M_3	M–D	5	13·5±0·81	1·82	10·4–14·9	
	Tri B	5	12·9±0·31	0·70	11·9–13·7	*
	Tal B	5	12·0±0·61	1·37	9·6–12·8	
	Hypo-conulid					

* $P<0·05$
† $P<0·01$

Table 201
Pongo pygmaeus female

Maxillary teeth		n	$\bar{X}\pm$s.e.	s.d.	Range
I^1	M–D	4	13·8±0·56	1·04	12·9–15·3
	B–L	9	12·4±0·26	0·77	11·7–14·0
I^2	M–D	4	8·4±0·13	0·25	8·1–8·7
	B–L	5	8·3±0·33	0·74	7·5–9·3
C	M–D	4	12·5±0·78	1·56	10·2–13·7
	B–L	5	10·8±0·22	0·49	10·5–11·7
P^3	M–D	8	9·6±0·33	0·94	8·0–11·2
	B–L	6	11·8±0·29	0·72	11·0–12·9
P^4	M–D	8	9·3±0·27	0·77	8·3–10·1
	B–L	8	12·1±0·23	0·64	11·2–13·2
M^1	M–D	10	11·9±0·24	0·75	10·4–13·4
	Ant B	10	12·2±0·30	0·93	10·2–13·3
	Post B	9	11·7±0·26	0·79	10·2–12·9
M^2	M–D	9	12·1±0·27	0·79	11·2–13·4
	Ant B	9	12·9±0·38	1·13	10·3–13·8
	Post B	9	11·6±0·28	0·84	10·2–12·7
M^3	M–D	5	11·5±0·20	0·46	11·0–12·0
	Ant B	5	12·6±0·18	0·40	12·2–13·0
	Post B	4	10·3±0·33	0·66	9·5–10·9

Table 202
Pongo pygmaeus female

Mandibular teeth		n	$\bar{X}\pm$s.e.	s.d.	Range
I_1	M–D	9	8·5±0·13	0·38	8·0–9·1
	B–L	8	9·4±0·28	0·80	8·3–10·3
I_2	M–D	9	8·4±0·19	0·58	7·5–9·4
	B–L	9	9·7±0·26	0·78	8·5–10·5
C	M–D	3	9·7±1·48	2·55	7·8–12·6
	B–L	6	11·9±0·46	1·13	10·0–12·8
P_3	M–D	7	13·8±0·38	1·01	12·5–15·0
	B–L	7	9·3±0·21	0·55	8·4–10·1
P_4	M–D	7	10·5±0·30	0·80	9·4–11·6
	B–L	6	10·4±0·21	0·53	9·7–11·2
M_1	M–D	10	11·4±0·85	2·68	10·4–13·2
	Tri B	10	11·1±0·19	0·60	10·3–12·1
	Tal B	10	11·3±0·19	0·62	10·5–12·1
M_2	M–D	8	13·3±0·21	0·58	12·4–14·3
	Tri B	8	12·0±0·34	0·96	10·2–13·0
	Tal B	8	12·1±0·26	0·72	10·9–13·0
M_3	M–D	4	12·9±0·40	0·79	11·7–13·5
	Tri B	5	11·8±0·28	0·62	11·1–12·5
	Tal B	5	11·3±0·33	0·73	10·6–12·4
	Hypoconulid				

Table 203
Gorilla gorilla male

Maxillary teeth		n	$\bar{X}\pm$s.e.	s.d.	Range	P
I^1	M–D	3	14·7±0·15	0·25	14·5–15·0	†
	B–L	2	11·6±0·00	0·00	11·6–11·6	*
I^2	M–D	2	10·8±0·30	0·42	10·5–11·1	†
	B–L	4	10·1±0·38	0·75	9·4–11·0	
C	M–D	4	21·4±0·83	1·66	19·6–23·3	†
	B–L	4	16·0±0·39	0·78	14·9–16·6	†
P^3	M–D	4	12·4±0·35	0·71	11·7–13·3	
	B–L	3	15·8±0·35	0·61	15·1–16·2	*
P^4	M–D	3	11·2±0·17	0·30	10·9–11·5	
	B–L	3	15·2±0·40	0·69	14·8–16·0	*
M^1	M–D	5	15·7±0·35	0·79	14·5–16·4	*
	Ant B	4	15·4±0·60	1·19	14·2–17·0	
	Post B	4	15·0±0·62	1·25	13·2–16·0	*
M^2	M–D	4	17·5±0·11	0·22	17·3–17·8	†
	Ant B	3	16·9±0·27	0·46	16·4–17·3	*
	Post B	4	16·0±0·60	1·20	14·4–17·2	*
M^3	M–D	4	16·1±0·68	1·35	14·1–16·9	
	Ant B	4	15·8±0·40	0·80	14·6–16·3	*
	Post B	4	13·7±1·00	1·99	10·8–15·4	

* $P<0·05$
† $P<0·01$

Table 204
Gorilla gorilla male

Mandibular teeth		n	$\bar{X}\pm$s.e.	s.d.	Range	P
I_1	M–D	4	7·9±0·15	0·30	7·6–8·3	
	B–L	4	9·7±0·44	0·89	8·5–10·4	*
I_2	M–D	3	10·2±0·06	0·10	10·1–10·3	
	B–L	3	10·9±0·29	0·50	10·4–11·4	
C	M–D	4	16·9±1·10	2·19	14·8–20·0	
	B–L	5	18·1±0·48	1·08	16·2–19·0	†
P_3	M–D	4	16·9±0·39	0·77	15·9–17·6	
	B–L	5	11·9±0·14	0·31	11·6–12·4	*
P_4	M–D	5	11·8±0·37	0·82	11·1–13·0	
	B–L	4	14·6±0·13	0·26	14·3–14·9	*
M_1	M–D	6	16·2±0·32	0·79	14·7–17·0	
	Tri B	3	13·7±0·59	1·00	12·6–14·5	
	Tal B	3	13·3±0·84	1·46	11·8–14·7	
M_2	M–D	5	18·3±0·29	0·65	17·2–18·9	*
	Tri B	4	16·6±0·09	0·17	16·4–16·8	*
	Tal B	4	16·2±0·41	0·83	15·0–16·8	*
M_3	M–D	5	18·4±0·52	1·16	16·7–19·6	†
	Tri B	5	16·1±0·39	0·88	14·6–16·8	†
	Tal B	3	14·8±0·32	0·55	14·2–15·2	*
	Hypo-conulid					

* $P<0·05$
† $P<0·01$

Table 205
Gorilla gorilla female

Maxillary teeth		n	$\bar{X}\pm$s.e.	s.d.	Range
I^1	M–D	8	14·0±0·14	0·39	13·4–14·6
	B–L	8	10·3±0·16	0·46	9·6–10·9
I^2	M–D	8	9·9±0·36	1·03	8·7–11·7
	B–L	8	9·4±0·29	0·83	8·1–10·4
C	M–D	7	15·0±0·21	0·56	14·3–15·6
	B–L	6	11·3±0·19	0·47	10·5–11·8
P^3	M–D	9	11·6±0·28	0·83	10·6–13·3
	B–L	9	14·8±0·13	0·38	13·8–15·1
P^4	M–D	7	11·0±0·21	0·55	10·3–11·8
	B–L	8	14·4±0·17	0·47	13·4–14·9
M^1	M–D	11	14·8±0·27	0·89	13·4–16·3
	Ant B	10	15·0±0·35	1·10	13·8–17·4
	Post B	10	14·0±0·22	0·68	13·0–15·4
M^2	M–D	10	16·0±0·28	0·89	14·0–17·3
	Ant B	9	15·8±0·25	0·75	14·3–16·7
	Post B	7	14·8±0·29	0·77	13·3–15·5
M^3	M–D	6	14·6±0·30	0·72	13·6–15·2
	Ant B	5	14·5±0·54	1·22	13·2–16·3
	Post B	3	12·9±0·93	1·61	11·6–14·7

Table 206
Gorilla gorilla female

Mandibular teeth		n	$\bar{X}\pm$s.e.	s.d.	Range
I_1	M–D	9	8·0±0·09	0·28	7·5–8·2
	B–L	10	8·8±0·07	0·21	8·4–9·0
I_2	M–D	8	10·2±0·34	0·97	9·0–11·9
	B–L	10	10·3±0·18	0·58	9·6–11·5
C	M–D	4	12·0±0·49	0·97	11·0–13·3
	B–L	7	13·0±0·30	0·78	11·6–13·8
P_3	M–D	6	14·8±0·30	0·72	14·0–16·1
	B–L	7	11·2±0·50	1·31	9·5–13·3
P_4	M–D	7	11·3±0·24	0·63	10·3–12·1
	B–L	8	13·3±0·37	1·06	11·5–14·7
M_1	M–D	11	15·4±0·25	0·83	14·0–16·6
	Tri B	9	13·2±0·21	0·63	12·3–14·4
	Tal B	8	13·1±0·18	0·50	12·3–13·7
M_2	M–D	8	17·0±0·31	0·88	15·7–18·2
	Tri B	8	15·1±0·28	0·79	13·6–16·3
	Tal B	8	14·8±0·32	0·90	13·1–15·7
M_3	M–D	4	15·6±0·31	0·62	14·9–16·4
	Tri B	6	14·4±0·26	0·62	13·6–15·3
	Tal B	5	12·9±0·32	0·71	12·3–14·1
	Hypo-conulid				

Table 207
Pan troglodytes male

Maxillary teeth		n	$\bar{X}\pm$s.e.	s.d.	Range	P
I^1	M–D	14	12·6±0·22	0·82	10·5–13·5	†
	B–L	15	10·1±0·20	0·76	9·0–11·3	*
I^2	M–D	15	9·3±0·16	0·61	7·9–10·3	
	B–L	14	9·2±0·16	0·58	8·4–10·8	
C	M–D	14	15·0±0·41	1·57	13·0–18·0	†
	B–L	14	12·0±0·24	0·93	10·5–13·8	†
P^3	M–D	17	8·2±0·20	0·84	6·3–9·5	
	B–L	17	10·5±0·18	0·75	9·4–12·0	*
P^4	M–D	16	7·2±0·14	0·55	6·2–7·9	
	B–L	17	10·5±0·18	0·76	9·3–12·2	†
M^1	M–D	19	10·3±0·13	0·56	9·3–11·2	
	Ant B	19	11·7±0·17	0·73	10·7–13·2	†
	Post B	16	11·0±0·16	0·63	10·0–12·3	
M^2	M–D	19	10·4±0·15	0·65	9·6–12·3	
	Ant B	18	12·0±0·17	0·70	10·6–13·3	†
	Post B	17	11·0±0·14	0·59	9·5–11·9	
M^3	M–D	17	10·0±0·15	0·60	9·0–11·3	†
	Ant B	15	11·6±0·21	0·80	10·0–12·8	†
	Post B	14	10·4±0·17	0·63	9·6–11·6	†

* $P<0·05$
† $P<0·01$

Table 208
Pan troglodytes male

Mandibular teeth		n	$\bar{X}\pm$s.e.	s.d.	Range	P
I_1	M–D	11	8·3±0·23	0·75	6·8–9·0	
	B–L	13	9·7±0·16	0·57	8·8–11·0	*
I_2	M–D	13	8·9±0·22	0·80	7·6–10·2	*
	B–L	17	10·0±0·18	0·75	8·6–12·0	
C	M–D	13	12·7±0·44	1·58	10·9–15·9	†
	B–L	15	13·0±0·31	1·19	10·4–14·7	†
P_3	M–D	19	10·2±0·27	1·17	8·1–12·4	
	B–L	18	9·1±0·23	0·99	7·5–11·0	†
P_4	M–D	17	7·8±0·17	0·72	6·6–9·2	
	B–L	16	9·1±0·21	0·87	7·4–10·5	*
M_1	M–D	18	11·0±0·14	0·59	10·0–11·9	
	Tri B	15	10·0±0·18	0·68	9·0–11·2	*
	Tal B	17	10·2±0·15	0·60	8·7–11·0	
M_2	M–D	19	11·4±0·20	0·89	9·5–12·9	*
	Tri B	17	10·9±0·18	0·75	9·9–12·5	†
	Tal B	19	10·7±0·15	0·66	9·2–11·8	*
M_3	M–D	17	10·8±0·20	0·84	9·4–12·3	
	Tri B	15	10·6±0·16	0·60	9·9–11·5	†
	Tal B	13	9·8±0·14	0·50	8·9–10·6	
	Hypo-conulid					

* $P<0·05$
† $P<0·01$

Table 209
Pan troglodytes female

Maxillary teeth		n	$\bar{X}\pm$s.e.	s.d.	Range
I[1]	M–D	51	11·9±0·12	0·88	10·0–13·4
	B–L	50	9·6±0·12	0·82	8·3–11·7
I[2]	M–D	43	8·8±0·13	0·87	6·9–10·9
	B–L	45	8·9±0·12	0·80	7·6–11·0
C	M–D	32	11·7±0·30	1·68	6·5–15·0
	B–L	30	9·5±0·28	1·53	7·4–13·9
P[3]	M–D	46	8·1±0·14	0·94	6·5–12·3
	B–L	45	9·9±0·15	0·98	7·1–11·8
P[4]	M–D	46	7·2±0·14	0·96	5·8–11·4
	B–L	46	9·8±0·12	0·84	6·4–11·4
M[1]	M–D	51	10·1±0·10	0·72	9·0–11·9
	Ant B	50	10·9±0·14	1·01	7·0–12·8
	Post B	49	10·5±0·15	1·02	6·1–12·2
M[2]	M–D	50	10·1±0·10	0·74	9·0–11·9
	Ant B	45	11·1±0·14	0·92	7·2–13·3
	Post B	48	10·3±0·15	1·05	5·5–12·2
M[3]	M–D	37	9·5±0·14	0·83	8·0–11·1
	Ant B	36	10·6±0·13	0·78	8·8–12·5
	Post B	33	9·4±0·15	0·83	7·5–11·6

Table 210
Pan troglodytes female

Mandibular teeth		n	$\bar{X}\pm$s.e.	s.d.	Range
I_1	M–D	42	8·0±0·12	0·78	5·5–9·6
	B–L	45	9·1±0·13	0·89	7·2-11·0
I_2	M–D	41	8·4±0·10	0·63	7·0–9·6
	B–L	41	9·5±0·14	0·91	7·9–11·5
C	M–D	30	10·4±0·27	1·47	8·2–14·0
	B–L	31	10·2±0·26	1·44	8·0–14·1
P_3	M–D	45	9·9±0·14	0·96	8·4–12·4
	B–L	45	8·1±0·15	0·99	6·6–10·8
P_4	M–D	43	7·5±0·10	0·67	6·0–9·1
	B–L	43	8·5±0·12	0·76	7·2–10·0
M_1	M–D	45	10·8±0·09	0·63	9·8–12·7
	Tri B	42	9·4±0·12	0·79	7·4–11·0
	Tal B	45	9·6±0·11	0·77	8·1–11·4
M_2	M–D	47	11·0±0·09	0·65	9·5–12·2
	Tri B	46	10·2±0·13	0·86	8·3–12·3
	Tal B	46	10·2±0·13	0·86	8·5–13·1
M_3	M–D	40	10·4±0·13	0·82	9·0–12·2
	Tri B	36	9·7±0·12	0·74	8·0–10·7
	Tal B	37	9·4±0·13	0·78	8·0–10·8
	Hypo-conulid				

Table 211
Pan paniscus paniscus

Maxillary teeth		n	$\bar{X} \pm$ s.e.	s.d.	Range
I^1	M–D	15	10·3	0·9	8·9–11·9
	B–L	15	7·9	0·6	7·2–9·2
I^2	M–D	13	7·9	0·7	6·9–9·2
	B–L	13	7·3	0·6	6·7–8·5
C	M–D	15	11·1	0·9	9·7–13·3
	B–L	15	8·8	0·8	7·6–10·7
P^3	M–D	17	7·4	0·6	6·6–8·4
	B–L	17	9·3	0·6	8·4–10·3
P^4	M–D	16	6·3	0·5	5·7–7·6
	B–L	16	9·0	0·6	8·0–10·3
M^1	M–D	7	8·5	0·6	7·9–9·4
	Ant B	6	9·5	0·5	9·2–10·4
	Post B				
M^2	M–D	7	8·4	0·7	7·6–9·6
	Ant B	7	10·2	0·4	9·6–10·6
	Post B				
M^3	M–D	6	7·8	0·3	7·5–8·2
	Ant B	6	9·8	0·4	9·4–10·3
	Post B				

D. C. Johanson, 1974

Table 212
Pan paniscus male

Mandibular teeth		n	$\bar{X} \pm$ s.e.	s.d.	Range
I_1	M–D	16	7·4	0·7	6·1–8·7
	B–L	15	7·0	0·5	6·3–8·1
I_2	M–D	17	7·5	0·6	6·3–8·3
	B–L	17	7·1	0·4	6·7–8·4
C	M–D	16	10·0	0·7	8·7–11·4
	B–L	16	7·6	0·4	6·8–8·5
P_3	M–D	17	8·1	0·5	7·2–8·9
	B–L	17	7·4	1·2	5·8–9·7
P_4	M–D	18	7·1	0·4	6·1–7·7
	B–L	18	7·8	0·7	6·5–9·2
M_1	M–D	5	9·1	0·7	8·2–9·9
	Tri B	5	8·8	0·5	8·1–9·4
	Tal B	5	9·0	0·7	7·9–9·8
M_2	M–D	7	9·9	0·4	9·4–10·6
	Tri B	7	9·1	0·6	8·4–10·0
	Tal B	7	8·8	0·3	8·3–9·2
M_3	M–D	7	9·2	0·5	8·1–9·4
	Tri B	7	8·4	0·5	7·8–9·2
	Tal B	7	7·9	0·2	7·5–8·6
	Hypo-conulid				

D. C. Johanson, 1974

Table 213
Pan paniscus female

Maxillary teeth		n	$\bar{X}\pm$s.e.	s.d.	Range
I^1	M–D	20	10·4	0·7	9·0–11·5
	B–L	21	7·6	0·4	6·8–8·5
I^2	M–D	20	7·9	0·7	7·1–10·1
	B–L	20	7·1	0·4	6·4–7·7
C	M–D	18	9·0	0·5	8·2–9·9
	B–L	18	6·9	0·4	6·3–7·6
P^3	M–D	26	7·2	0·4	6·2–7·8
	B–L	27	9·2	0·4	8·3–10·2
P^4	M–D	23	6·1	0·5	5·0–6·6
	B–L	24	8·8	0·4	7·7–9·8
M^1	M–D	6	8·3	0·4	7·6–8·8
	Ant B	6	9·7	0·4	9·3–10·4
	Post B				
M^2	M–D	7	8·4	0·7	7·6–9·6
	Ant B	7	10·2	0·4	9·6–10·6
	Post B				
M^3	M–D	4	8·2	0·2	8·0–8·4
	Ant B	3	10·2	0·3	9·9–10·5
	Post B				

D. C. Johanson, 1974

Table 214
Pan paniscus female

Mandibular teeth		n	$\bar{X}\pm$s.e.	s.d.	Range
I_1	M–D	22	7·2	0·7	5·6–8·5
	B–L	22	6·8	0·3	6·2–7·5
I_2	M–D	24	7·3	0·8	5·2–9·0
	B–L	24	6·9	0·3	6·4–7·5
C	M–D	20	8·8	0·7	7·5–10·9
	B–L	20	6·5	0·7	5·8–8·9
P_3	M–D	24	8·2	0·4	7·5–9·1
	B–L	24	7·0	0·9	5·2–8·4
P_4	M–D	24	7·0	0·8	5·4–9·1
	B–L	24	7·6	0·6	5·6–8·5
M_1	M–D	5	9·5	0·4	9·0–10·1
	Tri B	4	8·9	0·2	8·7–9·2
	Tal B	3	8·8	0·5	8·4–9·3
M_2	M–D	5	9·7	0·3	9·3–10·0
	Tri B	5	9·2	0·4	8·8–9·7
	Tal B	5	8·9	0·3	8·7–9·3
M_3	M–D	5	9·2	0·5	8·4–9·8
	Tri B	5	8·7	0·4	8·2–9·1
	Tal B	5	8·2	0·3	7·8–8·6
	Hypo-conulid				

D. C. Johanson, 1974

References

Adloff, P. (1908). *Das Gebiss des Menschen und der Anthropomorphen.* Berlin: Julius Springer.

Ashton, E. H. and Zuckerman, S. (1950). Some quantitative dental characteristics of the chimpanzee, gorilla and orang-utan. *Phil. Trans. Roy. Soc. Lond.* **234B**: 471–484.

Bennejeant, C. (1936). *Anomalies et variations dentaires chez les Primates.* Paris: Clermont–Ferrand. Imprimeries Paul Vallier.

Biggerstaff, R. H. (1966). Metric and taxonomic variations in the dentitions of two Asian cercopithecoid species: *Macaca mulatta* and *Macaca speciosa. Am. J. Phys. Anthrop.* **24**: 231–238.

Booth, A. H. (1956). The distribution of primates on the Gold Coast. *J. West Afr. Sci. Assoc.* **2**: 122–133.

Buettner-Janusch, J. and Andrew, R. J. (1962). The use of the incisors by *Primates* in grooming. *Am. J. Phys. Anthrop.* **20**: 127–129.

Butler, P. M. (1939). Studies on the mammalian dentition: differentiation of the postcanine dentition. *Proc. Zool. Soc. Lond.* Ser. B **109**: 1–36.

Butler, P. M. (1956). The ontogeny of molar patterns. *Biol. Rev.* **31**: 30–70.

Cabrera, A. (1957). *Catalogo de los mamiferos de America del Sur.* IV: No. I, Buenos Aires y Peru. Instituto Nacional de Investigacion de la Ciencias Naturales, Ciencio Zoologicia.

Carpenter, C. R. (1940). A field study in Siam of the behavior and social relations of the gibbon *(Hylobates lar). Comp. Psychol. Monogr.* **16**: 1–212.

Cartmill, M. (1972). *Daubentonia,* woodpeckers and klinorhynchy. *Am. J. Phys. Anthrop.* **37**: 432. Abst.

Colyer, F. (1936). *Variations and diseases of the teeth of animals.* London: John Bale Sons and Danielsson.

Cooke, H. B. S. (1968). The fossil mammal fauna of Africa. *Quart. Rev. Biol.* **43**: 234–264.

Dahlberg, A. A. (1949). The dentition of the American Indian. Papers on the physical anthropology of the American Indian. Fourth Viking Fund Seminar: 138–176. New York: The Viking Fund, Inc.

Dahlberg, A. A. (1950). The evolutionary significance of the protostylid. *Am. J. Phys. Anthrop.* **8**: 15–25.

Davis, D. D. (1962). Mammals of the lowland rain-forest of North Borneo. *Bull. Singapore Nat. Mus.* **31**: 5–129.

Day, M. H. (1965). *Guide to fossil man: a handbook of human paleontology.* Cleveland and New York: The World Publishing Co.

Delson, E. (1973). Fossil colobine monkeys of the Circum-Mediterranean region. Ann Arbor, Michigan: University Microfilms.

Dietz, V. H. (1944). A common dental morphotropic factor, the Carabelli cusp. *J. Am. Dent. Assoc.* **31**: 784–789.

Dollman, G. (1937). Exhibition of skins of marmosets and tamarins. *Abstr. Proc. Zool. Soc. Lond.* pp. 64-65.

Ellefson, J. O. (1968). Territorial behavior in the common white-handed gibbon, *Hylobates lar* Linn. In *Primates*: 180, (Ed. Jay, P. C.). New York: Henry Holt and Co.

Erikson, G. E. (1963). Brachiation in the New World monkeys. In *The Primates:* 135-164. (Eds Napier, J. and Barnicot, N. A.). Symp. Zool. Soc. Lond. Vol. 10. London and New York: Academic Press.

Every, R. G. (1970). Sharpness of teeth in man and other primates. *Postilla* **143**: 1–30.

Fooden, J. (1972). Breakup of Pangaea and isolation of relict mammals in Australia, South America and Madagascar. *Science* **175**: 894–898.

Forbes, H. O. (1894). *A hand-book to the Primates* Vol. I (Ed. Sharpe, R. Bowdler). London: W. C. Allen and Co. Ltd. Allen's Naturalist's Library.

Freedman, L. (1957). The fossil Cercopithecoidea of South Africa. *Ann. Trans. Mus.* pt **2**, 1–121.

Friant, M. (1935). Description et interpretation de la dentition d'un jeune *Indris. C.r. Ass. Anat.* **30**: 205–213.

Friant, M. (1942). Persistance d'un caractère archaique fondamental des molaires supérieures chez un Singe Platyrhinien, le Mycetes. *Bull. Mus. Nat. Hist. Naturelle, Paris* **14**: 106–108.

Frisch, J. E. (1963). Dental variability in a population of gibbons (*Hylobates lar*). In *Dental anthropology* Vol. 5: 15–28. (Ed. Brothwell, D. R.). New York: The Macmillan Co.

Frisch, J. E. (1965). Trends in the evolution of the Hominoid dentition. *Bibl. Primatol.* **3**: 1–130. Basel: Karger.

Garn, S. M., Kerewsky, R. S. and Swindler, D. R. (1966). Canine "field" in sexual dimorphism of tooth size. *Nature, Lond.* **212**: 1501–1502.

George, R. M. (1973). The musculature of the limbs of *Urogale everetti* with reference to the comparative anatomy of the Tupaiidae. Ph.D. Thesis, University of Washington, Seattle.

Gingerich, P. D. (1972). Molar occlusion and jaw mechanics of the Eocene primate *Adapis. Am. J. Phys. Anthrop.* **36**: 359–368.

Goldstein, M. S. (1948). Dentition of Indian crania from Texas. *Am. J. Phys. Anthrop.* **6**: 63–84.

Goodall, J. (1963). Feeding behavior of wild chimpanzees. In *The Primates*: 39–47, (Eds Napier, J. and Barnicot, N. A.). Symp. Zool. Soc. Lond. Vol. 10. London and New York: Academic Press.

Greene, D. L. (1973). Gorilla dental sexual dimorphism and early hominid taxonomy. In *Craniofacial biology of Primates* Vol. 3: 82–100, (Ed. Zingeser, M. R.). Basel: Karger.

Gregory, W. K. (1916). Studies on the evolution of the primates. *Bull. Am. Mus. Nat. Hist.* **35**: 239–355.
Gregory, W. K. (1922). *The origin and evolution of the human dentition.* Baltimore: Williams and Wilkins Co.
Gregory, W. K. and Hellman, M. (1926). The dentition of *Dryopithecus* and the origin of man. *Anthrop. Pap. Am. Mus. Nat. Hist.* **28**: 9–117.
Groves, C. P. (1970). The forgotten leaf-eaters, and the phylogeny of the Colobine. In *Old World monkeys: evolution, systematics and behaviour*: 557–587, (Eds Napier, J. R. and Napier, P. H.). New York and London: Academic Press.
Hall, K. R. L. (1963). Variations in the ecology of the Chacma baboon, *Papio ursinus.* In *The Primates*: 1–28, (Eds Napier, J. and Barnicot, N.A.). Symp. Zool. Soc. Lond. Vol. 10. London and New York: Academic Press.
Hellman, M. (1928). Racial characters in human dentition. *Proc. Am. Phil. Soc.* **67**: 157–174.
Hershkovitz, P. (1966). Taxonomic notes on Tamarins, genus *Saguinus* (Callithricidae, Primates), with descriptions of four new forms. *Folia primat.* **4**: 381–395.
Hershkovitz, P. (1971). Basic crown patterns and cusp homologies of mammalian teeth. In *Dental morphology and evolution*: 95–150, (Ed. Dahlberg, A. A.). Chicago: University of Chicago Press.
Hiiemae, K. M. and Crompton, A. W. (1971). A cinefluorographic study of feeding in the American opossum. In *Dental morphology and evolution*: 299–334, (Ed. Dahlberg, A. A.). Chicago: University of Chicago Press.
Hiiemae, K. M. and Kay, R. F. (1973). Evolutionary trends in the dynamics of primate mastication. In *Craniofacial biology of Primates* **3**: 28–64, (Ed. Zingeser, M. R.). Basel: Karger.
Hill, J. P. (1965). On the placentation of *Tupaia. J. Zool. Lond.* **146**: 278–304.
Hill, W. C. O. (1953). *Primates comparative anatomy and taxonomy, Vol. I: Strepsirhini.* New York: Interscience Publishers, Inc.
Hill, W. C. O. (1955). *Primates comparative anatomy and taxonomy, Vol. II: Haplorhini, Tarsioidea.* Edinburgh: The University Press.
Hill, W. C. O. (1957). *Primates comparative anatomy and taxonomy, Vol. III: Pithecoidea, Platyrrhini.* New York: Interscience Publishers, Inc.
Hill, W. C. O. (1959). Anatomy of *Callimico goeldii. Trans. Am. Phil. Soc.* **49**: 3–116.
Hill, W. C. O. (1960). *Primates comparative anatomy and taxonomy, Vol. IV: Part A, Cebidae.* New York: Interscience Publishers, Inc.
Hill, W. C. O. (1970). *Primates comparative anatomy and taxonomy, Vol. VIII: Cynopithecinae.* New York: Wiley–Interscience, Inc.
Hooijer, D. A. (1948). Prehistoric teeth of man and the orang-utan from central Sumatra, with notes on the fossil orang-utan from Java and southern China. *Zoolog-Mededelingen* **29**: 173–301.
Hornbeck, P. V. and Swindler, D. R. (1967). Morphology of the lower fourth premolar of certain Cercopithecidae. *J. Dent. Res.* **46**: 5, suppl: 979–983.
Hrdlička, A. (1920). Shovel shaped teeth. *Am. J. Phys. Anthrop.* **3**: 429–465.
Hurme, V. O. and Van Wagenen, G. (1961). Basic data on the emergence of permanent teeth in the rhesus monkey (*Macaca mulatta*). *Proc. Am. Philos. Soc.* **105**: 105–140.
Itani, J. (1958). On the acquisition and propagation of a new food habit in the troop of Japanese monkeys at Takasakiyama. *Primates* **1**: 131–148.

James, W. W. (1960). *The jaws and teeth of Primates.* London: Pitman Medical Publishing Co. Ltd.

Jane, J. A., Campbell, C. B. G. and Yashon, D. (1969). The origin of the corticospinal tract of the tree shrew (*Tupaia glis*) with observations on its brain stem and spinal laminations. *Brain Behav. Evol.* **2**: 160–182.

Johanson, D. C. (1974). Some metric aspects of the permanent and deciduous dentition of the pygmy chimpanzee (*Pan paniscus*). *Am. J. Phys. Anthrop.* **41**: 39–48.

Jolly, A. (1966). *Lemur behavior.* Chicago: University of Chicago Press.

Jolly, C. J. (1970a). The large African monkeys as an adaptive array. In *Old World monkeys*: 141–174. (Ed. Napier, J. R. and Napier, P. H.) New York and London: Academic Press.

Jolly, C. J. (1970b). The seed-eaters: a new model of hominid differentiation based on a baboon analogy. *Man* **5**: 5–26.

Kälin, J. (1962). Über *Moeripithecus marcgrafi* Schlosser und die phyletischen vorstuffen der Bilophodontie der Cercopithecidae. *Bibl. Primatol.* **1**: 32–42.

Kay, R. F. and Hiiemae, K. M. (1974). Jaw movement and tooth use in recent and fossil primates. *Am. J. Phys. Anthrop.* **40**: 227–256.

Kern, J. A. (1964). Observations on the habits of the proboscis monkey, *Nasalis larvatus* (Wurmb), made in Brunei Bay area, Borneo. *Zoologica* **49**: 183–191.

Kinzey, W. G. (1971). Evolution of the human canine tooth. *Am. Anthrop.* **73**: 680–694.

Kinzey, W. G. (1972). Canine teeth of the monkey, *Callicebus moloch*: lack of sexual dimorphism. *Primates* **13**: 365–369.

Kinzey, W. G. (1973). Reduction of the cingulum in Ceboidea. In *Craniofacial biology of primates* **3**: 101–127, (Ed. Zingeser, M. R.). Basel: Karger.

Kitahara-Frisch, J. (1973). Taxonomic and phylogenetic uses of the study of variability in the hylobatid dentition. In *Craniofacial biology of primates* **3**: 128–147, (Ed. Zingeser, M. R.). Basel: Karger.

Korenhof, C. A. W. (1960). *Morphogenetical aspects of the human upper molar.* Druk: Uitgeversmaatschappij Neerlandia-Utrecht.

Kraus, B. S. (1951). Carabelli's anomaly of the maxillary molar teeth. *Am. J. Hum. Gen.* **3**: 348–355.

Kraus, B. S. and Furr, M. L. (1953). Lower first premolars, Part I: a definition and classification of discrete morphologic traits. *J. Dent. Res.* **32**: 554–564.

Le Gros Clark, W. E. (1971). *The antecedents of man: an introduction to the evolution of the Primates* Third edition. Chicago: Quadrangle Books.

Leutenegger, W. (1971). Metric variability of the postcanine dentition in Colobus monkeys. *Am. J. Phys. Anthrop.* **35**: 91–100.

Long, J. O. and Cooper, R. W. (1968). Physical growth and dental eruption in captive-bred squirrel monkeys, *Saimiri sciureus* (Leticia, Columbia). In *The squirrel monkey*: 193–205. New York and London: Academic Press.

Ludwig, F. J. (1957). The mandibular second premolars: morphologic variation and inheritance. *J. Dent. Res.* **36**: 263–273.

Lyon, M. W., Jr. (1913). Treeshrews: an account of the mammalian family Tupaiidae. *Proc. U.S. Natn. Mus.* **45**: 1–188. Washington, D.C.

Martin, R. D. (1975). Ascent of the primates. *Natural History* **84**: 52–61.

Mills, J. R. E. (1955). Ideal dental occlusion in the primates. *Dent. Practitioner* **1**: 47–61.

Mills, J. R. E. (1963). Occlusion and malocclusion of the teeth of primates. In *Dental anthropology* **5**: 29–51, (Ed. Brothwell, D. R.). New York: Macmillan Co.

Mivart, St. George (1867). Notes on the osteology of the Insectivora. *J. Anat. Lond. Primates:* 65–81, (Ed. Zingeser, M. R.). Basel: Karger.

Mivart, St. George (1867). Notes on the osteology of the Insectivora. *J. Anat. Lond.* **1**: 281–312.

Moorrees, C. F. A. (1957). *The aleut dentition.* Cambridge, Massachusetts: Harvard University Press.

Napier, J. R. (1961). Prehensility and opposability in the hands of primates. In *Vertebrate locomotion*: 115–132, (Ed. Harris, J. E.). Symp. Zool. Soc. Lond. Vol. 5. London and New York: Academic Press.

Napier, J. R. and Napier, P. H. (1967). *A handbook of living Primates.* London and New York: Academic Press.

Napier, J. R. and Walker, A. C. (1967). Vertical clinging and leaping, a newly recognized category of locomotor behaviour among primates. *Folia primat.* **6**: 180–203.

Nelson, C. T. (1938). The teeth of the Indians of Pecos Pueblo. *Am. J. Phys. Anthrop.* **23**: 261–293.

Ockerse, T. (1959). The eruption sequence and eruption times of the teeth of the vervet monkey. *J. Dent. Assoc. S. Afr.* **14**: 422–423.

Orlosky, F. J. (1973). Comparative dental morphology of extant and extinct Cebidae. Ph.D. Thesis, University of Washington, Seattle.

Orlosky, F. J., Swindler, D. R. and McCoy-Beck, H. A. (1974). Metric trends of the anterior teeth in African monkeys. *Human Biol.* **46**: 647–661.

Osborn, H. F. (1907). *Evolution of mammalian molar teeth to and from the triangular type,* (Ed. Gregory, W. K.). New York: The Macmillan Co.

Owen, R. (1840–1845). *Odontography* 2 Vols. London: Hippolyte Bailliere.

Pedersen, P. O. (1949). The east Greenland Eskimo dentition. *Meddelelser om Grønland* **142**: 1–256.

Petter, J. J. (1962). Ecological and behavioral studies of Madagascar lemurs in the field. **102**: 181–514. In *The relatives of Man: modern studies of the relation of the evolution of the nonhuman Primates to human evolution,* (Ed. Buettner-Janusch, J.). New York: Annals of the New York Academy of Sciences.

Pilbeam, D. R. (1969). *Tertiary Pongidae of East Africa: evolutionary relationships and taxonomy.* Peabody Mus. Nat. Hist., Bull. **31**. Yale University, New Haven, Connecticut.

Pocock, R. I. (1925a). Additional notes on the external characters of some platyrrhine monkeys. *Proc. Zool. Soc. Lond.*, pp. 27–47.

Pocock, R. I. (1925b). Notes on Ceropithecine genera, *Rhinostigma* and *Miopithecus. Ann Mag. Nat. Hist.* **16**: 264–268.

Remane, A. (1921). Beiträge zur Morphologie des Anthropoidengebisses. *Wiegmann. Archiv. für Naturgeschichte* **87**: 1–179.

Remane, A. (1960). Zähne und Gebiss. *Primatologia* **3**: 637–846.

Robinson, J. T. (1956). *The dentition of the Australopithecinae.* Transvaal Mus. Memoir No. 9: 1–179. Pretoria, South Africa.

Saheki, M. (1966). Morphological studies of *Macaca fuscata.* IV: Dentition. *Primates* **7**: 407–422.

Sauer, E. G. F. and Sauer, E. M. (1963). The South-West African bush-baby of the *Galago senegalensis* group. *J. S. W. Afr. Sci. Soc.* **16**: 5–35.

Schultz, A. H. (1935). Eruption and decay of the permanent teeth in primates. *Am. J. Phys. Anthrop.* **19**: 489–581.

Schultz, A. H. (1944). Age changes and variability in gibbons. *Am. J. Phys. Anthrop.* **2**: 1–129.

Schultz, A. H. (1958). Cranial and dental variability in Colobus monkeys. *Proc. Zool. Soc. Lond.* **130**: 79–105.

Schultz, A. H. (1960). Age changes and variability in the skulls and teeth of the central American monkeys *Alouatta, Cebus* and *Ateles*. Proc. Zool. Soc. Lond. **133**: 337–390.

Schuman, E. L. and Brace, C. L. (1955). Metric and morphologic variations in the dentition of the Liberian chimpanzee. *Human Biol.* **26**: 239–268.

Schwartz, J. H. (1974a). Dental development in the prosimians and its bearing on their evolution. University Microfilms, Ann Arbor, Michigan.

Schwartz, J. H. (1974b). Observations on the dentitions of the Indriidae. *Am. J. Phys. Anthrop.* **41**: 107–114.

Schwarz, E. (1931). A revision of the genera and species of Madagascar Lemuridae. *Proc. Zool. Soc. Lond.*, pp. 399–428.

Serra, della O. (1951). Variacoes do articulado dos dentes incisivos nos macacos do genero *Alouatta* Lac. 1799. *Papéis avulsos do Dep. Zool. S Paulo* **10**: 139–146.

Serra, della, O. (1952a). A seqüência eruptiva dos dentes definitivos nos simios Platyrrhina e sua interpretačao filogenética. *Anais Fac. Farm. Odont. Univ. S Paulo* **10**: 215–296.

Serra, della O. (1952b). O tuberculo intermedianio posterior (metaconulo) dos Platyrrhina do genero *Alouatta* Lac. *Anais Fac. Farm. Odont. Univ. S Paulo* **10**: 297–301.

Simpson, G. G. (1945). The principles of classification and a classification of mammals. *Bull. Am. Mus. Nat. Hist.* **85**: 1–350.

Simpson, G. G. (1955). The Phenacolemuridae, new family of early primates. *Bull. Amer. Mus. Nat. Hist.* **105**: 411–442.

Sirianni, J. E. (1974). Dental variability in African Cercopithecidae: A morphologic, metric and discriminant analysis. Ph.D. Thesis, University of Washington, Seattle.

Skinner, E. W. (1954). *The science of dental materials.* Philadelphia: W. B. Saunders Co.

Sokal, R. R. and Rohlf, F. J. (1969). *Biometry: the principles and practice of statistics in biological research.* San Francisco: W. H. Freeman and Co.

Steele, D. G. (1973). Dental variability in the tree shrews (Tupaiidae). In *Craniofacial biology of Primates,* **3**: 154–179, (Ed. Zingeser, M. R.). Basel: Karger.

Steuerwald, E. A. (1969). Review of the phylogenetic position of the tree shrew (*Tupaia glis* Diard) with new observations on the Arteria Carotis Interna. Ph.D. Thesis, Michigan State University, East Lansing.

Swindler, D. R. (1968). The maxillary incisors and evolution of Old World monkeys. In *Taxonomy and phylogeny of Old World Primates with references to the origin of Man*: 57–67, (Ed. Chiarelli, B.). Torino, Italy: Rosenberg and Sellier.

Swindler, D. R., Gaven, J. A. and Turner, W. M. (1963). Molar tooth size variability in African monkeys. *Human Biol.* **35**: 104–122.

Swindler, D. R., McCoy, H. A. and Hornbeck, P. V. (1967). The dentition of the baboon (*Papio anubis*). In *The baboon in medical research* Vol. II: 133–150, (Ed. Vagtborg, H). Austin: University of Texas Press.

Swindler, D. R. and Orlosky, F. J. (1972). Metric and morphological variability in the dentition of colobine monkeys. *J. Human Evol.* **3**: 135–160.

Swindler, D. R. and Sirianni, J. E. (1969). Variability of maxillary incisors among primates. *Proc. VIIIth Congr. Anthrop. Ethnol. Sci. S-2, Dental Aspects,* pp. 308–311.

Szalay, F. S. (1969). Mixodectidae, Microsyopidae, and the insectivore-primate transition. *Bull. Am. Mus. Nat. Hist.* **140**: 193–330.

Tattersall, I. (1973). Cranial anatomy of the Archaeolemurinae (Lemuroidea, Primates). *Anthrop. Pap. Am. Mus. Nat. Hist.*, New York, pp. 5–110.

Todd, T. W. (1918). *An introduction to the mammalian dentition.* London: Henry Kimpton.

Tuttle, R. H. (1967). Knuckle walking and the evolution of hominoid hands. *Am. J. Phys. Anthrop.* **26**: 171–182.

Vandebroek, G. (1961). The comparative anatomy of the teeth of lower and nonspecialized mammals. Internat. Colloq. on the evolution of lower and non-specialized mammals. Kon VI. *Acad. Ventensch Lett. Sch. Kunsten Belgie, Brussels. Kl. Wetensch.* **1**: 215–313.

Van Valen, L. (1965). Tree shrews, primates, and fossils. *Evolution* **19**: 137–151.

Van Valen, L. (1966). Deltatheridia, a new order of mammals. *Bull. Am. Mus. Nat. Hist.* **132**: 1–126.

Van Valen, L. and Sloan, R. E. (1965). The earliest primates. *Science* **150**: 743–745.

Voruz, C. (1970). Origine des dents bilophodonts des Cercopithecoidea. *Mammalia* **34**: 269–293.

Warrer, E. (1952). Metrisk analyse of gibsmodeller. *Tandlaegebladet* **56**: 95–105.

Washburn, S. L. and DeVore, I. (1961). The social life of baboons. *Sci. Amer.* **204**: 62–71.

Weidenreich, F. (1937). The dentition of *Sinanthropus pekinensis*: a comparative odontography of the hominids. *Palaeont. Sin.* n.s. D, *no.* **1** (whole series no. 101). Text and Atlas.

Wharton, C. H. (1950). The tarsier in captivity. *J. Mammal.* **31**: 260–268.

Zingeser, M. R. (1966). Occlusofacial relationships in the mature howler monkey (*Alouatta caraya*). *Am. J. Phys. Anthrop.* **24**: 171–180.

Zingeser, M. R. (1967). Odontometric characteristics of the howler monkey (*Alouatta caraya*). *J. Dent. Res.* Suppl., part 1:975–978. (Ed. Dahlberg, A. A.) Int. Symp. Dent. Morphol.

Zingeser, M. R. (1968a). Characteristics of the masticatory system. In *Biology of the howler monkey*: 141–150. (Ed. Malinow, M. R.) Basel: Karger.

Zingeser, M. R. (1968b). Functional and phylogenetic significance of integrated growth and form in occluding monkey canine teeth (*Alouatta caraya* and *Macaca mulatta*). *Am. J. Phys. Anthrop.* **28**: 263–270.

Zingeser, M. R. (1969). Cercopithecoid canine tooth honing mechanisms. *Am. J. Phys. Anthrop.* **31**: 205–214.

Zingeser, M. R. (1970). The morphological basis of the underbite trait in langurs (*P. melalophus* and *T. cristatus*) with an analysis of adaptive and evolutionary implications. *Am. J. Phys. Anthrop.* **32**: 179–186.

Zingeser, M. R. (1973a). Occlusofacial morphological integration (*Homo sapiens, Alouatta caraya, Cebus capucinus*). In *Craniofacial biology of Primates*: 241–257. (Ed. Zingeser, M. R.). Basel: Karger.

Zingeser, M. R. (1973b). Dentition of *Brachyteles arachnoides* with reference to Alouattine and Atelinine affinities. *Folia primat.* **20**: 351–390.

Taxonomic Index